AF352802

Genetics of Halophilic Microorganisms

Genetics of Halophilic Microorganisms

Special Issue Editors

Rafael Montalvo-Rodríguez
Julie A. Maupin-Furlow

MDPI • Basel • Beijing • Wuhan • Barcelona • Belgrade • Manchester • Tokyo • Cluj • Tianjin

Special Issue Editors

Rafael Montalvo-Rodríguez
University of Puerto Rico
USA

Julie A. Maupin-Furlow
University of Florida
USA

Editorial Office
MDPI
St. Alban-Anlage 66
4052 Basel, Switzerland

This is a reprint of articles from the Special Issue published online in the open access journal *Genes* (ISSN 2073-4425) (available at: https://www.mdpi.com/journal/genes/special_issues/halophilic_microorganisms).

For citation purposes, cite each article independently as indicated on the article page online and as indicated below:

LastName, A.A.; LastName, B.B.; LastName, C.C. Article Title. *Journal Name* **Year**, *Article Number, Page Range.*

ISBN 978-3-03928-955-4 (Hbk)
ISBN 978-3-03928-956-1 (PDF)

Cover image courtesy of Puerto Rico Desde el Aire.

Contents

About the Special Issue Editors

Rafael Montalvo-Rodríguez is a Professor with the Biology Department, University of Puerto Rico-Mayaguez. The Extremophiles Laboratory studies the diversity, physiology, and genetics of extremophiles in Puerto Rico and the Caribbean.

Julie A. Maupin-Furlow is a Professor with the Department of Microbiology and Cell Science, University of Florida. Her laboratory studies extremophiles, including halophiles of the domain Archaea, to advance our understanding of systems important to cell biology and to develop useful tools for biotechnology.

 *genes*

Editorial

Insights through Genetics of Halophilic Microorganisms and Their Viruses

Rafael Montalvo-Rodríguez [1,*] and Julie A. Maupin-Furlow [2,*]

[1] Department of Biology, University of Puerto Rico, Box 9000, Mayagüez, PR 00681, USA

[2] Department of Microbiology and Cell Science, Institute of Food and Agricultural Sciences, University of Florida, Gainesville, FL 32611, USA

* Correspondence: rafael.montalvo@upr.edu (R.M.-R.); jmaupin@ufl.edu (J.A.M.-F.)

Received: 30 March 2020; Accepted: 1 April 2020; Published: 2 April 2020

Abstract: Halophilic microorganisms are found in all domains of life and thrive in hypersaline (high salt content) environments. These unusual microbes have been a subject of study for many years due to their interesting properties and physiology. Study of the genetics of halophilic microorganisms (from gene expression and regulation to genomics) has provided understanding into mechanisms of how life can occur at high salinity levels. Here we highlight recent studies that advance knowledge of biological function through study of the genetics of halophilic microorganisms and their viruses.

Keywords: halophile; archaea; bacteria; fungi; virus; genomics; metagenomics; stress; DNA methylation; DNA recombination

1. Metagenomics

The employment of metagenomics to study microbial diversity and discovering genes with novel functions have proven to be a very powerful tool in microbiology. Considering that halophilic microorganisms present a challenge for this kind of study (mainly because they are understudied and they have a high G+C content), Uritskiy and DiRuggiero [1] present several proposals for the application of these techniques to halophilic microorganisms. The authors explore the limitations and challenges these methodologies currently have and present outlines on how to create better pipelines to study halophilic microbiomes.

Couto-Rodríguez and Montalvo-Rodríguez [2] used metagenomics to perform a comprehensive temporal study of the microbial community present at the solar salters of Cabo Rojo, Puerto Rico. Their findings revealed that the microbial diversity at genus level of this thalassohaline environment is stable through time, dominated by members of the *Euryarchaeota*, followed by *Bacteroidetes* and *Proteobacteria*. Functional annotation analysis of metagenomic sequences showed a diversity of metabolic genes related to nitrogen fixation, ammonia oxidation, sulfate reduction, sulfur oxidation, and phosphate solubilization. Binning methods allowed the reconstruction of four putative genomes belonging to novel species of Archaea and Bacteria.

2. Viruses Genomics

Viruses of halophilic archaea have been a subject of research for years. The obtained knowledge not only provides insights on how infection occurs at high salinity environments, but also it can be useful to develop genetic systems to study halophilic microorganisms. On that line, Dyall-Smith et al. [3] describes the novel myovirus *ChaoS9* which host is *Halobacterium salinarum*. The viral genome consists of a linear dsDNA with approximatedly 55 kb in length. This novel halovirus showed some relationship to PhiH1 and PhiCh1. The genome annotation and organization is also presented. On this direction, Dyall-Smith et al. [4] determined the genome sequence of the halovirus PhiH1 (ΦH1). Eventhough this myohalovirus was discovered in 1982, there is little information about its genome composition. The authors sequenced the genome and present the annotation of the 97 protein coding putative ORFs.

3. Transcriptomics

Transcriptomics studies are extremely useful in understanding the expression and repression of genes in an organism. Tafer et al. [5] combined the tools of genomics and transcriptomics to study the halophilic fungus *Aspergillus salisburgensis*. This organism was isolated at the salt mines of Austria where several extreme conditions exists (high salinity, low nutrient availability and darkness). This fungus was compared to *Aspergillus sclerotialis* which is a halotolerant strain of the genus. Genomic comparison showed several differences specifically at transport-related genes. Differences at gene expression and regulation at the transcriptomic level were also found. The work provides insights on what strategies fungi have develop to grow at extreme conditions specially at high salinity.

Antisense RNA (asRNA) can function in gene regulation in cells. de Almeida et al. [6] used a transcriptomic approach to map the primary antisense transcriptome of *Halobacterium salinarum* (sp. NRC-1). The researchers found that around 21% of the genes in *Halobacterium salinarum* contain asRNA. A further description of genes possesing this feature is presented as well as comparisons with *Haloferax volcanii* are established.

4. Recombination and DNA Modification Systems

Genome sequencing is useful in revealing species with geographic subpopulations, habitat specialization or high frequencies of recombination. With that in mind, Sun et al. [7] analyzed the genome sequences of 25 strains of *Wallemia mellicola*, a xerotolerant and halotolerant fungal species of widespread distribution in indoor and outdoor habitats. From Slovenian chocolate to the hypersaline waters of Spain, the researchers found the *W. mellicola* genome sequences to be relatively homogenous with no apparent clusters of strains based on habitat or geographic location. The authors suggest that *W. mellicola* strains undergo a reasonable amount of recombination shuffling between genomes of individual organisms and likely do this via sexual reproduction. This suggestion is based on phylogenetic analysis of core Benchmarking Universal Single-Copy Orthologues (BUSCOs), the density of single nucleotide polymorphisms (SNPs) and the identification of putative mating-type loci.

DNA methyltransferases (MTases) and restriction modification (RM) systems are important in a variety of functions including restricting foreign genomes and host DNA repair by recombination. In this special issue, Fullmer et al. [8] provide a survey of the distribution of RM system and orphan MTase gene homologs among halophilic Archaea of the class *Halobacteria*. One striking result was the irregular distribution of RM system candidate genes among the orders, genera, species, and even communities and populations of the *Halobacteria*. Based on this patchy distribution, the authors suggest that the RM systems are selfish genetic elements that undergo frequent horizontal gene transfer and gene loss. By contrast, the orphan MTase gene homologs were highly conserved and, thus, appeared functionally constrained among the *Halobacteria* lineages. Under-(CTAG) and over-(GATC) represented motifs were also identified in the genome sequences that may be targets of the MTase and RM systems.

5. Metabolism and Stress Responses

Halophilic microorganisms are considered a resource for industrial catalysts that function in organic solvent, high salt or other extreme conditions that most organisms cannot tolerate. Liao et al. [9] provide insight into an anaerobic haloalkaliphilic bacterium, *Alkalitalea saponilacus*, that uses xylan as a sole carbon and energy source and produces propionic acid as a major product. Microbes and the enzymes that hydrolze xylan (xylanases) are useful in the biobleaching of wood pulp as well as in the depolymerization of lignocellulosic biomass to generate renewable fuels and chemicals. In this special issue, the authors [9] find *A. saponilacus* secretes an extracellular fraction that hydrolyzes xylan in high salt and, through genomic sequencing, identify gene homologs relating to the pathway for complete xylan degradation. One future aim of this work is to develop a method to recover the xylanase for use in biobleaching wood pulp. Now that the genome sequence is available, genetic engineering may be an option for enhancing production of this halotolerant xylanase.

Halophilic archaea are masters at handling stress as these microbes thrive in hypersaline environments that promote hyperosmotic shock, desiccation, high UV exposure and other extreme factors. By screening a transposon mutant library of *Haloferax volcanii*, Gomez et al. [10] probed the molecular factors responsible for oxidative stress response. Transposon mutants hypertolerant of oxidant were isolated and found to have insertions at loci associated with post-translational modification, transport, polyamine biosynthesis, electron transfer and other cellular processes. As follow-up by markerless deletion, the authors demonstrated that cells producing 20S proteasomes of $\alpha 2$ and β (and not $\alpha 1$) subunits were more tolerant of oxidative stress than wild type. Thus, modulation of the subunit composition of one of the central proteolytic systems of these microbes (i.e., proteasomes) appears important in stress response.

Halophile genetics can provide understanding into the function of unusual groups of proteins as evidenced by the work of Nagel et al. [11]. In this work, the researchers examined the function of ORFs predicted to encode small proteins of less than 100 amino acids that harbor a zinc finger motif (Cys/His pattern of two Cys or His residues separated by two to three intermediate amino acids). Through systematic and targeted deletion, the researchers identified 12 ORFs encoding putative zinc finger proteins that were correlated with the ability of cells to adapt to stress, form biofilms and/or swarm. This type of approach offers a strong foundation for future studies to reveal how these zinc finger proteins may interact with DNA, RNA, proteins, lipids, and/or small molecules to alter the biological function of the cell.

Funding: J.M.F. work was supported in part by the U.S. Department of Energy, Office of Basic Energy Sciences, Division of Chemical Sciences, Geosciences and Biosciences, Physical Biosciences Program (DOE DE-FG02-05ER15650) to advance understanding of bioenergy, the National Institutes of Health (NIH R01 GM57498) to examine archaeal systems important to human health, and the National Science Foundation (MCB-1642283) for global analysis of halophile biology.

Conflicts of Interest: The authors declare that there is no conflict of interest concerning this work.

References

1. Uritskiy, G.; DiRuggiero, J. Applying genome-resolved metagenomics to deconvolute the halophilic microbiome. *Genes* **2019**, *10*, 220. [CrossRef] [PubMed]
2. Couto-Rodríguez, R.L.; Montalvo-Rodríguez, R. Temporal analysis of the microbial community from the crystallizer ponds in Cabo Rojo, Puerto Rico, using metagenomics. *Genes* **2019**, *10*, 422. [CrossRef] [PubMed]
3. Dyall-Smith, M.; Palm, P.; Wanner, G.; Witte, A.; Oesterhelt, D.; Pfeiffer, F. *Halobacterium salinarum* virus *ChaoS9*, a novel halovirus related to PhiH1 and PhiCh1. *Genes* **2019**, *10*, 194. [CrossRef] [PubMed]
4. Dyall-Smith, M.; Pfeifer, F.; Witte, A.; Oesterhelt, D.; Pfeiffer, F. Complete genome sequence of the model halovirus PhiH1 (ΦH1). *Genes* **2018**, *9*, 493. [CrossRef] [PubMed]
5. Tafer, H.; Poyntner, C.; Lopandic, K.; Sterflinger, K.; Piñar, G. Back to the salt mines: Genome and transcriptome comparisons of the halophilic fungus. *Genes* **2019**, *10*, 381. [CrossRef] [PubMed]
6. de Almeida, J.P.P.; Vêncio, R.Z.N.; Lorenzetti, A.P.R.; Caten, F.T.; Gomes-Filho, J.V.; Koide, T. The primary antisense transcriptome of *Halobacterium salinarum* NRC-1. *Genes* **2019**, *10*, 280. [CrossRef] [PubMed]
7. Sun, X.; Gostinčar, C.; Fang, C.; Zajc, J.; Hou, Y.; Song, Z.; Gunde-Cimerman, N. Genomic evidence of recombination in the basidiomycete *Wallemia mellicola*. *Genes* **2019**, *10*, 427. [CrossRef] [PubMed]
8. Fullmer, M.S.; Ouellette, M.; Louyakis, A.S.; Papke, R.T.; Gogarten, J.P. The patchy distribution of restriction modification system genes and the conservation of orphan methyltransferases in halobacteria. *Genes* **2019**, *10*, 233. [CrossRef] [PubMed]
9. Liao, Z.; Holtzapple, M.; Yan, Y.; Wang, H.; Li, J.; Zhao, B. Insights into xylan degradation and haloalkaline adaptation through whole-genome analysis of *Alkalitalea saponilacus*, an anaerobic haloalkaliphilic bacterium capable of secreting novel halostable xylanase. *Genes* **2018**, *10*, 1. [CrossRef] [PubMed]

10. Gomez, M.; Leung, W.; Dantuluri, S.; Pillai, A.; Gani, Z.; Hwang, S.; McMillan, L.J.; Kiljunen, S.; Savilahti, H.; Maupin-Furlow, J.A. Molecular factors of hypochlorite tolerance in the hypersaline archaeon *Haloferax volcanii*. *Genes* **2018**, *9*, 562. [CrossRef] [PubMed]

11. Nagel, C.; Machulla, A.; Zahn, S.; Soppa, J. Several one-domain zinc finger μ-proteins of *Haloferax volcanii* are important for stress adaptation, biofilm formation, and swarming. *Genes* **2019**, *10*, 361. [CrossRef] [PubMed]

Review

Applying Genome-Resolved Metagenomics to Deconvolute the Halophilic Microbiome

Gherman Uritskiy and Jocelyne DiRuggiero *

Department of Biology, Johns Hopkins University, Baltimore, MD 21218, USA; guritsk1@jhu.edu
* Correspondence: jdiruggiero@jhu.edu

Received: 15 February 2019; Accepted: 11 March 2019; Published: 14 March 2019

Abstract: In the past decades, the study of microbial life through shotgun metagenomic sequencing has rapidly expanded our understanding of environmental, synthetic, and clinical microbial communities. Here, we review how shotgun metagenomics has affected the field of halophilic microbial ecology, including functional potential reconstruction, virus–host interactions, pathway selection, strain dispersal, and novel genome discoveries. However, there still remain pitfalls and limitations from conventional metagenomic analysis being applied to halophilic microbial communities. Deconvolution of halophilic metagenomes has been difficult due to the high G + C content of these microbiomes and their high intraspecific diversity, which has made both metagenomic assembly and binning a challenge. Halophiles are also underrepresented in public genome databases, which in turn slows progress. With this in mind, this review proposes experimental and analytical strategies to overcome the challenges specific to the halophilic microbiome, from experimental designs to data acquisition and the computational analysis of metagenomic sequences. Finally, we speculate about the potential applications of other next-generation sequencing technologies in halophilic communities. RNA sequencing, long-read technologies, and chromosome conformation assays, not initially intended for microbiomes, are becoming available in the study of microbial communities. Together with recent analytical advancements, these new methods and technologies have the potential to rapidly advance the field of halophile research.

Keywords: extremophiles; halophilic microorganisms; hypersaline habitats; metagenomics; microbiome; shotgun sequencing; genome assembly and binning; functional annotation

---

1. Introduction

Microbial life is one of the most diverse and bioenergetically dominant forces in the earth's ecosphere [1], making microbiome research a critical component of modern ecology. The unparalleled taxonomic and functional diversity of microbial communities has allowed them to populate all locations on the planet [2,3], including environments unfit for colonization by other life forms. In hypersaline environments, unique environmental pressures have forced microbiota to evolve with specific survival adaptations, resulting in highly resilient communities that push the boundaries of life's limit (Figure 1). Halophiles have been found to play important roles in soil bioenergetic processes [4] and food storage and preservation [5,6], and have also been detected in the human gut microbiota [7]. Additionally, studying halophilic life forms has revealed many fundamental aspects of life's survival limits and strategies, including the potential to endure the harsh environments we are most likely to find on other planets [8,9]. Prior to the introduction of high-throughput sequencing, our understanding of halophile genomics was limited to studying cultured organisms [10,11]. While next-generation sequencing technologies have become commonplace in microbiology, the halophile field lacks a critical analysis of prospects and potential applications of these technologies in halophilic microbiomes.

Figure 1. Photographs of commonly studied hypersaline environments: (**A**) saltern flats, (**B**) halite nodules, (**C**) hypersaline microbial mats, (**D**) hypersaline lakes, (**E**) underwater haloclines, and (**F**) hypersaline soils. * Sources for images (free-to-use sources): https://commons.wikimedia.org/wiki/File:Salterns,_salt_making_fields,_tamil_nadu_-_panoramio.jpg, https://en.wikipedia.org/wiki/Phototrophic_biofilm#/media/File:Microbial_mat_section.jpg, https://commons.wikimedia.org/wiki/File:Saline_Lake_at_Ras_Mohamed_National_Park.jpg, https://commons.wikimedia.org/wiki/File:Halocline.png, https://pxhere.com/en/photo/1132612.

In this review, we discuss key aspects of halophile community composition and function that metagenomics has revealed and provide examples of studies in various hypersaline environments for a perspective on analytical progress. We then examine the advantages and limitations of applying shotgun metagenomic sequencing in uncovering the structure and function of halophilic microbiomes. We outline the factors and characteristics that make the deconvolution of halophilic metagenomes a major challenge and propose analytical adjustments to be made when investigating these complex communities. Both experimental design and computation analysis approaches that are appropriate in halophilic metagenomics are summarized. Finally, we discuss novel sequencing technologies that show promise in further propelling the halophile metagenomic field.

2. Shotgun Sequencing in Metagenomics

Rapid developments in high-throughput DNA sequencing technologies since the early 2000s have propelled our understanding of not only single-organism genetics, but also microbiome community structure and function [12]. Marker gene (particularly the *16S rRNA* gene) amplicon sequencing has revealed the taxonomic composition of a given community through sequencing a small target of the community's DNA. In contrast, whole-metagenomic sequencing (WMGS) theoretically allows for reconstruction of the entire microbial community's DNA content. This has led to a number of important findings in microbiome research [12–14], as biologists have been able to thoroughly investigate microbial communities at the genetic level without the need for culturing [15].

However, while sequencing technologies are rapidly developing, producing complete genomes of all the microorganisms found in a community is currently unattainable due to low sequencing coverage of the less abundant organisms. Additionally, sequence repeats and regions of homology between organisms limits genome recovery from short-read data, resulting in incomplete assemblies. Instead, long contiguous pieces (contigs) of genomes are produced, ranging in length from 1 Kbp to 1 Mbp [16,17]. These contigs then need to be grouped based on the genome they belong to, a process known as binning. It is only recently that binning has become reliable enough to produce reasonably high-quality metagenome-assembled genomes (MAGs). The ability to produce high-quality MAGs has in turn led to the discovery of thousands of novel organisms and has thus enabled many breakthroughs in characterizing the taxonomic and functional components of microbiomes [18–20].

Shotgun metagenomics offers tremendous advantages in recovering taxonomic and functional potential components of microbial communities, but sequencing costs deter some researchers from deploying this approach in their studies. The high average read coverage required for the assembly of a genome from shotgun reads [21] presents a major challenge for the assembly of less-abundant organisms in a metagenomic context. These highly diverse but underrepresented taxa often constitute significant proportions of microbial communities and play important roles in biome functioning [22]. Despite these challenges, WMGS carries tremendous benefits, empowering researchers to study previously unknown aspects of microbiomes. In particular, WMGS allows for the reconstruction of a given community's gene content, which has enabled ecologists to predict the functional potential of entire communities. This new angle of microbiome analysis has enabled the prediction of metabolic processes potentially present in communities and the study of community natural selection at the functional level [23,24]. The possibility of studying the functional potential of any organism in a community means that our understanding of microbial genetics, dynamics, evolution, and function is no longer limited to cultured organisms. In many fields, such as human microbiome research, this has hailed a new era for research [25,26].

2.1. Halophilic Microbiome Research Powered by Shotgun Metagenomics

Numerous breakthroughs in halophilic microbiome research have been enabled by WMGS [11] (Table 1). This sequencing approach reveals the taxonomic structure of microbiomes in high-salt environments with significantly less taxonomy-based biases than conventional ribosomal amplicon sequencing. Indeed, in conventional *16S rDNA* amplicon sequencing, primer choices can have a substantial impact on taxonomic distribution, and it is difficult to reliably amplify multiple domains of life, e.g., Bacteria and Archaea, with the same primer set [27]. While WMGS still has biases associated with G + C content, taxonomic annotation of shotgun reads usually results in more accurate and robust taxonomic profiles than amplicon sequencing [28]. This is particularly important in high-salt environments, where both Archaea and Bacteria are found in high abundance. For example, shotgun sequencing has provided more comprehensive taxonomic profiles of an endolithic halite community (Figure 1B) and the discovery that a unique algae was present in this community, in addition to Halobacteria, Cyanobacteria, and other heterotrophic bacteria [29]. In the study of a hypersaline lake (Figure 1D), the use of shotgun sequencing revealed the functional redundancy between taxonomically dissimilar communities constituted of both bacteria and archaea along a salinity gradient [30]. WMGS also provides DNA sequences that are not targeted by *16S rDNA* amplification, including eukaryotic genomes, DNA viruses, and extrachromosomal DNA, such as plasmids. For example, in a study investigating the community composition of saltern ponds (Figure 1A) along a salinity gradient, the use of metagenomics allowed access to both the cellular and viral components of the community within the same sequencing datasets, revealing increased virus abundance at higher salt concentrations [31].

The reconstruction of viral genomes from hypersaline environments [32] using WMGS has resulted in improved characterization of this major component of halophilic microbiomes. Viruses take on the vital role of predators in many microbiomes and contribute to nutrient turnover with their lytic activity [33,34]. While nonshotgun approaches have been used previously to characterize halophilic metaviromes [35,36], high-throughput sequencing has empowered a more streamlined and unbiased recovery and annotation of viral sequences from various types of high-salt environments (Table 1). For example, an investigation of the metavirome in deep-sea haloclines (Figure 1E) through nontargeted shotgun sequencing revealed the stratification of virus lineages along the salinity gradient of the haloclines, likely associated with their host specificity [37]. In WMGS from solar salterns (Figure 1A), perfect alignments between the CRISPR spacers of microorganisms and viral sequences have been used together with di- and trinucleotide frequencies to predict and validate host specificity among halophilic phages across several locations [38]. Another study looking at halophilic *Cyanobacteria* in endolithic communities (Figure 1B) used virus sequences encoded in CRISPR arrays as a high-sensitivity strain signature, which allowed for the tracking of strain dispersal in the region [39].

As previously mentioned, one of the biggest strengths of WMGS is the ability to reconstruct the functional potential of a microbial community. With WMGS, hypersaline water [8,40], soil [4], and endolithic [41] microbiomes have been characterized in terms of their metabolic function, particularly their ability to use a wide range of energy sources. In particular, building on previous culture-dependent methods, systematic functional analysis of halophilic metagenomes has led to major improvements in our understanding of halophile osmotic adaptation and evolution [42]. For example, longitudinal analysis of halite endolith (Figure 1B) microbiota after a heavy rainfall revealed metaproteome adaptations to the temporarily decreased salt concentrations [41]. Functional annotation of longitudinal studies of halophiles from saltern, hypersaline lake, and salt mineral environments has also led to the characterization of horizontal gene transfers, evolutionary dynamics, and functional adaptations across time and space [40,41,43,44]. Functional potential profiling has also uncovered selective pressures and community functional dynamics that were not possible to investigate through taxonomy alone due to high functional redundancy. For example, the investigation of metagenomes from hypersaline soils (Figure 1F) has allowed researchers to uncover core differences in the functioning of their communities compared to more homogeneous aquatic hypersaline environments, which stems from nutrient scarcity, limited mobility, and niche stratification [4]. In a metagenomic study of phototropic hypersaline microbial mats (Figure 1C), functional annotation and pathway quantitation led to a better understanding of energy and nutrient capture and cycling between layers of the mats [45]. In particular, identification of MAGs with complementary parts of nitrogen and sulfur metabolism pathways suggested a dependence on the metabolite exchange between community members. A functional potential investigation of microbial communities of solar saltern ponds (Figure 1A) revealed a higher prevalence of DNA replication and repair machinery in communities found in saturated brine compared to subsaturated saline environments [31]. With WMGS analysis rapidly improving and halophile databases rapidly growing [46], more breakthroughs will follow.

Another major aspect of metagenomics facilitated by WMGS is the reconstruction of novel individual genomes of halophiles. This is particularly important because extreme halophiles, and extremophiles in general, have been difficult to isolate due to specific growth condition requirements, symbiotic relationships, and cross-species functional pathways [47]. The binning of metagenomics assemblies has enabled researchers to recover hundreds of halophilic MAGs in the past decade [46], with many belonging to previously unknown orders, or even phyla [48]. For example, metagenomic binning of WMGS data from Lake Tyrel resulted in the recovery of near-complete genomes from a new clade of Nanohaloarchaea [49]. Similarly, metagenomic binning of solar saltern metagenomes uncovered several novel lineages of Euryarchaeota, Nanohaloarchaea, and Gammaproteobacteria. Functional annotation of these novel lineages allowed researchers to infer their metabolic functions within the microbiome [50]. In a halite endolith (Figure 1B) longitudinal study following a rare rain, community composition at the strain level was interrogated by genome-resolved metagenomics, leading to a general model of fine-scale taxonomic rearrangement of microbial communities following acute perturbations [41]. In addition to these individual discoveries, the rapidly increasing number of annotated reference halophile genomes allows for more accurate taxonomic and functional annotation in halophilic microbiomes, propelling the field in a positive-feedback loop [46].

Table 1. Studies that have contributed novel aspects of halophilic microbial communities through whole-metagenomic sequencing (WMGS) in hypersaline environments (list is not exhaustive). MAG: metagenome-assembled genome.

Environment	Longitudinal Dynamics	MAG Discovery	Functional Potential	Virus Analysis
Hypersaline lakes	Andrade [51], Tschitschko [44], Podell [52]	Narasingarao [49]	Vavourakis [53], Naghoni [30]	Emerson [54], Tschitschko [44], Ramos-Barbero [55]
Salterns	Plominsky [2]	Ramos-Barbero [56], Ghai [50]	Plominsky [31], Ghai [50]	Moller [38], Di Meglio [57]
Hypersaline microbial mats	Mobberley [45], Berlanga [58]	Mobberley [45]	Mobberley [45], Ruvindy [59], Wong [60]	White [61]
Haloclines	N/A	Speth [62]	Guan [63], Pachiadaki [64]	Antunes [37]
Halite endoliths	Uritskiy [41], Finstad [39]	Finstad [39], Uritskiy [41],	Crits-Christoph [65], Uritskiy [41]	Crits-Christoph [65]
Hypersaline soils	Narayan [66]	Vera-Gargallo [4]	Vera-Gargallo [4], Pandit [67]	NA

2.2. Limitations of Shotgun Metagenomics in Halophile Research

In contrast to human and synthetic microbiomes, the reconstruction of environmental metagenomes has been complicated by their sheer diversity and microdiversity. This is especially true in high-salt environments, which often host microbial communities with low taxonomic diversity but very high intraspecific diversity and characteristically high G + C content [68,69]. The presence of a large number of highly similar strains presents major challenges for deconvoluting their DNA content during metagenomic assembly and binning. This is particularly problematic in many halophiles that have genomic island regions of high inter-strain variability stemming from horizontal gene transfer [70,71]. On the other hand, the high G + C content of many dominant halophiles reduces the fraction of unique sequences in the samples [56,72], posing another challenge at the assembly stage. For example, halophilic endolith communities are typically dominated by Halobacteria and Salinibacter, but their high strain diversity and G + C content (over 60%) leads to relatively poor assembly and MAG quality [32]. In contrast, other community members that are less abundant and have low G + C content, such as Cyanobacteria, Actinobacteria, and Gammaproteobacteria, have yielded high-quality MAGs [41].

Due to the previously mentioned difficulties in culturing a diversity of halophiles, there are a relatively small number of genomes available. In 2018, there were just 942 complete halophile genomes available in NCBI databases [46], a tiny number in the era of high-throughput sequencing, which thus far has yielded over 200,000 prokaryotic complete genomes [73]. This leaves MAG extraction from environmental sequencing data the primary method for obtaining genomes of halophilic organisms, which has been difficult because of their metagenomic properties. In a negative feedback loop, this in turn has further stalled the progress of halophilic microbiome research, as the lack of available reference genomes has made taxonomic and functional annotation difficult. As WMGS becomes commonplace in microbiome research, it is crucial that the halophile field takes full advantage of the new technology and the use of newly available bioinformatic tools to further its understanding of microbial community assembly and function. Since 2014–2015, improvements in analytical methods and assembly software such as metaSPAdes [74], binning software such as metaBAT2 [75], and processing pipelines such as metaWRAP [18] have allowed for effective deconvolution of WMGS data from even the most complex microbiomes. These new analytical methods will greatly benefit the halophile research field, if applied effectively.

3. Experimental Design Considerations for Sequencing Halophilic Metagenomes

Obtaining MAG-level resolution in a metagenome enables more accurate and meaningful functional pathway and taxonomic annotation and allows for detailed analysis of specific members of the community. With this in mind, the end goal of many microbiome studies is accurate and complete binning of sequence data. There are two general approaches to metagenomic sequencing and analysis for this purpose: (1) co-assembly of multiple shallowly sequenced samples or (2) individual processing of a few deeply sequenced samples. Both approaches have their benefits and limitations, depending on the microbiome that is sequenced and the biological question to answer.

In the first approach, samples are sequenced with relatively low-read coverage, and reads from all samples are combined during metagenomic assembly (Figure 2A). In research projects that demand a large number of samples, such as longitudinal studies, this results in low sequencing costs per sample, while also producing high-quality MAGs from the co-assembly by leveraging differential abundances of the contigs across samples [18,75]. The taxonomic and functional composition of individual samples can be investigated by linking the taxonomic and functional annotations of each contig with its abundance in each sample, allowing for easy comparisons between large numbers of samples [41,43]. Finally, co-assembling data from multiple samples enhances the recovery of genomes from low-abundance organisms, which is not possible from individual samples due to low coverage [49]. However, the use of co-assembly in metagenomics comes with significant drawbacks [56], including the high computational costs of co-assembling large data and the high level of microdiversity introduced by each new biological replicate. This latter point might be counterintuitive, but it leads to poor assemblies of very abundant taxa because accumulated mismatches from strain heterogeneity complicate the De Bruijn graph during assembly. This is particularly problematic with halophilic microbiomes, which are often dominated by highly diverse groups of Euryarchaeota and Bacteroidetes [48]. The high population microdiversity of these taxa is exacerbated when using multiple biological replicates, which results in poor, fragmented, or chimeric assemblies [56]. This in turn translates into poor-quality MAGs. However, when a broad capture of community diversity across many samples is the intent of the study, these limitations should then be considered in data interpretation.

An alternative approach to co-assembly is to sequence a small number of samples with deep coverage and process them individually (Figure 2B). Because of the reduced microdiversity, individual assemblies produce larger contigs, given a comparable sequencing depth [76]. After binning each sample separately, MAGs can be combined into a single set through dereplication, removing duplicate MAGs that share a high nucleotide identity [77]. As with the co-assembly approach, differential contig coverage across samples may be used to improve binning results [40]. While this method is superior in highly heterogeneous communities such as halophilic microbiomes, it comes with a major increase in sequencing cost per sample. For most metagenomes, a meaningful assembly (N50 > 5 Kbp) requires 25–50 Gbp of sequencing data per sample, which limits the number of samples that can be multiplexed on a sequencing run. In turn, the limited replication reduces the effectiveness of binning, which leverages differential coverage of contigs across many samples to increase binning accuracy [78]. For many studies that require a large number of replicates, such as longitudinal studies, the cost of this approach may become prohibitively expensive.

A. Co-assembly binning

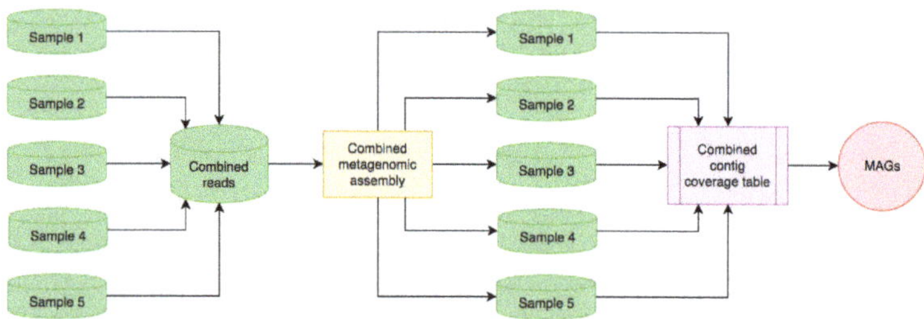

B. Individual binning

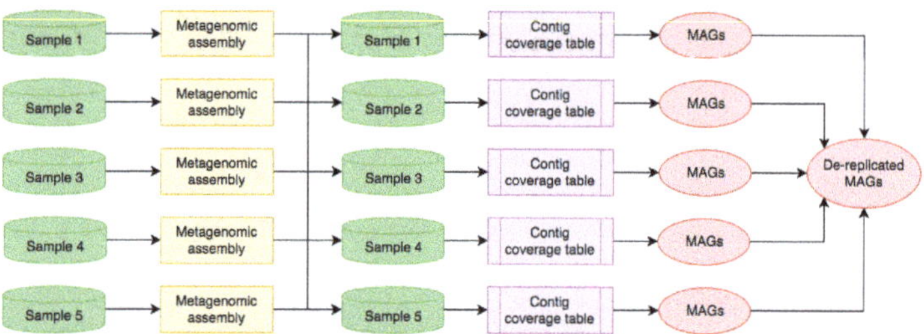

Figure 2. Flowcharts showing two common experimental designs and analysis workflows: (**A**) co-assembly and (**B**) individual sample processing and binning.

An additional consideration in choosing a strategy for metagenomic sequencing and analysis is that of intersample community diversity. Communities in aquatic biomes, such as hypersaline lakes or brine ponds, are often more homogenous, harboring the same microorganisms with different relative abundances at different sampling locations. Under those conditions, a co-assembly strategy for metagenomics, as discussed above, is often preferred [43,49,79]. In contrast, in terrestrial microbiomes with limited dispersal, such as halite nodules in salars of the Atacama Desert, which contain unique taxonomic compositions, an individual assembly approach is more advantageous [29,39]. Hybrid approaches are also possible in many cases, as binning of the individual and grouped assemblies may be combined and dereplicated to obtain the most robust MAGs of both rare and abundant species [80]. Regardless of the experimental design, it is critical to process samples, generate libraries, and sequence samples together to avoid batch effects [81]. If more than one flow cell is required to achieve the desired read depth, it is usually better to sequence the pooled libraries on several flow cells than to sequence each sample on its own flow cell [81]. For library preparation, it is recommended to use protocols that produce minimal G + C biases in coverage, particularly in halophilic communities that have high G + C content variation in their metagenomes [82,83].

The take-home message is that, when conducting a halophile metagenomic study, it is especially important to design a sampling and sequencing scheme with statistical questions in mind. Because of the high strain-level diversity typically found in halophilic microbiomes, an experimental design should avoid adding unnecessary replicates into the study, as each added biological replicate will introduce more microdiversity into the data, further complicating the assembly and binning stages of the analysis [56]. In practical terms, unless the intent of the study is to capture maximum diversity, the experimental design should include the minimum number of biological replicates that will allow for the intended statistical analysis downstream.

3.1. Best Bioinformatics Practices for Halophilic Metagenome Analysis

When processing halophilic metagenome sequencing data, it is important to adjust existing pipelines to accommodate for high intraspecific diversity, G + C content diversity, and underrepresentation in most sequence databases. While this section does not provide step-by-step instructions for bioinformatics analysis, it outlines core considerations and adjustments that should be made when processing halophilic metagenomes. Automated metagenomic analysis pipelines such as metaWRAP [18] or SqueezeM [84] may be used to streamline and simplify analysis: However, pipelines that are specifically designed for animal microbiomes, such as gut microbiota, should be avoided. Indeed, these latter pipelines rely strongly on pre-existing taxonomic and functional databases of closely related organisms, as the majority of organisms found in host-associated microbiomes have been sequenced and characterized.

The preprocessing of WMGS data, which typically includes read trimming, duplicate read removal, and metagenomic assembly, is standard for most types of metagenomes. We encourage testing a variety of software and comparing the results to evaluation programs such as FastQC [85] (for read quality) and MetaQUAST [86] (for assembly quality), as some methods may be more suited to specific microbial community types [87]. For metagenomic assembly, metaSPAdes [74] is currently considered to be the best overall, while MegaHIT [88] is a better solution when resources are a limiting factor, as it is significantly faster and requires less memory [89]. Thanks to recent improvements in assembly software, it is no longer necessary to subsample reads during this stage, as contig quality no longer drops off with increased read depth [89]. However, higher-quality assemblies of abundant organisms can be achieved through individual or grouped sample assembly, as described above.

In contrast to assembly, the annotation of halophilic metagenomes for taxonomies and functions can be somewhat compromised because halophiles have extremely limited representation in standard-distribution taxonomic databases [90,91], which introduces significant biases in sequence annotation. As of 2018, there were only 942 published complete halophilic genomes available in NCBI [46], the main database used as a reference in most taxonomic and functional annotation software. Regarding methods for taxonomic profiling, general alignment-based methods such as MegaBLAST [92] are usually too specific for annotating non-assembled halophilic DNA sequences because they rely on high sequence similarity and skew the annotation toward taxa that are better-represented in the database. To produce more balanced taxonomic annotations given the limited databases, it is recommended to assign taxonomies to assembled contigs based on the genes that they carry and then infer the taxonomy of reads based on their alignment with the contigs. If the intent is to obtain the most accurate taxonomic distribution profile of the community, extracting and annotating marker genes (such as *16S rRNA* genes) with EMIRGE is usually the best alternative [93], as rRNA gene databases are more established and encompass greater taxonomic diversity [94].

Functional annotation (the functional categorization of genes) in halophile metagenomes is also severely limited by existing databases, especially compared to human microbiomes. Because many halophilic genes are not annotated in NCBI databases, metagenome-inclusive custom or specific databases are preferred because they contain a greater variety of noncultured organisms. In particular, services such as the "Integrated Microbial Genomes and Microbiomes" systems from JGI [95] include taxonomic and functional annotation models that are trained on user-submitted metagenomic data, including high-quality MAGs. The annotation sensitivity resulting from using the newest metagenomic data is extremely valuable for both functional and taxonomic annotation in relatively understudied systems, such as halophilic microbiomes. Regardless of the database being used, it is important to regularly update to the most recent release, as new organisms are constantly being sequenced. Annotation pipelines geared toward human microbiomes such as HUMANN2 [96] should be avoided, as they rely on the presence of closely related organisms in databases.

For many metagenomic studies, an important objective is the genome-resolved description of the microbiome of interest, since the analysis of individual MAGs opens up many avenues for more accurate and meaningful functional pathway annotations and strain-level comparative metagenomics. To that extent, the success of metagenomic binning of assemblies depends greatly on software choice, as binning programs perform differently with various data types [18]. Additionally, many popular binning software programs, such as metaBAT1, are trained on gut microbiome data [75], potentially limiting their efficacy in complex halophilic communities. Furthermore, benchmarking of such algorithms is often done on real or synthetic gut microbial communities [87]. Because of this, it is recommended to bin the metagenomic assembly with a variety of the most recent binning software, such as metaBAT2 [75] and CONCOCT [97], and to use a binning consolidation tool, such as metaWRAP or DAS_Tool, to produce the best final bin set [18,98]. When estimating the read coverage of the contigs in a given sample to be fed into the binning algorithms, it is important to remember that they represent collapsed averages of a number of strains. Given the high intraspecific diversity of halophilic microbiomes [56], more accurate abundance estimations could potentially be obtained with slightly relaxed read alignment parameters, allowing for more approximate matches.

Considering the overwhelming number of metagenomic bioinformatics tools coming out each year, it is difficult to keep up to date with the best analytical methods. In general, we advise testing and benchmarking multiple software programs for each analytical step to determine the best option, as many conventionally used software programs behave unpredictably with halophilic sequence data. For annotation, emphasis should be placed on high sensitivity rather than high precision, given the database limitations.

3.2. The Future of Halophilic Metagenomics

Beyond shotgun sequencing of a microbiome's DNA content, there exist a number of other sequencing technologies that have become available and may further our understanding of halophilic ecosystems. Studies applying these technologies to more developed microbial fields, such as human gut microbiomes, have shown their great promise and their potential applications in halophilic microbial communities in the near future.

Conventional Illumina sequencing is limited to short DNA fragments (50 bp–250 bp), as errors accumulate rapidly at higher read lengths. However, read length, together with sequencing coverage, is undoubtedly a major limiting factor for metagenomics sequence assembly. Longer reads result in more accurate assembly and reduced chimeras, while they improve the contiguity of the assembly by allowing the assembly of repetitive DNA elements [99]. Recent sequencing technologies (minION from Oxford Nanopore and SMRT from PacBio sequencing) produce longer DNA fragments compared to Illumina. PacBio is able to consistently produce long reads (N50 up to 10 Kbp) with a relatively high degree of accuracy [100,101], while Nanopore sequencing produces even longer reads (N50 up to 100 Kbp), but with some sacrifices in accuracy [102,103]. Read lengths from these technologies enable not only the sequencing of complete ribosomal genes for improved taxonomic annotation, but also a significant improvement in the accuracy of metagenomics assembly and binning [101,104]. In highly diverse halophilic communities, long reads can help assemble ambiguous regions resulting from taxonomic heterogeneity, drastically improving the quality of the metagenome assembly [104]. Pseudo-single-cell technology from 10X Genomics, which tags each read with a barcode unique to the cell it came from, also shows great promise in halophilic microbiome deconvolution, as it is able to produce strain-specific synthetic long reads originating from single cells [105]. With reported maximum read lengths of over 1 Mbp from Nanopore, long-read technology is rapidly approaching the point where sequencing complete genomes in a single read is theoretically possible [106]. When this becomes reality, it will propel the field of metagenomics into a new post-assembly era. However, the recovery of less abundant taxa will remain a concern given the relatively low throughput of these methods.

Chromosome conformation capture with Hi-C is another technology that shows great promise in the field of halophilic metagenomics. A Hi-C assay crosslinks DNA based on spatial proximity: The chimeric segments resulting from the crosslink events are then sequenced, revealing sections of DNA that are proximal to each other. Conventionally used to indirectly measure the proximity between sections of a genome, Hi-C was successfully applied in 2017 to microbiomes to improve binning predictions [107]. Considering the difficulty of binning halophilic metagenomes due to their heterogeneity, Hi-C could significantly improve halophile MAG extraction. Hi-C-based binning also enables the recovery of extrachromosomal elements such as viral and plasmid DNA, which so far has been difficult to accomplish [108]. Hi-C can also be used to produce DNA proximity maps in individual MAGs for the study of chromatin conformation in prokaryotes at the metagenomic and single-cell scale [108].

Finally, genome-resolved metatranscriptomics (the analysis of a microbial community's RNA content) has been widely used in a variety of microbiomes to interrogate microbial transcriptional activities [25,109]. Metatranscriptomics has been used in halophile research to characterize carbon cycling in saline soils [110] and has been extensively used to characterize activity in other soil microbiomes [111,112]. However, it remains a largely underdeployed tool in many other high-salt systems, partly due to the difficulty in depleting ribosomal sequences in archaeal RNA. Another major deterrent has been the difficulty in standardizing transcript expression to the abundance of each individual organism in a sample. In other words, if a transcript is more abundant in a given sample, it can be difficult to determine if the organism carrying it is more abundant in that sample, or if it is truly highly expressed. However, with rapid improvements in genome-resolved metagenomic analysis of halophile communities, it is possible that the metatranscriptomic problem can be simplified down to more conventional transcriptome analysis by investigating the transcriptomes of individual MAGs.

4. Conclusion

Successful applications of whole metagenomics in halophilic communities has already led to numerous breakthroughs in our understanding of their functional composition, virus–host interactions, and strain diversity and dispersal, and has allowed for the genome extraction of previously unknown halophiles. However, the genomic qualities and composition characteristics of halophilic communities have made them difficult to deconvolute in a metagenomic context, limiting the information that can be extracted from halophilic shotgun metagenomes. Combined with relatively low numbers of cultures of halophiles, this has led to their underrepresentation in existing taxonomical and functional databases, which has further complicated analysis. While in silico deconvolution of halophilic metagenomes is a challenge, it can be accomplished with analysis workflows that account for the specific characteristics of halophile communities. With proper tuning, rapidly advancing sequencing technology has the potential to reconstruct the complete nucleic acid content of halophilic communities, allowing the halophile field to focus on microbial functional activity and interactions.

Acknowledgments: This work was supported by grants NNX15AP18G from NASA and DEB1556574 from the NSF. We thank Sarah Preheim for useful suggestions in the writing of this manuscript.

Conflicts of Interest: The authors declare no conflicts of interest.

References

1. Graham, E.B.; Knelman, J.E.; Schindlbacher, A.; Siciliano, S.; Breulmann, M.; Yannarell, A.; Beman, J.M.; Abell, G.; Philippot, L.; Prosser, J.; et al. Microbes as Engines of Ecosystem Function: When Does Community Structure Enhance Predictions of Ecosystem Processes? *Front. Microbiol.* **2016**, *7*, 111. [CrossRef] [PubMed]

2. Kallmeyer, J.; Pockalny, R.; Adhikari, R.R.; Smith, D.C.; D'Hondt, S. Global distribution of microbial abundance and biomass in subseafloor sediment. *Proc. Natl. Acad. Sci. USA* **2012**, *109*, 16213–16216. [CrossRef] [PubMed]

3. Whitman, W.B.; Coleman, D.C.; Wiebe, W.J. Prokaryotes: The unseen majority. *Proc. Natl. Acad. Sci. USA* **1998**, *95*, 6578–6583. [CrossRef] [PubMed]

4. Vera-Gargallo, B.; Ventosa, A. Metagenomic Insights into the Phylogenetic and Metabolic Diversity of the Prokaryotic Community Dwelling in Hypersaline Soils from the Odiel Saltmarshes (SW Spain). *Genes* **2018**, *9*, 152. [CrossRef] [PubMed]

5. Gibtan, A.; Park, K.; Woo, M.; Shin, J.-K.; Lee, D.-W.; Sohn, J.H.; Song, M.; Roh, S.W.; Lee, S.-J.; Lee, H.-S. Diversity of Extremely Halophilic Archaeal and Bacterial Communities from Commercial Salts. *Front. Microbiol.* **2017**, *8*, 631. [CrossRef] [PubMed]

6. Henriet, O.; Fourmentin, J.; Delincé, B.; Mahillon, J. Exploring the diversity of extremely halophilic archaea in food-grade salts. *Int. J. Food Microbiol.* **2014**, *191*, 36–44. [CrossRef] [PubMed]

7. Seck, E.H.; Dufour, J.-C.; Raoult, D.; Lagier, J.-C. Halophilic & halotolerant prokaryotes in humans. *Future Microbiol.* **2018**, *13*, 799–812. [PubMed]

8. Oren, A. Halophilic archaea on Earth and in space: Growth and survival under extreme conditions. *Philos. Trans. R. Soc. A Math. Phys. Eng. Sci.* **2014**, *372*, 20140194. [CrossRef] [PubMed]

9. Ma, Y.; Galinski, E.A.; Grant, W.D.; Oren, A.; Ventosa, A. Halophiles 2010: Life in Saline Environments. *Appl. Environ. Microbiol.* **2010**, *76*, 6971–6981. [CrossRef] [PubMed]

10. Rinke, C.; Schwientek, P.; Sczyrba, A.; Ivanova, N.N.; Anderson, I.J.; Cheng, J.-F.; Darling, A.; Malfatti, S.; Swan, B.K.; Gies, E.A.; et al. Insights into the phylogeny and coding potential of microbial dark matter. *Nature* **2013**, *499*, 431–437. [CrossRef] [PubMed]

11. Hedlund, B.P.; Dodsworth, J.A.; Murugapiran, S.K.; Rinke, C.; Woyke, T. Impact of single-cell genomics and metagenomics on the emerging view of extremophile "microbial dark matter.". *Extremophiles* **2014**, *18*, 865–875. [CrossRef] [PubMed]

12. Riesenfeld, C.S.; Schloss, P.D.; Handelsman, J. METAGENOMICS: Genomic Analysis of Microbial Communities. *Annu. Rev. Genet.* **2004**, *38*, 525–552. [CrossRef] [PubMed]

13. Ranjan, R.; Rani, A.; Metwally, A.; McGee, H.S.; Perkins, D.L. Analysis of the microbiome: Advantages of whole genome shotgun versus 16S amplicon sequencing. *Biochem. Biophys. Res. Commun.* **2015**, *469*, 967–977. [CrossRef] [PubMed]

14. Tessler, M.; Neumann, J.S.; Afshinnekoo, E.; Pineda, M.; Hersch, R.; Velho, L.F.M.; Segovia, B.T.; Lansac-Toha, F.A.; Lemke, M.; DeSalle, R.; et al. Large-scale differences in microbial biodiversity discovery between 16S amplicon and shotgun sequencing. *Sci Rep.* **2017**, *7*, 6589. [CrossRef] [PubMed]

15. Quince, C.; Walker, A.W.; Simpson, J.T.; Loman, N.J.; Segata, N. Corrigendum: Shotgun metagenomics, from sampling to analysis. *Nat. Biotechnol.* **2017**, *35*, 1211. [CrossRef] [PubMed]

16. Ghurye, J.S.; Cepeda-Espinoza, V.; Pop, M. Metagenomic Assembly: Overview, Challenges and Applications. *Yale J. Biol Med.* **2016**, *89*, 353–362. [PubMed]

17. Olson, N.D.; Treangen, T.J.; Hill, C.M.; Cepeda-Espinoza, V.; Ghurye, J.; Koren, S.; Pop, M. Metagenomic assembly through the lens of validation: Recent advances in assessing and improving the quality of genomes assembled from metagenomes. *Brief. Bioinform.* **2017**. [CrossRef] [PubMed]

18. Uritskiy, G.V.; DiRuggiero, J.; Taylor, J. MetaWRAP—a flexible pipeline for genome-resolved metagenomic data analysis. *Microbiome* **2018**, *6*, 158. [CrossRef] [PubMed]

19. Tully, B.J.; Graham, E.D.; Heidelberg, J.F. The reconstruction of 2,631 draft metagenome-assembled genomes from the global oceans. *Sci. Data* **2018**, *5*, 170203. [CrossRef] [PubMed]

20. Sangwan, N.; Xia, F.; Gilbert, J.A. Recovering complete and draft population genomes from metagenome datasets. *Microbiome* **2016**, *4*, 197. [CrossRef] [PubMed]

21. Sims, D.; Sudbery, I.; Ilott, N.E.; Heger, A.; Ponting, C.P. Sequencing depth and coverage: Key considerations in genomic analyses. *Nat. Rev. Genet.* **2014**, *15*, 121–132. [CrossRef] [PubMed]

22. Zaheer, R.; Noyes, N.; Polo, R.O.; Cook, S.R.; Marinier, E.; Van Domselaar, G.; Belk, K.E.; Morley, P.S.; McAllister, T.A. Impact of sequencing depth on the characterization of the microbiome and resistome. *Sci. Rep.* **2018**, *8*, 5890. [CrossRef] [PubMed]

23. Sharifi, F.; Ye, Y. From Gene Annotation to Function Prediction for Metagenomics. *Alcohol* **2017**, *1611*, 27–34.

24. Zhang, Y.; Wang, J.; Qi, J.; Zhao, F.; He, S.; Wei, S. Metagenomic sequencing reveals microbiota and its functional potential associated with periodontal disease. *Sci. Rep.* **2013**, *3*, 1843.

25. Wang, W.-L.; Xu, S.-Y.; Ren, Z.-G.; Tao, L.; Jiang, J.-W.; Zheng, S.-S. Application of metagenomics in the human gut microbiome. *WJG* **2015**, *21*, 803–814. [CrossRef] [PubMed]

26. Quince, C.; Walker, A.W.; Simpson, J.T.; Loman, N.J.; Segata, N. Shotgun metagenomics, from sampling to analysis. *Nat. Biotechnol.* **2017**, *35*, 833–844. [CrossRef] [PubMed]

27. Poretsky, R.; Rodríguez-R, L.M.; Luo, C.; Tsementzi, D.; Konstantinidis, K.T. Strengths and Limitations of 16S rRNA Gene Amplicon Sequencing in Revealing Temporal Microbial Community Dynamics. *PLoS ONE* **2014**, *9*, e93827. [CrossRef] [PubMed]

28. White, J.R.; Nagarajan, N.; Pop, M. Statistical Methods for Detecting Differentially Abundant Features in Clinical Metagenomic Samples. *PLoS Comput. Biol.* **2009**, *5*, e1000352. [CrossRef] [PubMed]

29. Crits-Christoph, A.; Gelsinger, D.R.; Ma, B.; Wierzchos, J.; Ravel, J.; Davila, A.; Casero, M.C.; DiRuggiero, J. Functional interactions of archaea, bacteria, and viruses in a hypersaline endolithic community. *Environm. Microbiol.* **2016**, *18*, 2064–2077. [CrossRef] [PubMed]

30. Naghoni, A.; Emtiazi, G.; Amoozegar, M.A.; Cretoiu, M.S.; Stal, L.J.; Etemadifar, Z.; Fazeli, S.A.S.; Bolhuis, H. Microbial diversity in the hypersaline Lake Meyghan, Iran. *Sci. Rep.* **2017**, *7*, 11522. [CrossRef] [PubMed]

31. Plominsky, A.M.; Henríquez-Castillo, C.; Delherbe, N.; Podell, S.; Ramirez-Flandes, S.; Ugalde, J.A.; Santibañez, J.F.; Engh, G.V.D.; Hanselmann, K.; Ulloa, O.; et al. Distinctive Archaeal Composition of an Artisanal Crystallizer Pond and Functional Insights Into Salt-Saturated Hypersaline Environment Adaptation. *Front. Microbiol.* **2018**, *9*, 1800. [CrossRef] [PubMed]

32. Roux, S.; Enault, F.; Ravet, V.; Colombet, J.; Bettarel, Y.; Auguet, J.; Bouvier, T.; Lucas-Staat, S.; Vellet, A.; Prangishivili, D.; et al. Analysis of metagenomic data reveals common features of halophilic viral communities across continents. *Environ. Microbiol.* **2015**, *18*, 889–903. [CrossRef] [PubMed]

33. Pedrós-Alió, C.; Calderón-Paz, J.I.; MacLean, M.H.; Medina, G.; Marrasé, C.; Gasol, J.M.; Guixa-Boixereu, N. The microbial food web along salinity gradients. *FEMS Microbiol. Ecol.* **2000**, *32*, 143–155. [CrossRef]

34. Guixa-Boixareu, N.; Calderón-Paz, J.; Heldal, M.; Bratbak, G.; Pedrós-Alió, C. Viral lysis and bacterivory as prokaryotic loss factors along a salinity gradient. *Aquat. Microb. Ecol.* **1996**, *11*, 215–227. [CrossRef]

35. Santos, F.; Meyerdierks, A.; Peña, A.; Amann, R.; Antón, J.; Rosselló-Mora, R.; Rosselló-Mora, R.; Rosselló-Móra, R. Metagenomic approach to the study of halophages: The environmental halophage 1. *Environ. Microbiol.* **2007**, *9*, 1711–1723. [CrossRef] [PubMed]

36. Santos, F.; Yarza, P.; Parro, V.; Briones, C.; Antón, J. The metavirome of a hypersaline environment. *Environ. Microbiol.* **2010**, *12*, 2965–2976. [CrossRef] [PubMed]

37. Antunes, A.; Alam, I.; Simões, M.F.; Daniels, C.; Ferreira, A.J.; Siam, R.; El-Dorry, H.; Bajic, V.B. First Insights into the Viral Communities of the Deep-sea Anoxic Brines of the Red Sea. *Genomics Prot. Bioinform.* **2015**, *13*, 304–309.

38. Moller, A.G.; Liang, C. Determining virus-host interactions and glycerol metabolism profiles in geographically diverse solar salterns with metagenomics. *PeerJ* **2017**, *5*, e2844. [CrossRef] [PubMed]

39. Finstad, K.M.; Probst, A.J.; Thomas, B.C.; Andersen, G.L.; Demergasso, C.; Echeverría, A.; Amundson, R.G.; Banfield, J.F. Microbial Community Structure and the Persistence of Cyanobacterial Populations in Salt Crusts of the Hyperarid Atacama Desert from Genome-Resolved Metagenomics. *Front. Microbiol.* **2017**, *8*, 1435. [CrossRef] [PubMed]

40. Kimbrel, J.A.; Ballor, N.; Wu, Y.-W.; David, M.M.; Hazen, T.C.; Simmons, B.A.; Singer, S.W.; Jansson, J.K. Microbial Community Structure and Functional Potential Along a Hypersaline Gradient. *Front. Microbiol.* **2018**, *9*, 1492. [CrossRef] [PubMed]

41. Uritskiy, G.; Getsin, S.; Munn, A.; Gomez-Silva, B.; Davila, A.; Glass, B.; Taylor, J.; DiRuggiero, J. Response of extremophile microbiome to a rare rainfall reveals a two-step adaptation mechanism. *bioRxiv* **2018**, *bioRxiv*, 442525.

42. Becker, E.A.; Seitzer, P.M.; Tritt, A.; Larsen, D.; Krusor, M.; Yao, A.I.; Wu, D.; Madern, D.; Eisen, J.A.; Darling, A.E.; et al. Phylogenetically Driven Sequencing of Extremely Halophilic Archaea Reveals Strategies for Static and Dynamic Osmo-response. *PLoS Genet.* **2014**, *10*, e1004784. [CrossRef] [PubMed]

43. DeMaere, M.Z.; Williams, T.J.; Allen, M.A.; Brown, M.V.; Gibson, J.A.E.; Rich, J.; Lauro, F.M.; Dyall-Smith, M.; Davenport, K.W.; Woyke, T.; et al. High level of intergenera gene exchange shapes the evolution of haloarchaea in an isolated Antarctic lake. *Proc. Natl. Acad. Sci. USA* **2013**, *110*, 16939–16944. [CrossRef] [PubMed]

44. Tschitschko, B.; Erdmann, S.; Roux, S.; Panwar, P.; Brazendale, S.; Cavicchioli, R.; DeMaere, M.Z.; A Allen, M.; Williams, T.J.; Hancock, A.M.; et al. Genomic variation and biogeography of Antarctic haloarchaea. *Microbiome* **2018**, *6*, 113. [CrossRef] [PubMed]

45. Mobberley, J.M.; Lindemann, S.R.; Bernstein, H.C.; Moran, J.J.; Renslow, R.S.; Babauta, J.; Hu, D.; Beyenal, H.; Nelson, W.C. Organismal and spatial partitioning of energy and macronutrient transformations within a hypersaline mat. *FEMS Microbiol. Ecol.* **2017**, *93*. [CrossRef] [PubMed]

46. Loukas, A.; Kappas, I.; Abatzopoulos, T.J. HaloDom: A new database of halophiles across all life domains. *J. Biol. Res. Thessalon.* **2018**, *25*, 2. [CrossRef] [PubMed]

47. Solden, L.; Lloyd, K.; Wrighton, K. The bright side of microbial dark matter: Lessons learned from the uncultivated majority. *Curr. Opin. Microbiol.* **2016**, *31*, 217–226. [CrossRef] [PubMed]

48. Ventosa, A.; De La Haba, R.R.; Sánchez-Porro, C.; Papke, R.T. Microbial diversity of hypersaline environments: A metagenomic approach. *Curr. Opin. Microbiol.* **2015**, *25*, 80–87. [CrossRef] [PubMed]

49. Narasingarao, P.; Podell, S.; A Ugalde, J.; Brochier-Armanet, C.; Emerson, J.B.; Brocks, J.J.; Heidelberg, K.B.; Banfield, J.F.; E Allen, E. De novo metagenomic assembly reveals abundant novel major lineage of Archaea in hypersaline microbial communities. *ISME J.* **2011**, *6*, 81–93. [CrossRef] [PubMed]

50. Pašić, L.; Fernández, A.B.; Martin-Cuadrado, A.-B.; Papke, R.T.; Rodriguez-Brito, B.; Ghai, R.; Mizuno, C.M.; McMahon, K.D.; Stepanauskas, R.; Rohwer, F.; et al. New Abundant Microbial Groups in Aquatic Hypersaline Environments. *Sci. Rep.* **2011**, *1*, 135.

51. Andrade, K.; Logemann, J.; Heidelberg, K.B.; Emerson, J.B.; Comolli, L.R.; A Hug, L.; Probst, A.J.; Keillar, A.; Thomas, B.C.; Miller, C.S.; et al. Metagenomic and lipid analyses reveal a diel cycle in a hypersaline microbial ecosystem. *ISME J.* **2015**, *9*, 2697–2711. [CrossRef] [PubMed]

52. Podell, S.; Ugalde, J.A.; Narasingarao, P.; Banfield, J.F.; Heidelberg, K.B.; Allen, E.E. Assembly-Driven Community Genomics of a Hypersaline Microbial Ecosystem. *PLoS ONE* **2013**, *8*, e61692. [CrossRef] [PubMed]

53. Vavourakis, C.D.; Andrei, A.-S.; Mehrshad, M.; Ghai, R.; Sorokin, D.Y.; Muyzer, G. A metagenomics roadmap to the uncultured genome diversity in hypersaline soda lake sediments. *Microbiome* **2018**, *6*, 168. [CrossRef] [PubMed]

54. Emerson, J.B.; Andrade, K.; Thomas, B.C.; Norman, A.; Allen, E.E.; Heidelberg, K.B.; Banfield, J.F. Virus-host and CRISPR dynamics in Archaea-dominated hypersaline Lake Tyrrell, Victoria, Australia. *Archaea* **2013**, 370871. [CrossRef] [PubMed]

55. Ramos-Barbero, M.D.; Martínez, J.M.; Almansa, C.; Rodríguez, N.; Villamor, J.; Gomariz, M.; Escudero, C.; Rubin, S.; Antón, J.; Martínez-García, M.; et al. Prokaryotic and viral community structure in the singular chaotropic salt lake salar de uyuni. *Environ. Microbiol.* **2019**. [CrossRef] [PubMed]

56. Ramos-Barbero, M.D.; Martin-Cuadrado, A.-B.; Viver, T.; Santos, F.; Martinez-Garcia, M.; Antón, J. Recovering microbial genomes from metagenomes in hypersaline environments: The Good, the Bad and the Ugly. *Syst. Appl. Microbiol.* **2019**, *42*, 30–40. [CrossRef] [PubMed]

57. Di Meglio, L.; Santos, F.; Gomariz, M.; Almansa, C.; López, C.; Antón, J.; Nercessian, D. Seasonal dynamics of extremely halophilic microbial communities in three Argentinian salterns. *FEMS Microbiol. Ecol.* **2016**, *92*, fiw184. [CrossRef] [PubMed]

58. Berlanga, M.; Palau, M.; Guerrero, R. Functional Stability and Community Dynamics during Spring and Autumn Seasons Over 3 Years in Camargue Microbial Mats. *Front. Microbiol.* **2017**, *8*, 2619. [CrossRef] [PubMed]

59. Ruvindy, R.; White, R.A., 3rd; Neilan, B.A..; Burns, B.P. Unravelling core microbial metabolisms in the hypersaline microbial mats of Shark Bay using high-throughput metagenomics. *ISME J.* **2016**, *10*, 183–196. [CrossRef] [PubMed]

60. Wong, H.L.; White, R.A.; Visscher, P.T.; Charlesworth, J.C.; Vázquez-Campos, X.; Burns, B.P. Disentangling the drivers of functional complexity at the metagenomic level in Shark Bay microbial mat microbiomes. *ISME J.* **2018**, *12*, 2619–2639. [CrossRef] [PubMed]

61. White, R.A.; Wong, H.L.; Ruvindy, R.; Neilan, B.A.; Burns, B.P.; Iii, R.A.W. Viral Communities of Shark Bay Modern Stromatolites. *Front. Microbiol.* **2018**, *9*, 1223. [CrossRef] [PubMed]

62. Speth, D.R.; Lagkouvardos, I.; Wang, Y.; Qian, P.-Y.; Dutilh, B.E.; Jetten, M.S.M. Draft Genome of Scalindua rubra, Obtained from the Interface Above the Discovery Deep Brine in the Red Sea, Sheds Light on Potential Salt Adaptation Strategies in Anammox Bacteria. *Microb. Ecol.* **2017**, *74*, 1–5. [CrossRef] [PubMed]

63. Guan, Y.; Hikmawan, T.; Antunes, A.; Ngugi, D.; Stingl, U. Diversity of methanogens and sulfate-reducing bacteria in the interfaces of five deep-sea anoxic brines of the Red Sea. *Res. Microbiol.* **2015**, *166*, 688–699. [CrossRef] [PubMed]

64. Pachiadaki, M.G.; Yakimov, M.M.; Lacono, V.; Leadbetter, E.; Edgcomb, V. Unveiling microbial activities along the halocline of Thetis, a deep-sea hypersaline anoxic basin. *ISME J.* **2014**, *8*, 2478–2489. [CrossRef] [PubMed]

65. Crits-Christoph, A.; Robinson, C.K.; Ma, B.; Ravel, J.; Wierzchos, J.; Ascaso, C.; Artieda, O.; Souza-Egipsy, V.; Casero, M.C.; DiRuggiero, J. Phylogenetic and Functional Substrate Specificity for Endolithic Microbial Communities in Hyper-Arid Environments. *Front. Microbiol.* **2016**, *7*, 301. [CrossRef] [PubMed]

66. Narayan, A.; Patel, V.; Singh, P.; Patel, A.; Jain, K.; Karthikeyan, K.; Shah, A.; Madamwar, D. Response of microbial community structure to seasonal fluctuation on soils of Rann of Kachchh, Gujarat, India: Representing microbial dynamics and functional potential. *Ecol. Genet. Genomics* **2018**, *6*, 22–32. [CrossRef]

67. Pandit, A.S.; Joshi, M.N.; Bhargava, P.; Shaikh, I.; Ayachit, G.N.; Raj, S.R.; Saxena, A.K.; Bagatharia, S.B. A snapshot of microbial communities from the Kutch: One of the largest salt deserts in the World. *Extremophiles* **2015**, *19*, 973–987. [CrossRef] [PubMed]

68. Cuadros-Orellana, S.; Martin-Cuadrado, A.-B.; Legault, B.; D'Auria, G.; Zhaxybayeva, O.; Papke, R.T.; Rodriguez-Valera, F. Genomic plasticity in prokaryotes: The case of the square haloarchaeon. *ISME J.* **2007**, *1*, 235–245. [CrossRef] [PubMed]

69. Papke, R.T.; Koenig, J.E.; Rodriguez-Valera, F.; Doolittle, W.F. Frequent Recombination in a Saltern Population of Halorubrum. *Science* **2004**, *306*, 1928–1929. [PubMed]

70. Pašić, L.; Rodriguez-Mueller, B.; Martin-Cuadrado, A.-B.; Rodriguez-Valera, F.; Mira, A.; Rohwer, F. Metagenomic islands of hyperhalophiles: The case of Salinibacter ruber. *BMC Genomics* **2009**, *10*, 570. [CrossRef] [PubMed]

71. Martin-Cuadrado, A.-B.; Pašić, L.; Rodriguez-Valera, F. Diversity of the cell-wall associated genomic island of the archaeon Haloquadratum walsbyi. *BMC Genomics* **2015**, *16*, 13950. [CrossRef] [PubMed]

72. Chen, Y.-C.; Liu, T.; Yu, C.-H.; Chiang, T.-Y.; Hwang, C.-C. Effects of GC bias in next-generation-sequencing data on de novo genome assembly. *PLoS ONE* **2013**, *8*, e62856. [CrossRef] [PubMed]

73. Haft, D.H.; DiCuccio, M.; Badretdin, A.; Brover, V.; Chetvernin, V.; O'Neill, K.; Li, W.; Chitsaz, F.; Derbyshire, M.K.; Gonzales, N.R.; et al. RefSeq: An update on prokaryotic genome annotation and curation. *Nucleic Acids Res.* **2017**, *46*, D851–D860. [CrossRef] [PubMed]

74. Nurk, S.; Meleshko, D.; Korobeynikov, A.; Pevzner, P.A. metaSPAdes: A new versatile metagenomic assembler. *Genome Res.* **2017**, *27*, 824–834. [CrossRef] [PubMed]

75. Kang, D.D.; Froula, J.; Egan, R.; Wang, Z.; Rahmann, S. MetaBAT, an efficient tool for accurately reconstructing single genomes from complex microbial communities. *PeerJ* **2015**, *3*, 1165. [CrossRef] [PubMed]

76. Haro-Moreno, J.M.; López-Pérez, M.; De La Torre, J.R.; Picazo, A.; Camacho, A.; Rodriguez-Valera, F. Fine metagenomic profile of the Mediterranean stratified and mixed water columns revealed by assembly and recruitment. *Microbiome* **2018**, *6*, 128. [CrossRef] [PubMed]

77. Olm, M.R.; Brown, C.T.; Brooks, B.; Banfield, J.F. dRep: A tool for fast and accurate genomic comparisons that enables improved genome recovery from metagenomes through de-replication. *ISME J.* **2017**, *11*, 2864–2868. [CrossRef] [PubMed]

78. Goodrich, J.K.; Di Rienzi, S.C.; Poole, A.C.; Koren, O.; Walters, W.A.; Caporaso, J.G.; Knight, R.; Ley, R.E. Conducting a microbiome study. *Cell* **2014**, *158*, 250–262. [CrossRef] [PubMed]

79. Vavourakis, C.D.; Ghai, R.; Rodríguez-Valera, F.; Sorokin, D.Y.; Tringe, S.G.; Hugenholtz, P.; Muyzer, G. Metagenomic Insights into the Uncultured Diversity and Physiology of Microbes in Four Hypersaline Soda Lake Brines. *Front. Microbiol.* **2016**, *7*, 533. [CrossRef] [PubMed]

80. Stewart, R.D.; Auffret, M.D.; Warr, A.; Wiser, A.H.; Press, M.O.; Langford, K.W.; Liachko, I.; Snelling, T.J.; Dewhurst, R.J.; Walker, A.W.; et al. Assembly of 913 microbial genomes from metagenomic sequencing of the cow rumen. *Nat. Commun* **2018**, *9*, 870. [CrossRef] [PubMed]

81. Gibbons, S.M.; Duvallet, C.; Alm, E.J. Correcting for batch effects in case-control microbiome studies. *PLoS Comput. Biol.* **2018**, *14*, e1006102. [CrossRef] [PubMed]

82. Paul, S.; Bag, S.K.; Das, S.; Harvill, E.T.; Dutta, C. Molecular signature of hypersaline adaptation: Insights from genome and proteome composition of halophilic prokaryotes. *Genome Biol.* **2008**, *9*. [CrossRef] [PubMed]

83. Jones, M.B.; Highlander, S.K.; Anderson, E.L.; Li, W.; Dayrit, M.; Klitgord, N.; Fabani, M.M.; Seguritan, V.; Green, J.; Pride, D.T.; et al. Library preparation methodology can influence genomic and functional predictions in human microbiome research. *Proc. Natl. Acad. Sci. USA* **2015**, *112*, 14024–14029. [CrossRef] [PubMed]

84. Tamames, J.; Puente-Sanchez, F. SqueezeM, a highly portable, fully automatic metagenomic analysis pipeline. *Bioinformatics* **2018**, *bioRxiv*, 347559.

85. Brown, J.; Pirrung, M.; McCue, L.A. FQC Dashboard: Integrates FastQC results into a web-based, interactive, and extensible FASTQ quality control tool. *Bioinformatics* **2017**, *33*, 3137–3139. [CrossRef] [PubMed]

86. Mikheenko, A.; Saveliev, V.; Gurevich, A. MetaQUAST: Evaluation of metagenome assemblies. *Bioinformatics* **2015**, *32*, 1088–1090. [CrossRef] [PubMed]

87. Sczyrba, A.; Hofmann, P.; Belmann, P.; Koslicki, D.; Janssen, S.; Dröge, J.; Gregor, I.; Majda, S.; Fiedler, J.; Dahms, E.; et al. Critical Assessment of Metagenome Interpretation-a benchmark of metagenomics software. *Nat. Methods* **2017**, *14*, 1063–1071. [CrossRef] [PubMed]

88. Li, D.; Luo, R.; Liu, C.-M.; Leung, C.-M.; Ting, H.-F.; Sadakane, K.; Yamashita, H.; Lam, T.-W. MEGAHIT v1.0: A fast and scalable metagenome assembler driven by advanced methodologies and community practices. *Methods* **2016**, *102*, 3–11. [CrossRef] [PubMed]

89. Vollmers, J.; Wiegand, S.; Kaster, A.-K. Comparing and Evaluating Metagenome Assembly Tools from a Microbiologist's Perspective—Not Only Size Matters! *PLoS ONE* **2017**, *12*, e0169662. [CrossRef] [PubMed]

90. Wheeler, D.L. Database resources of the National Center for Biotechnology Information. *Nucleic Acids Res.* **2001**, *29*, 11–16. [CrossRef] [PubMed]

91. O'Leary, N.A.; Wright, M.W.; Brister, J.R.; Ciufo, S.; Haddad, D.; McVeigh, R.; Rajput, B.; Robbertse, B.; Smith-White, B.; Ako-Adjei, D.; et al. Reference sequence (RefSeq) database at NCBI: Current status, taxonomic expansion, and functional annotation. *Nucleic Acids Res.* **2015**, *44*, D733–D745. [CrossRef] [PubMed]

92. Chen, Y.; Ye, W.; Zhang, Y.; Xu, Y. High speed BLASTN: An accelerated MegaBLAST search tool. *Nucleic Acids Res.* **2015**, *43*, 7762–7768. [CrossRef] [PubMed]

93. Miller, C.S.; Baker, B.J.; Thomas, B.C.; Singer, S.W.; Banfield, J.F. EMIRGE: Reconstruction of full-length ribosomal genes from microbial community short read sequencing data. *Genome Biol.* **2011**, *12*, R44. [CrossRef] [PubMed]

94. Quast, C.; Pruesse, E.; Yilmaz, P.; Gerken, J.; Schweer, T.; Yarza, P.; Peplies, J.; Glockner, F.O. The SILVA ribosomal RNA gene database project: Improved data processing and web-based tools. *Nucleic Acids Res.* **2012**, *41*, D590–D596. [CrossRef] [PubMed]

95. Chen, I.A.; Markowitz, V.M.; Chu, K.; Palaniappan, K.; Szeto, E.; Pillay, M.; Ratner, A.; Huang, J.; Andersen, E.; Huntemann, M.; et al. IMG/M: Integrated genome and metagenome comparative data analysis system. *Nucleic Acids. Res.* **2017**, *45*, D507–D516. [CrossRef] [PubMed]

96. Abubucker, S.; Segata, N.; Goll, J.; Schubert, A.M.; Izard, J.; Cantarel, B.L.; Rodriguez-Mueller, B.; Zucker, J.; Thiagarajan, M.; Henrissat, B.; et al. Metabolic Reconstruction for Metagenomic Data and Its Application to the Human Microbiome. *PLoS Comput. Biol.* **2012**, *8*, e1002358. [CrossRef] [PubMed]

97. Alneberg, J.; Bjarnason, B.S.; De Bruijn, I.; Schirmer, M.; Quick, J.; Ijaz, U.Z.; Lahti, L.; Loman, N.J.; Andersson, A.F.; Quince, C. Binning metagenomic contigs by coverage and composition. *Nat. Methods* **2014**, *11*, 1144–1146. [CrossRef] [PubMed]

98. Sieber, C.M.K.; Probst, A.J.; Sharrar, A.; Thomas, B.C.; Hess, M.; Tringe, S.G.; Banfield, J.F. Recovery of genomes from metagenomes via a dereplication, aggregation and scoring strategy. *Nat. Microbiol.* **2018**, *3*, 836–843. [CrossRef] [PubMed]

99. Wommack, K.E.; Bhavsar, J.; Ravel, J. Metagenomics: Read length matters. *Appl. Environ. Microbiol.* **2008**, *74*, 1453–1463. [CrossRef] [PubMed]

100. Rhoads, A.; Au, K.F. PacBio Sequencing and Its Applications. *Genomics Prot. Bioinform.* **2015**, *13*, 278–289.

101. Frank, J.A.; Pan, Y.; Tooming-Klunderud, A.; Eijsink, V.G.H.; McHardy, A.C.; Nederbragt, A.J.; Pope, P.B.; Nederbragt, A.; Pope, P. Improved metagenome assemblies and taxonomic binning using long-read circular consensus sequence data. *Sci. Rep.* **2016**, *6*, 25373. [CrossRef] [PubMed]

102. Brown, B.L.; Watson, M.; Minot, S.S.; Rivera, M.C.; Franklin, R.B. MinION™ nanopore sequencing of environmental metagenomes: A synthetic approach. *GigaScience* **2017**, *6*, 1–10. [CrossRef] [PubMed]

103. Rang, F.J.; Kloosterman, W.P.; De Ridder, J. From squiggle to basepair: Computational approaches for improving nanopore sequencing read accuracy. *Genome Biol.* **2018**, *19*, 90. [CrossRef] [PubMed]

104. Driscoll, C.B.; Otten, T.G.; Brown, N.M.; Dreher, T.W. Towards long-read metagenomics: Complete assembly of three novel genomes from bacteria dependent on a diazotrophic cyanobacterium in a freshwater lake co-culture. *Stand. Genomic Sci.* **2017**, *12*, 9. [CrossRef] [PubMed]

105. Moss, E.; Bishara, A.; Tkachenko, E.; Kang, J.B.; Andermann, T.M.; Wood, C.; Handy, C.; Ji, H.; Batzoglou, S.; Bhatt, A.S. De novo assembly of microbial genomes from human gut metagenomes using barcoded short read sequences. *bioRxiv* **2017**. [CrossRef]

106. Jain, M.; Koren, S.; Miga, K.H.; Quick, J.; Rand, A.C.; A Sasani, T.; Tyson, J.R.; Beggs, A.D.; Dilthey, A.T.; Fiddes, I.T.; et al. Nanopore sequencing and assembly of a human genome with ultra-long reads. *Nat. Biotechnol* **2018**, *36*, 338–345. [CrossRef] [PubMed]

107. Press, M.O.; Wiser, A.H.; Kronenberg, Z.N.; Langford, K.W.; Shakya, M.; Lo, C.-C.; Mueller, K.A.; Sullivan, S.T.; Chain, P.S.G.; Liachko, I. Hi-C deconvolution of a human gut microbiome yields high-quality draft genomes and reveals plasmid-genome interactions. *Genomics* **2017**, *bioRxiv*, 198713.

108. Burton, J.N.; Liachko, I.; Dunham, M.J.; Shendure, J. Species-Level Deconvolution of Metagenome Assemblies with Hi-C–Based Contact Probability Maps. *G3* **2014**, *4*, 1339–1346. [CrossRef] [PubMed]

109. Lavelle, A.; Sokol, H. Gut microbiota: Beyond metagenomics, metatranscriptomics illuminates microbiome functionality in IBD. *Nat. Rev. Gastroenterol. Hepatol.* **2018**, *15*, 193–194. [CrossRef] [PubMed]

110. Ren, M.; Zhang, Z.; Wang, X.; Zhou, Z.; Chen, D.; Zeng, H.; Zhao, S.; Chen, L.; Hu, Y.; Zhang, C.; et al. Diversity and Contributions to Nitrogen Cycling and Carbon Fixation of Soil Salinity Shaped Microbial Communities in Tarim Basin. *Front. Microbiol.* **2018**, *9*, 431. [CrossRef] [PubMed]

111. Garoutte, A.; Cardenas, E.; Tiedje, J.; Howe, A. Methodologies for probing the metatranscriptome of grassland soil. *J. Microbiol. Methods* **2016**, *131*, 122–129. [CrossRef] [PubMed]

112. Jiang, Y.; Xiong, X.; Danska, J.; Parkinson, J. Metatranscriptomic analysis of diverse microbial communities reveals core metabolic pathways and microbiome-specific functionality. *Microbiome* **2016**, *4*, 13780. [CrossRef] [PubMed]

 *genes*

Article

Temporal Analysis of the Microbial Community from the Crystallizer Ponds in Cabo Rojo, Puerto Rico, Using Metagenomics

Ricardo L. Couto-Rodríguez [1,2] **and Rafael Montalvo-Rodríguez** [2,*]

[1]   Department of Microbiology and Cell Science, University of Florida, Gainesville, FL 32603, USA; r.coutorodriguez@ufl.edu
[2]   Biology Department, Box 9000, University of Puerto Rico, Mayagüez, PR 00681, USA
*   Correspondence: rafael.montalvo@upr.edu; Tel.: +1-787-832-4040 X2421

Received: 1 April 2019; Accepted: 29 May 2019; Published: 31 May 2019

Abstract: The Cabo Rojo solar salterns are a hypersaline environment located in a tropical climate, where conditions remain stable throughout the year. These conditions can favor the establishment of steady microbial communities. Little is known about the microbial composition that thrives in hypersaline environments in the tropics. The main goal of this study was to assess the microbial diversity present in the crystallizer ponds of Cabo Rojo, in terms of structure and metabolic processes across time using metagenomic techniques. Three samplings (December 2014, March and July 2016) were carried out, where water samples (50 L each) were filtered through a Millipore pressurized filtering system. DNA was subsequently extracted using physical–chemical methods and sequenced using paired end Illumina technologies. The sequencing effort produced three paired end libraries with a total of 111,816,040 reads, that were subsequently assembled into three metagenomes. Out of the phyla detected, the microbial diversity was dominated in all three samples by *Euryarchaeota*, followed by *Bacteroidetes* and *Proteobacteria*. However, sample MFF1 (for Muestreo Final Fraternidad) exhibited a higher diversity, with 12 prokaryotic phyla detected at 34% NaCl (w/v), when compared to samples MFF2 and MFF3, which only exhibited three phyla. Precipitation events might be one of the contributing factors to the change in the microbial community composition through time. Diversity at genus level revealed a more stable community structure, with an overwhelming dominance of the square archaeon *Haloquadratum* in the three metagenomes. Furthermore, functional annotation was carried out in order to detect genes related to metabolic processes, such as carbon, nitrogen, and sulfur cycles. The presence of gene sequences related to nitrogen fixation, ammonia oxidation, sulfate reduction, sulfur oxidation, and phosphate solubilization were detected. Through binning methods, four putative novel genomes were obtained, including a possible novel genus belonging to the *Bacteroidetes* and possible new species for the genera *Natronomonas*, *Halomicrobium*, and *Haloquadratum*. Using a metagenomic approach, a 3-year study has been performed in a Caribbean hypersaline environment. When compared to other salterns around the world, the Cabo Rojo salterns harbor a similar community composition, which is stable through time. Moreover, an analysis of gene composition highlights the importance of the microbial community in the biogeochemical cycles at hypersaline environments.

Keywords: metagenomics; hypersaline; halophilic archaea; Puerto Rico; Caribbean

---

1. Introduction

Marine solar salterns are classified as hypersaline environments due to their high NaCl concentrations (above 3 M) [1]. The organisms that frequently predominate in these ecosystems are known as halophiles, which can thrive at around 10–15% NaCl (w/v) or above [2]. Halophiles possess a wide variety of applications, including enzymes used for food processing and biosynthetic processes like hydrolases, such as amylases, lipases, and proteases [3–5]. Hypersaline habitats have been extensively studied worldwide, especially in template locations such as Turkey, where the microbial diversity was determined from six hypersaline lakes across the country; the Santa Pola salterns in Alicante, Spain (which have been the most extensively studied); and the Dead Sea [6–10]. The extreme conditions of these environments make them model ecosystems for understanding microbial community dynamics [11]. Despite these findings, relatively few studies have been performed in tropical environments, where conditions normally remain relatively stable throughout the year, with temperatures ranging from 35 to 40 °C year-round, as well as low precipitation rates and no drastic weather events in seasons (fall, winter, autumn, summer). Drastic changes can be introduced by hurricanes (like hurricane Maria, in September 2017) but this phenomenon does not happen every year.

The Cabo Rojo solar salterns are a tropical hypersaline environment that have been the subject of numerous diversity studies and novel microbes have been described. *Halogeometricum borinquense* was first isolated from these salterns [12] and subsequent microbial diversity surveys from the crystallizers and surrounding areas (mainly *Avicennia germinans* forests) yielded additional novel organisms (*Haloterrigena thermotolerans*, *Halomonas avicenneae* (now *Kushneria avicenniae*), *Halobacillus mangrovii*, *Kushneria aurantia*, as well as two recent novel isolates proposed as '*Haloarcula rubripromontorii*' and '*Halorubrum tropicale*') [13–18]. These findings suggest a more diverse community than that reported in other solar salterns worldwide. However, the aforementioned studies were performed using culture-dependent methods, where approximately 0.1% of the diversity can be isolated in pure culture [19]. Metagenomics has emerged as an answer to these limitations and has been applied for studies in habitats with high salinity. The most predominant studies have been undertaken in the Santa Pola salterns in Alicante, Spain, where the overwhelming dominance of the square archaeon, *Haloquadratum walsbyi*, has been confirmed at salinities of 30% NaCl (w/v) and above, as well as the presence of novel microbial groups previously undescribed in these environments [7,8]. Furthermore, ecological processes have also been described, using metagenomic techniques in the Santa Pola salterns. Other studies around the world include the Lake Tyrell, where the novel group of *Nanohaloarchaea* was first detected, and the Atacama Desert, where metagenomic analysis of endolithic halite microbial communities from the Salar Grande returned a novel genome also belonging to the *Nanohaloarchaea* [20,21]. These studies have shown that metagenomic methods in hypersaline environments successfully provide more comprehensive answers to community composition, as well as possible functions within these communities. The discovery of novel microbial groups has shifted our understanding of hypersaline environments towards new directions. This development will continue as new techniques become available and more data are released, which will in turn provide more tools towards characterizing novel taxa, as well as novel biocatalysts.

In this study, the aim was to obtain a comprehensive assessment of microbial and functional gene diversity at the crystallizer ponds of the solar salterns in Cabo Rojo, by performing a temporal metagenomic analysis. With this information, we intend to establish comparisons in terms of functional and microbial diversity with other metagenomes available from solar salterns around the world. The microbial diversity in the Cabo Rojo salterns was determined by means of culture independent methods, by using the pyrosequencing of partial 16S rRNA genes from a previous study [22]. However, to our knowledge, a full scale temporal metagenomic approach to assess gene diversity has not yet been performed in an extreme environment in the Caribbean.

2. Materials and Methods

2.1. Sampling and DNA Purification

Samplings (50 L per sample) were carried out in months of rainy (December 2014), and dry (March and July 2016) seasons. Water samples were taken from the crystallizer ponds at the Cabo Rojo salterns

(17°57′25.2″ N, 67°11′58.0″ W). This crystallizer system is served by the hypersaline Fraternidad Lagoon (Figure 1). Five-liter samples (from between the surface to 10 cm depth) of ten crystallizers were taken (Figure 1) and pooled to obtain a 50 L total volume per sampling. Samples were named MFF1 for the first sampling, MFF2 for the second sampling, and MFF3 for the third sampling (MFF stands for "Muestra Final Fraternidad", which means Fraternidad Final Sample). Temperature and salinity were taken for each sample (using a Fisherbrand™ salinity refractometer, Fisher Scientific, Pittsburgh, PA, USA) and averaged. Samples were then transported back to the laboratory. Each 50 L sample of saltern water was differentially filtered using a Millipore®pressurized filtering system, consisting of two nitrocellulose membrane filters (EMD Millipore, Burlington, MA, USA) of different pore sizes. The first membrane possessed a pore size of 5.0 µm, which was intended to retain eukaryotic cells, whereas the second membrane, with a pore size of 0.22 µm, was used for the collection of prokaryotic cells. Metagenomic DNA extraction was performed on cells present on the 0.22 µm membrane using the physical–chemical methods described previously by Martín-Cuadrado et al. [23]. Concentration and purity of DNA were measured using a Nanodrop™ spectrophotometer (Thermo Scientific, Waltham, MA, USA). Furthermore, a 0.8% (w/v) agarose gel electrophoresis was carried out in order to corroborate DNA quality before sequencing. Metagenomic DNA was then stored at −20 °C until it was used for sequencing.

Figure 1. Aerial map of the Solar Salterns of Cabo Rojo (17°57′25.2″ N, 67°11′58.0″ W). Water samples (5 L) were obtained from ten crystallizers (circled in black) and pooled together into one sample (50 L). Three samplings were performed on the same crystallizers in rainy (December 2014), and dry (March and July 2016) seasons. Photo taken by "Puerto Rico desde el Aire" reproduced with permission.

2.2. DNA Sequencing and Metagenome Assembly

DNA sequencing was performed using an Illumina HiSeq 2500. Library preparation and sequencing was carried out by the Molecular Research DNA (MR DNA) facility in Shallowater, TX. The sequencing reads obtained were quality checked using FastQC [24]. Low quality reads were trimmed for assembly using BBDuk (Geneious, Newark, NJ, USA). Taxonomy was assigned by comparing raw reads to Ribosomal Database Project (RDP)using a minimum alignment length of 100 bp and a threshold of 97%. Afterwards, assembly of remaining reads was performed using a MetaSPAdes assembler [25], where the quality of metagenomes assemblies was compared based on N50 values (median length of contigs), total contigs obtained, as well as the largest contig. Taxonomic and functional annotation of the assembled metagenome was carried out using the

MG-RAST [26] pipeline, which aligned sequences to reference databases, such as the KEGG, eggNOG, COG, and SEED subsystems [27–29].

2.3. Binning for Putative Genomes

Following the assembly of the metagenomes, the original reads were mapped back to the assembly in order to obtain a coverage, using the Burrows–Wheeler Aligner [30]. Subsequently, the coverage files, along with the final assembly, were binned for putative genomes using MetaBAT software [31]. Quality of the genomes, including completeness and contamination, was assessed using the CheckM tool [32]. The taxonomy of the quality bins obtained was assessed by means of amino acid identity (AAI), using the Microbial Genome Atlas (MiGA) web interphase [33]. Taxonomic novelty was determined by the maximum average amino acid identity (AAI) found against the genomes in the database. The p-value for this was estimated from the empirical distribution observed in all the reference genomes of NCBI's RefSeq at each taxonomic level, and indicates the probability of the observed AAI between genomes in the same taxon. Phylogeny using AAI was also determined using MiGA. Trees were generated using iTOL software [34]. Annotation of all genomes was carried out using the Rapid Annotation using Subsystems Technology (RAST) pipeline [35].

3. Results and Discussion

3.1. Sampling Site Conditions and Sequencing Analysis

The area of study was the solar salterns of Cabo Rojo, Puerto Rico. They are located at the coordinates 17°57′25.2″ N, 67°11′58.0″ W, which represent the southwestern part of the Island (Figure 1). This artisanal solar saltern has 508 years of continous operation. Water from the Fraternidad Lagoon (salinity approximately 14–19% w/v) was pumped to the crystallizers and the evaporation cycle took approximately 60 days. Samplings were performed at the middle to the end of the evaporation cycle (between 35 and 45 days). The crystallizers sampled on the three different occasions exhibited an average temperature of 31.1 °C and a NaCl concentration of 34% (w/v).

After DNA sequencing, assessment for the quality of assembly for the metagenome was based on number of contigs, longest contig length, and N50 (the minimum contig length in the set of contigs that comprises over half of the assembly) (Table 1). A lower number of contigs and high contig length and N50 are ideal for high-quality assemblies [36]. Table 1 also details the assembly statistics using MetaQUAST [37] for the three metagenomes, where the millions of reads were condensed to one hundred thousand contigs, a significant reduction. Additionally, the N50 values obtained surpassed the values obtained in other metagenomic studies performed in hypersaline environments [21].

Table 1. Sequencing results for the three metagenomes along with assembly statistics using MetaSPAdes assembler. Number of sequencing reads obtained through Illumina Sequencing, GC (Guanine Cytosine) content, range of read lengths are shown (in base pairs), number of assembled contigs, N50 and longest contig length are shown (in base pairs).

Sample	Number of Reads	GC Content (%)	Read Length (bp)	Number of contigs	N50 (bp)	Longest Contig (bp)
MFF1	29,432,758	56	35–251	318,469	3888	619,112
MFF2	41,746,817	57	35–151	420,402	4748	469,957
MFF3	40,634,465	58	35–151	379,415	4854	388,630

3.2. Microbial Community Composition

The taxonomical assignment of the sequencing reads containing the 16S rRNA gene is shown in Figure 2. In terms of the microbial diversity present, the phylum Euryarchaeota predominated in all three metagenomes, with more than 70% of the reads. This was also observed when the analysis was performed with annotated reads with predicted proteins and ribosomal RNA genes (Figure S1).

This abundance is expected, since this group contains the halophilic representatives from the Archaea domain. The Bacteroidetes group was the second in abundance, with about 20% of the reads. This group possesses one extremely halophilic representative in *Salinibacter ruber*. The third most predominant group was the Proteobacteria, with 1–3% of the reads. The presence of Proteobacteria was also expected, since it contains halophilic/halotolerant bacteria, such as the genera *Halomonas*, *Halovibrio*, and *Rhodovibrio*, among others [2]. Furthermore, taxonomic hits in MFF1 (as illustrated in Figure 2) reveal a diverse representation of other prokaryotic phyla. This representation is markedly different from the results found by Ghai et al. and Rhodes et al. [7,38], where only *Euryarchaeota*, *Bacteroidetes*, and *Proteobacteria* were encountered at a similar salinity of about 34% (w/v) or above. However, MFF2 and MFF3 were less diverse, with only three prokaryotic phyla detected, more consistent with the aforementioned results.

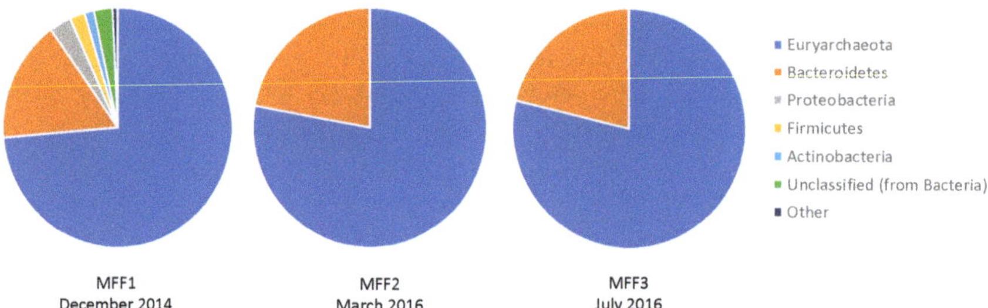

Figure 2. Taxonomic hits by phylum. Each slice indicates the number of reads with predicted 16SrRNA genes annotated to the indicated phylum. Samples were named MFF1 for the first sampling, MFF2 for the second sampling and MFF3 for the third sampling (MFF stands for "Muestra Final Fraternidad" which means Fraternidad Final Sample). Phylum *Euryarchaeota* were shown to be dominant in the three samples (73.69% for MFF1, 77.92% for MFF2, and 78.75% for MFF3), followed by *Bacteroidetes* (16.59, 22.03, 21.18%, respectively), and *Proteobacteria* (3.03, 0.05, 0.07%, respectively). Other groups include *Acidobacteria*, *Chlamydiae*, *Cyanobacteria*, *Deinococcus-Thermus*, *Fusobacteria*, *Planctomycetes*, and *Verrucomicrobia*, with less than 1% in MFF1.

When assessing microbial diversity at genus level (Table 2), the community structure showed high stability, with the same three dominant genera (*Haloquadratum*, *Salinibacter*, *Halorubrum*) through time. However, variations in the diversity of less frequent genera were observed across all three metagenomes. Other studies performed in hypersaline environments, with a few notable exceptions [39,40], have demonstrated that *Haloquadratum* usually predominates in salinities of 30% (w/v) and higher [1,8,9]. Our study shows that *Haloquadratum* is the dominant genera at the solar salterns of Cabo Rojo, and the first to show the predominance of this genus through time in tropical environments. Podell et al., [41] demonstrated that *Haloquadratum* abundance was positively correlated with high levels of potassium, magnesium, and sulfide, and negatively correlated with an increase in microbial diversity. Ionic composition data obtained by Rodríguez-García [22] in the Cabo Rojo salterns on June 2015 (1.0 inches of precipitation in the area) showed high concentrations of chloride ions (230 g/L), followed by magnesium (28.84 g/L) and potassium (11.22 g/L), typical for a thalasohaline environment. These data show, as in other marine solar salterns around the world, that the ionic composition in these crystallizer ponds is suitable for *Haloquadratum* predominance. Ionic composition could change over time due to precipitation effects. Data obtained from the National Weather Service's Advance Hydrologic Prediction Services (http://water.weather.gov/precip/) revealed that the amount of rainfall in the area of the Cabo Rojo salterns in November 2014 (first sampling was performed at the beginning of December 2014), March 2016, and July 2016 was 20.3 cm, 0.3 cm, and 7.6 cm of rain, respectively. These events were shown not to have an effect on salinity, as the salinity remained consistent at 34% (w/v) across all three samples, as well as the predominance of *Haloquadratum*. Precipitation is one of the many ways new microbes can be dispersed into new habitats [42]. Therefore, during a rain event, aquatic habitats can be recipients for new microorganisms. The data on precipitation suggest that the rainfall events in November could have contributed changes in ionic composition or a dilution effect on the water surface (causing cell lysis for haloarchaea) that directly affected microbial community structure and could explain the higher diversity of prokaryotic phyla in MFF1. Since Cabo Rojo is at a tropical location, Saharan dust could have also influenced precipitation and contributed to an increase in magnesium and potassium ions, which favored growth of *Haloquadratum* in all three samples [42]. Similarly, the amount of potassium found at this tropical saltern could favor an abundance of "salt-in" strategists, such as *Haloquadratum* and *Salinibacter*, and could perhaps be a contributing factor of their predominance in the three metagenomes.

Table 2. Taxonomic composition at genus level for each metagenome, using 16SrRNA gene sequences. The percentage of reads aligning with a minimum length of 100 bp and 97% identity at genus level are shown. *Unclassified sequences belong to phylum Euryarchaeota.

MFF1		MFF2		MFF3	
Genus	**Abundance**	**Genus**	**Abundance**	**Genus**	**Abundance**
Haloquadratum	53.77%	*Haloquadratum*	69.76%	*Haloquadratum*	62.47%
Salinibacter	16.02%	*Salinibacter*	12.99%	*Salinibacter*	21.17%
Halorubrum	8.65%	*Halorubrum*	7.55%	*Halorubrum*	4.33%
*Unclassified**	2.74%	*Halococcus*	1.99%	*Haloplanus*	4.12%
Haloplanus	2.39%	*Natronomonas*	1.39%	*Halococcus*	2.41%
Haloarcula	1.84%	*Halomicrobium*	1.08%	*Haloterrigena*	2.01%
Pseudomonas	1.74%	*Haloterrigena*	1.07%	*Natrinema*	1.08%
Halococcus	1.12%	*Haloplanus*	1.05%	*Natronomonas*	0.72%
Natronomonas	1.10%	*Haloferax*	0.45%	*Halovivax*	0.58%
Haloferax	0.94%	*Haloarcula*	0.43%	*Halobaculum*	0.51%
Halovivax	0.66%	*Halobaculum*	0.41%	*Haloferax*	0.35%
Dyella	0.61%	*Halobacterium*	0.41%	*Halomicrobium*	0.10%
Ruminococcus	0.46%	*Halogeometricum*	0.37%	*Haloarcula*	0.10%
Total sequences	19,228		37,156		31,202

Temporal studies in hypersaline environments at stable tropical climates are scarce. Most of these studies have been performed in variable climates, like the Lake Tebenquiche (Salar de Atacama) in Chile [43], and the Ocnei Lake in the Transylvanian Basin, Romania [44]. Even viruses have been the subject of temporal studies at solar salterns [45]. An interesting example of a temporal study performed at a less stable climate is the one at the Great Salt Lake in Utah, where the results showed that Salinibacter dominated as the main bacterial group through all samplings, whereas for the Archaea, Haloquadratum was also present in high numbers, although its abundance varied year-round. Other members of the haloarchaea found were *Halorubrum*, *Natronococcus*, and *Haloplanus*. The population changes observed in the Great Salt Lake were attributed to biotic factors, such as viruses and nutrients, and less to the seasonal temperature changes [46]. The presence of viruses could be a factor affecting microbial populations at salterns in tropical environments. Overall, the prokaryotic community structure of the crystallizers at the solar salterns in Cabo Rojo showed some variation at phylum-level but at genus-level seems to be stable over time.

3.3. Functional Annotation

In order to study the microbial processes that might be occurring in the Cabo Rojo solar salterns, contig sequences from the metagenomes were compared to the KEGG Orthology database and grouped into functional categories for further analysis.

Figure 3 details the predicted protein sequences of each metagenome into functional categories, according to the KEGG orthology database. Around 60% of the contigs obtained in all three samples were related to metabolism. This is consistent with other results in hypersaline environments, where a great number of metabolic processes, such as the primary production and degradation of organic compounds, are carried out [47]. Primary production is usually carried out at salinities of 25% NaCl and above, solely by the halophilic green algae *Dunaliella* [48,49]. However, it has been found that certain cyanobacteria are also capable of thriving at high salinities. For instance, cyanobacteria phylogenetically close to the genus *Halothece* have been reported in the Atacama Desert in Chile, with an average salinity of about 15% NaCl (w/v) [50]. Our results showed that cyanobacteria might be present at 34% NaCl in the crystallizers of Cabo Rojo. This was different from the Santa Pola salterns, where Ghai et al. reported that cyanobacteria were absent at salinities of 19% (w/v) and above [8].

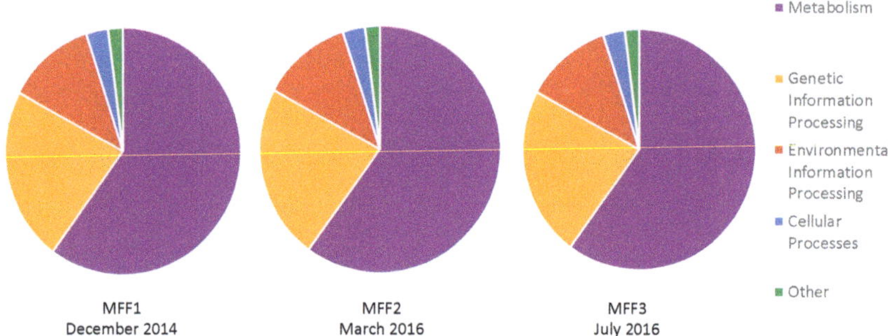

Figure 3. KEGG Orthology (KO) of functional genes obtained from samples MFF1, MFF2, and MFF3. Genes related to metabolism dominate in about 60% of the sequences in all three samples, followed by genetic information processing (23%), environmental information processing (12%), and cellular processes (3%). Other funtions are less than 1%. Samples were named MFF1 for the first sampling, MFF2 for the second sampling, and MFF3 for the third sampling (MFF stands for "Muestra Final Fraternidad" which means Fraternidad Final Sample).

Genes related to carbon fixation were encountered in all three metagenomes (Figure 4). Particularly, the gene encoding for the Ribulose-1,5-bisphosphate carboxylase/oxygenase enzyme, more commonly known as RuBisCO, was present. This enzyme is critical for carbon fixation because it catalyzes the very first step in the Calvin cycle. Cyanobacteria carry out carbon fixation using RuBisCO in their carboxysomes. Analysis of the 16SrRNA gene hits revealed the presence of *Cyanobacteria* in MFF1 but in low numbers (0.01%). However, when looking for predicted protein sequences attributed to RuBisCO, we could not find any related to *Cyanobacteria* in the metagenomes. RuBisCO sequences were attributed to *Natronomonas* and *Halomicrobium*, both of which have been reported to possess this enzyme and have also been found in other metagenomic studies [51]. We were able to obtain almost complete genome bins from a putative new species of *Natronomonas* (Bin RC33) and of *Halomicrobium* (Bin RC39) from the metagenomes. These two putative species might have an important role in carbon fixation at the solar salterns of Cabo Rojo. However, transcriptomic approaches would be necessary in order to determine if the genes present in both species are metabolically active at a salinity of 34% (w/v).

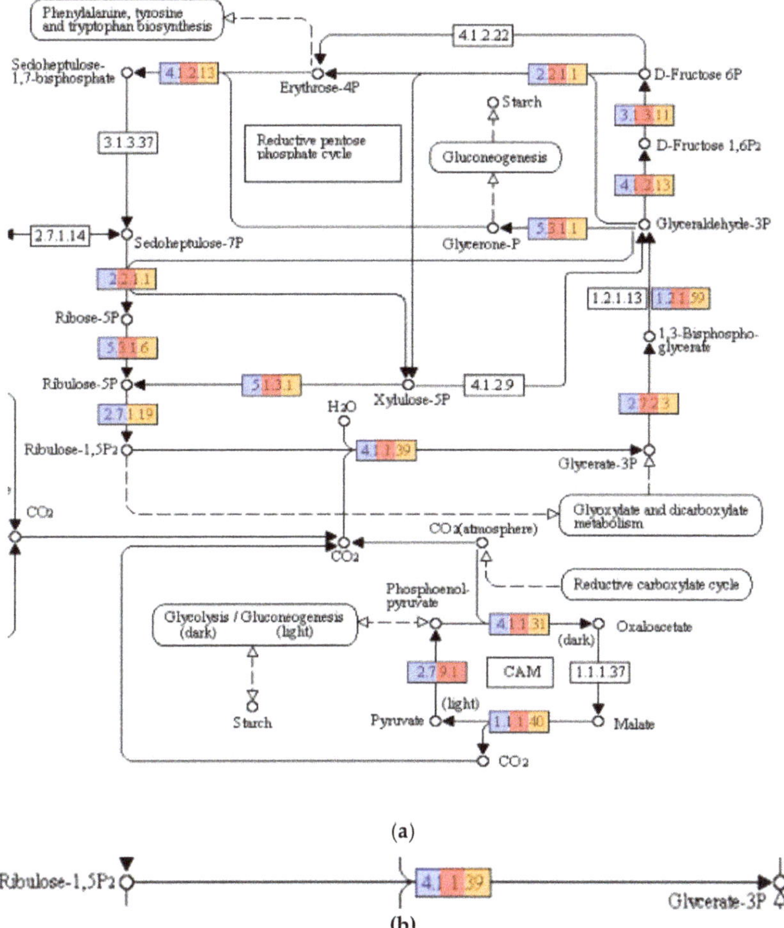

Figure 4. (**a**) Carbon metabolism pathways detected in the three metagenomes. Colored enzymes indicate the presence of the enzyme, light blue indicates presence in MFF1, red indicates presence in MFF2, and orange indicates presence in MFF3 (number of hits for each enzyme can be seen at Table S1 of supplementary figures). (**b**) Reaction catalyzed by the enzyme Ribulose-1,5-bisphosphate carboxylase/oxygenase (RuBisCO). The enzyme is present in all three metagenomes. Pathways obtained from KEGG pathways (*https://www.genome.jp/kegg/pathway.html*) with permission.

Microorganisms also play a pivotal role in the nitrogen cycle. Bacteria, in particular, are involved in all N-cycle pathways and their nitrogen metabolism has been studied extensively [52]. However, nitrogen metabolism in archaea is not well studied or understood. Archaea are known to be participants in all the reductive pathways of the N-cycle. Archaea inhabiting extreme environments are considered the principal driving force of the N-cycle [52,53]. Figure 5 illustrates the pathways concerning the nitrogen cycle and, as expected, genes encoding for enzymes related to the reductive pathways of the N-cycle were present. Nitrogen fixation, the process catalyzed by a nitrogenase, in which atmospheric nitrogen is converted to ammonia, is performed naturally by both bacteria and archaea. In Archaea, nitrogen fixation has been reported in the methanogenic representatives of the phylum *Euryarchaeota* [53].

Methanogenic archaea were not detected in this study using 16S rRNA genes, however, reads with predicted protein sequences showed similiarities to functions related to nitrogen fixation.

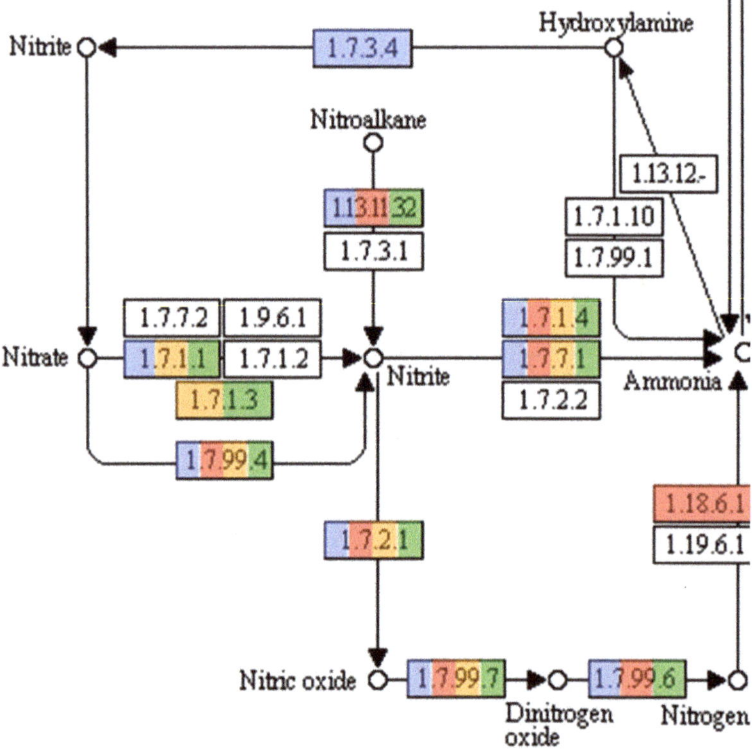

Figure 5. Nitrogen metabolism pathways present in the three metagenomes. Colored enzymes (Light blue and green for MFF1, red for MFF2, orange for MFF3) indicate the presence of hits related to the enzyme in the metagenome (number of hits for each enzyme can be seen at Table S1 of supplementary figures). Enzymes related to reductive pathways in the nitrogen cycle were encountered, including nitrate reductase (Enzyme Nomenclature database numbers (EC) 1.7.1.1, 1.7.1.3, 1.7.99.4), nitrite reductases (EC 1.7.2.1, 1.7.7.1, 1.7.1.4), and the nitrogenase needed for nitrogen fixation (EC 1.18.6.1). Pathways obtained from KEGG pathways (*https://www.genome.jp/kegg/pathway.html*) with permission.

Nitrification, the conversion of ammonia to nitrite and subsequently nitrate, is another pivotal process in the nitrogen cycle. Until recently, it was believed that this process was only undertaken by bacteria. However, several ammonia oxidizing archaea have been described, all representatives of the phylum *Thaumarchaeota* [54]. Protein sequences related to *Nitrosopumilus*, an ammonia oxidizing archaeon, were found in the three metagenomes [55,56]. However, as Figure 5 shows, there were no sequences matching the ammonia monooxygenase (AMO) enzyme. It has been argued that due to the high oxidation state of ammonia and the high energetic burden placed on halophilic organisms, this process would be too energy consuming for the amount of energy produced, and therefore not possible in high salinity environments [57]. This could suggest that either the sequences obtained are of an organism closely related to *Nitrosopumilus*, or that if the process is being performed, it could be the result of a novel less energy expensive pathway of ammonia oxidation. As sequencing technology improves, combined with better bioinformatics tools to analyze the enormous amounts of data, these gaps in information will be reduced.

As illustrated in Figure 6, hits matching sulfur metabolism were also found. The sulfur cycle is another prominent biogeochemical process undertaken in hypersaline environments. Both archaea and bacteria play a pivotal role in the cycling of sulfur. Sulfidogenesis, the production of H_2S from the reduction of elemental Sulfur (S^0), sulfate, thiosulfate, or sulfite, is a major step in the sulfur cycle. Bacteria possess sulfate reducing representatives in the phyla *Deltaproteobacteria* and *Firmicutes*. Sulfate-reducing bacteria in *Deltaproteobacteria* include the orders *Desulfobacteriales*, *Desulfovibrionales*, and *Syntrophobacteriales* [58], sequences of which were found in the three samples. Furthermore, the genus *Desulfotomaculum* from the *Firmicutes* was also present in our study. The archaeal genera known to reduce sulfate are *Archaeoglobus*, *Thermocladium*, and *Caldivirga*. However, hits related to these genera were not detected. This result is not surprising, because these organisms are not found at high salinity environments [59].

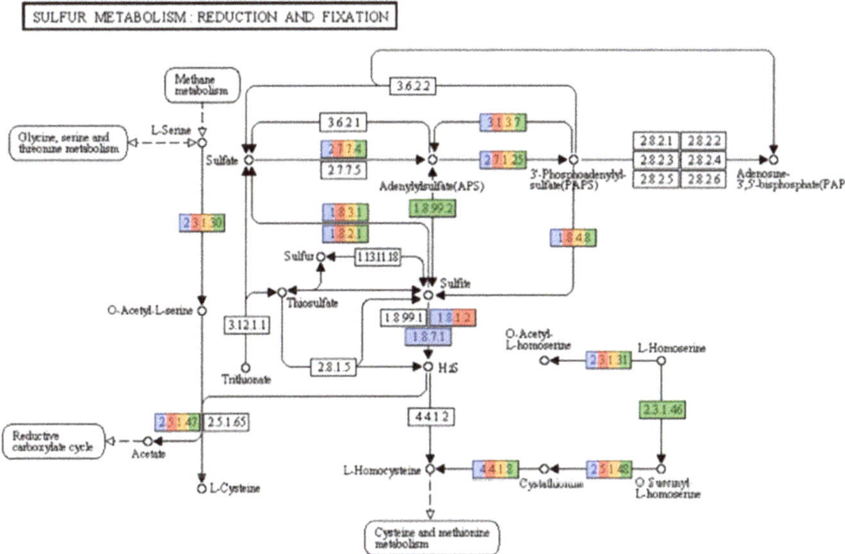

Figure 6. Sulfur metabolism pathways present in the metagenomes. Colored enzymes indicate the presence of hits related to the enzyme in the metagenome (number of hits for each enzyme can be seen at Table S1 of supplementary figures). Enzymes related to sulfate reduction (2.7.7.4, 1.8.99.2, 1.8.1.2, 1.8.7.1) were encountered. Pathways obtained from KEGG pathways (*https://www.genome.jp/kegg/pathway.html*) with permission.

The oxidation of H_2S is also another important pathway in the sulfur cycle, since hydrogen sulfide is toxic to plant and animal tissue. In hypersaline environments, representatives from the Gammaproteobacteria, such as *Halothiobacillus* and *Thiomicrospira*, among others, are classified as sulfur oxidizing bacteria (SOB) [60]. It is more common to find these types of bacteria in a hypersaline environment, due to the fact that their substrates are more reduced when compared to nitrifying organisms [57]. All three samples contained representatives from *Gammaproteobacteria*, including protein sequences matching those of *Halothiobacillus* and *Thiomicrospira*. Sulfur oxidizing archaea (SOA) have been poorly characterized and only two genera, Acidianus and Ferroglobus, are known to carry out sulfur oxidation [53]. Neither genus was encountered in our samples, nor were they expected, due to both being hyperthermophiles, growing optimally at temperatures above 60 °C.

These data show that microorganisms present at the solar salterns in Cabo Rojo might play an important role in the biogeochemical cycles, with most of the relevant pathways present in the metagenome. Furthermore, representatives known to perform processes in each of these cycles have been found. With further sampling, as well as the evolution of sequencing technologies, a more complete assessment can be carried out, as well as novel pathways being discovered that have not been described for the process at hypersaline environments [61,62].

3.4. Binning of Putative Novel Genomes

A reconstruction of putative genomes using binning techniques was performed, in order to determine if the predicted protein sequences found in the metagenomes could be assigned to specific organisms. Upon the assignment of taxonomic bins, it is important to avoid chimeric bins that might be produced which can lead to erroneous conclusions [19]. Caution should be taken before validating genomic bins, due to contaminating fragments. Of the software available, CheckM provides an accurate estimate of genome completeness and contamination [32]. A high number of taxonomic bins were obtained using binning methods, however most of these exhibited either a low degree of completeness or a high degree of completeness, but with high contamination. Nevertheless, we were able to obtain four genomic bins of significant quality from the three metagenomic libraries. All genomes presented a high amount of completeness and a low degree of contamination (Table 3).

Table 3. Details of genomic bins obtained.

Bin Name	Completeness (%)	Contamination (%)	GC Content (%)	Number of Contigs	N50 (bp)	Genome Size (Mb)	Predicted Proteins
RC33	94.40	2.40	66.22	11	392,903	1.5	1576
RC24	80.20	1.80	52.79	28	125,394	1.8	1533
RC39	96.00	3,20	68.18	75	54,228	2.4	2502
RC20	88.66	3.20	50.25	273	24,930	4.3	4449

Assigning taxonomy to uncultured organisms poses more of a challenge, due to the lack of phenotypic characterization. Therefore, the information available is based only on sequence data. The *Candidatus* status bypasses this limitation by assigning candidate names until phenotypic characters are appropriately characterized [63]. Several methods have been proposed to identify microbial species at the genome level. For cultured species, the DNA–DNA hybridization (DDH) has been the most traditional approach to differentiate closely related species with the 70% identity cutoff [64]. However, the average nucleotide identity (ANI) has been proposed as an alternate way of distinguishing bacterial and archaeal species. The cutoff for the ANI analyses is 95%, and has been employed successfully for the characterization of new microbial species [65–67]. Special caution should be taken, however, when describing species within a population, due to the members of a microbial population exhibiting gene differences of less than 5% of their total genes. Furthermore, ANI offers more robust resolution between genomes that share 80–100% ANI; organisms that show less than 80% ANI are too divergent to be compared based on this analysis [63]. Due to this, use of amino acid identity (AAI) is recommended to distinguish between more divergent organisms [66]. Organisms exhibiting an AAI of >85% are typically grouped within the same species, whereas those grouped in the genus-level exhibit an AAI of 60–80% [63,68]. Due to this, we used AAI to determine taxonomy for our four metagenomic bins (Table 4).

Table 4. Amino acid identity (AAI) of the four genomic bins obtained. The closest relative of each organism, along with its AAI and fraction of proteins shared, are listed.

Bin	Closest Relative	AAI	Fraction of Proteins Shared
RC33	*Natronomonas moolapensis*	59.74%	81.85%
RC24	*Pontibacter korlensis*	43.31%	66.02%
RC39	*Halomicrobium mukohataei*	62.81%	68.27%
RC20	*Haloquadratum walsbyi*	65.83%	73.46%

AAI results for RC33 revealed that *Natronomonas moolapensis* is its closest relative, with an identity of 59.74%. Phylogeny using AAI (Figure 7) suggests that this genome is a member of the family Halobacteriaceae and could represent a new species within the genus (*p* value = 0.0038). The high GC content, as well as the presence of proteins associated with hyperosmotic stress, indicate that this organism is halophilic, as a characteristic of organisms thriving in hypersaline environments. The strain might be non-motile due to the absence of motility genes, gas vesicle clusters, and chemotaxis genes. Furthermore, several enzymes from the glycerol utilization cluster, including glycerol kinase and glycerol-3-phosphate dehydrogenase, were detected. The presence of these enzymes suggests that this strain could possibly grow on media containing glycerol. We attempted to identify if this organism could be associated with any of the biogeochemical pathways previously mentioned. We found hits associated with ammonia assimilation and reduction (E.C. 1.7.7.1 in Figure 5). However, no hits associated with significant steps in pathways of carbon fixation (Figure 4) and sulfur metabolism (Figure 6) were encountered.

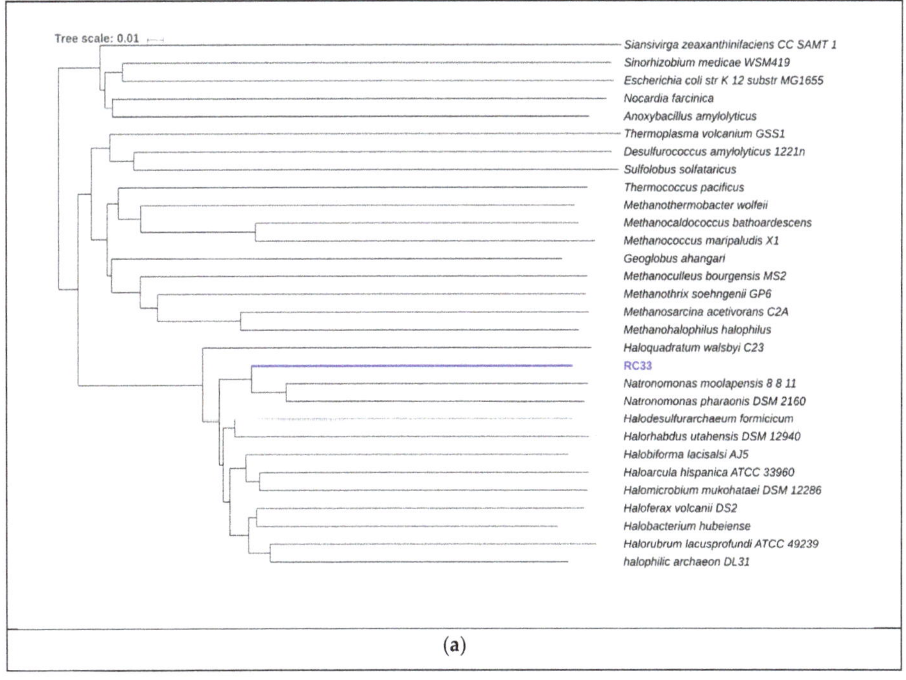

(a)

Figure 7. *Cont.*

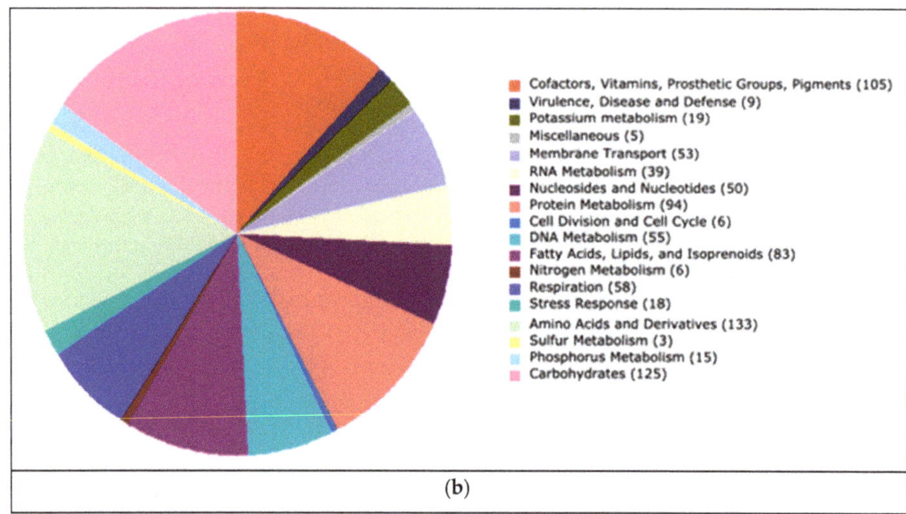

Figure 7. (**a**) Phylogeny for RC33 using amino acid identity. The scale represents change of amino acid substitution over time. (**b**) Subsystem category distribution for RC33. The graph represents the number of proteins that were grouped into a specific subsystems; 883 from a total 1570 coding sequences were identified to fit into subsystems. This chart was generated using Rapid Annotation System Technology (RAST).

Bin RC24 showed close relatedness to *Pontibacter korlensis* (Figure 8). Proteins encoding gram-negative cell walls were matched in the genome, therefore classifying this organism as gram-negative. No chemotaxis proteins or flagellar proteins were found, suggesting that this bacterium was non-motile. Ammonia assimilation genes, including ammonium transporter and nitrite reductase (EC 1.7.1.4 in Figure 5) genes, were detected. We could not identify any significant genes related to carbon fixation pathways (Figure 4) or sulfur metabolism pathways (Figure 6). Due to its presence in high salinity, as well as its relatedness to *Pontibacter*, it is suggested to be halotolerant. Nevertheless, AAI, as well as statistical analyses performed in MiGA, suggest that this organism is a new genus of the Bacteroidetes (p value = 0.0021), and it is suggested that it could represent a new family within the phylum (p value = 0.0051). Its low GC content is unusual compared to other organisms in hypersaline environments, however Ghai et al. [8] obtained similar results when they described a new genus of low GC *Actinobacteria* in the Santa Pola salterns through binning methods.

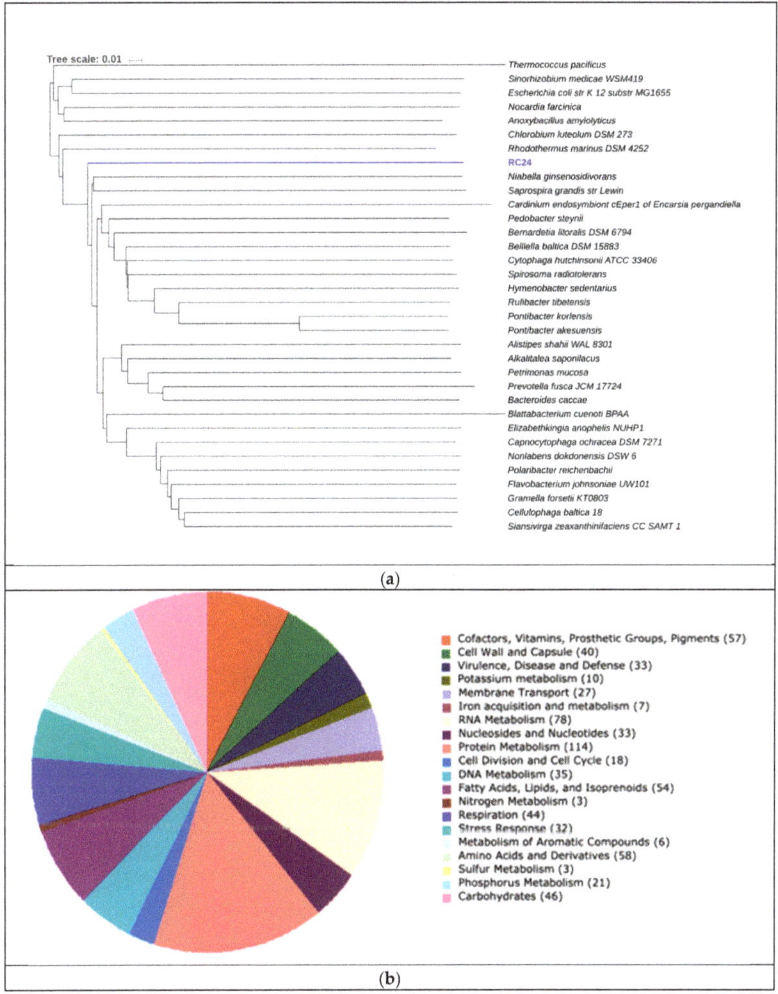

Figure 8. (**a**) Phylogeny for RC24 using AAI. The scale represents the number of amino acid substitutions over time. (**b**) Subsystem category distribution for RC24. The graph represents the number of proteins that were grouped into a specific subsystem; 723 of 1561 coding sequences were identified to fit into subsystems. This chart was generated using Rapid Annotation System Technology (RAST).

RC39 showed 62.81% similarity *to Halomicrobium mukohataei* (Figure 9). The organism is motile, with genes encoding for archaeal flagellar proteins. Furthermore, this organism possesses the genes necessary for ammonia assimilation, as well as nitrate and nitrite reductases (EC 1.7.99.4, 1.7.7.1, 1.7.1.4 respectively in Figure 5). *Halomicrobium mukohataei* has been described to be able to grow anaerobically under the presence of nitrate, as a terminal electron acceptor and forming nitrite as an end product in anaerobic respiration [69]. Similar growth has also been observed in other organisms, such as *Corynebacterium glutamicum*, where nitrate was used as an electron acceptor, producing nitrite as an end product [70]. We could not identify any essential genes related to carbon fixation pathways (Figure 4) or sulfur metabolism pathways (Figure 6). Once again, the presence of glycerol kinases and glycerol-3-phosphate dihydrogenase suggest that this organism can grow on media containing glycerol. Our results suggest this organism to be a novel species of the genus *Halomicrobium* (*p* value = 0.0046).

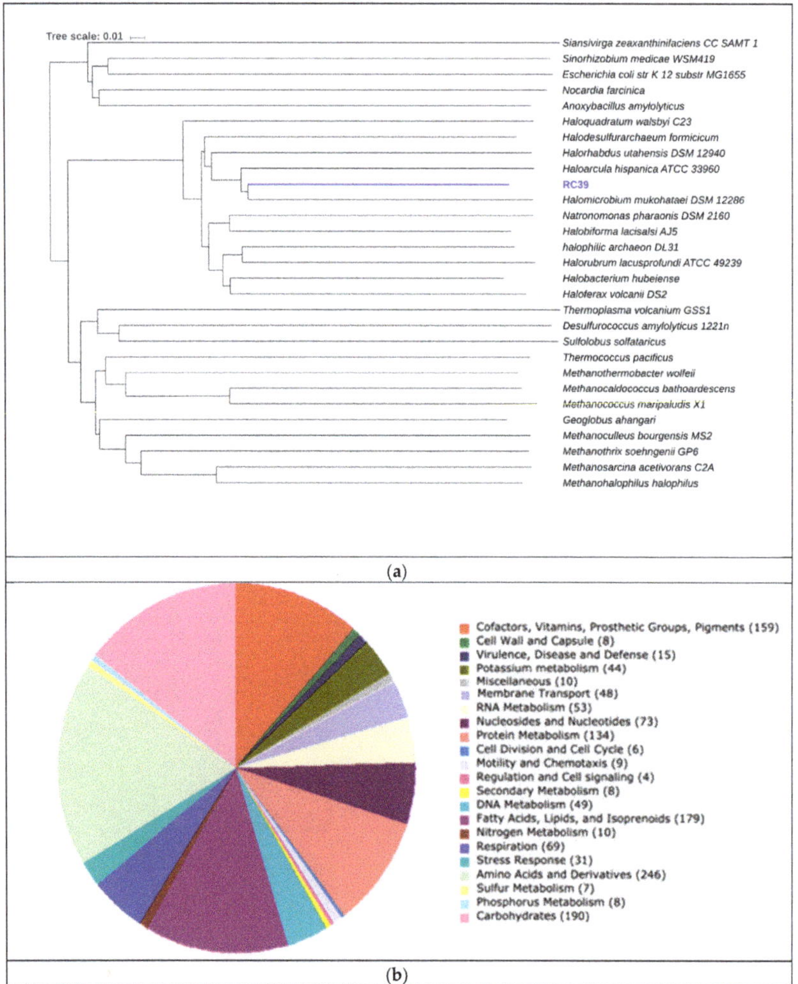

Figure 9. (**a**) Phylogeny for RC39 using AAI. The scale represents the number of amino acid substitutions over time. (**b**) Subsystem category distribution for RC39. The graph represents the number of proteins that were grouped into a specific subsystems; 1366 of 2439 coding sequences were identified to fit into subsystems. This chart was generated using Rapid Annotation System Technology (RAST).

Bin RC20 had an AAI of 65.83% with *Haloquadratum walsbyi* and MiGA analysis, suggesting it could represent a new species in the genus (*p* value = 0.0046) (Figure 10). The sequence of halomucin was blasted against the genome and was found. Halomucin, known as the largest archaeal protein, with 9159 amino acids, was described for the first time in *Haloquadratum walsbyi* [71]. Halomucin provides desiccation protection in saline environments to *Haloquadratum*, and is probably the secret to success for this organism in these environments. The presence of the gas vesicle cluster also coincides with the genome sequence of *Haloquadratum*. Bolhuis et al. [71] also described the presence of two bacteriorhodopsins and one halorhodopsin in *Haloquadratum*, which were also encountered here and is the reason they are able to grow phototrophically. This genome also encodes the presence of a TRAP (Tripartite ATP-independent periplasmic)-type C4-dicarboxylate transport system, two different ABC (ATP-Binding-Casette)-type sulfonate transport systems, and a phosphonate transport system, which are only described in *Haloquadratum walsbyi*. We detected genes related to nitrate/nitrite reductases (EC 1.7.99.4, 1.7.7.1, 1.7.1.4 in Figure 5). We also detected sulfate adenylyltransferase and adenylyl sulfate kinase (E.C. 2.7.7.4 and 2.7.1.25, respectively, in Figure 6) genes, both of which are important in sulfate reduction processes to sulfite [58]. No relevant genes related to carbon fixation pathways (Figure 4) were detected. The low GC content in this genome of 50.25% is also comparable to that of *Haloquadratum walsbyi* (47.9%). This low GC content is uncharacteristic of halophilic archaea due to their exposure to solar radiation. Due to the close relatedness of the genome with the *Haloquadratum walsbyi* genome, ANI was conducted in order to determine further resolution. ANI results showed an identity of 89.94%, indicating that RC20 is possibly a putative novel species of *Haloquadratum*.

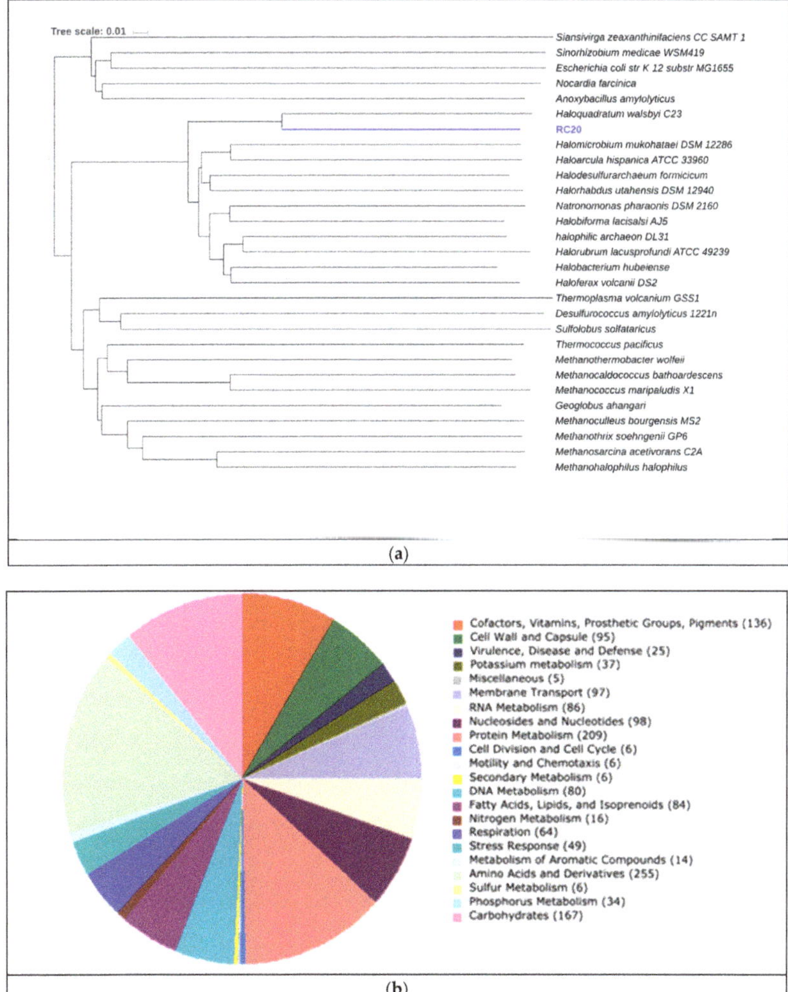

Figure 10. (**a**) Phylogeny for RC20 using AAI. The scale represents the amount of amino acid identity change over time. (**b**) Subsystem category distribution BIN20. The graph represents the number of proteins that were grouped into a specific subsystem; 1579 of 4980 coding sequences were identified to fit into subsystems. This chart was generated using Rapid Annotation System Technology (RAST).

Binning methods have uncovered previously undescribed microbes. As previously mentioned, Narasingarao et al. [20] recovered the recently proposed *Nanohaloarchaea* through binning methods. Furthermore, Ghai et al. [7] uncovered a novel group of low GC *Actinobacteria*, as well as a novel lineage of *Proteobacteria*, using metagenomic binning. Finally, through these methods, *Candidatus* "Nanopetramus" SG9 was also discovered [21]. Therefore, binning methods have been shown to be reliable in describing novel species in these environments, and may prove useful in providing a more non-biased assessment of the unculturable diversity in environmental samples. Moreover, a possible ecological role could be attributed to these reconstructed genomes. In our case, these putative genomes were shown to be associated to reductive pathways of the nitrogen cycle, as noted by the presence of nitrate and nitrite reductases in all four genomes. Additionally, we were able to identify a possible role of RC20 in sulfate reduction by the presence of related enzymes in sulfate adenylyltransferase and adenylyl sulfate kinase. One thing to note is that we were not able to detect any relevant genes related to carbon fixation, and it is therefore assumed that these organisms are not involved in these types of processes.

4. Conclusions

In this study, we used a deep sequencing strategy to obtain a considerable amount of sequence data (64 Gigabases) using a culture-independent approach, providing a more comprehensive perspective of microbial community structure and functional gene composition. By using a PCR unbiased culture-independent approach, a large microbial diversity has been discovered in the Cabo Rojo salterns. A diverse representation of prokaryotic phyla at a salinity of 34% (w/v) was shown, which has not been previously reported at this salinity. The microbial community structure at phylum-level could be influenced by weather fluctuations, which contribute to changes in the ionic composition of the crystallizer ponds. Additionally, using binning methods we have recovered four possible novel organisms that have been missed in our traditional culture-dependent surveys [12–18]. Although lack of phenotypic characterization of these strains makes validation of the organisms complicated, *Candidatus status* can be assigned to the four metagenomic bins. Further sampling may reveal more changes to this microbial community and can possibly unveil novel microbial groups. The importance of the organisms present in the Fraternidad crystallizer ponds is highlighted by the presence of essential genes related to the carbon, nitrogen, and sulfur cycles, with representatives from the diverse phyla encountered contributing to these cycles. Analysis of ionic composition of the crystallizer ponds following a precipitation event, as well as metatranscriptomics, may give us greater perspective regarding the active microbial community and their respective processes. Metagenomic analysis, although providing a wealth of information, does not distinguish active communities from dormant communities.

Supplementary Materials: All data can be found at the NCBI Sequence Read Archive under BioProject PRJNA343765 under accession numbers SRR8816317, SRR8816318 and SRR8816319 for samples MFF1, MFF2 and MFF3 respectively. The following are available online at http://www.mdpi.com/2073-4425/10/6/422/s1, Figure S1: Taxonomic hits by phylum of assembled contigs, Table S1: Kegg pathway hits for nitrogen metabolism, photosynthesis, and sulfur metabolism obtained from metagenomes MFF1, MFF2, and MFF3.

Author Contributions: R.M.-R. and R.C.-R. conceived and designed the experiments; R.C.-R. performed the experiments; R.M.-R. and R.C.-R. analyzed the data; R.M.-R. and R.C.-R. wrote the manuscript and approved the final version.

Funding: This study was supported by Howard Hughes Medical Institute (HHMI) "Enhancing Advanced Educational Opportunities in STEM Fields for Minority Students at UPRM" Grant Number 52007566.

Acknowledgments: The authors would like to thank members of the Extremophile Laboratory at UPRM for their help specially Daliana Campos and Nicole Feliberty. Special thanks to Dr. Carlos Rodríguez and Dr. Carlos Santos for their contributions and comments on this study.

Conflicts of Interest: The authors declare no conflict of interest.

References

1. Oren, A. Diversity of halophilic microorganisms: Environments, phylogeny, physiology, and applications. *J. Ind. Microbiol.* **2002**, *28*, 56–63.
2. Ventosa, A. Unusual micro-organisms from unusual habitats: Hypersaline environments. In *Symposia-Society for General Microbiology*; Cambridge University Press: Cambridge, UK, 2006; Volume 66, p. 223.
3. Sanchez-Porro, C.; Martı, S.; Mellado, E.; Ventosa, A. Diversity of moderately halophilic bacteria producing extracellular hydrolytic enzymes. *J. Appl. Microbiol.* **2003**, *94*, 295–300. [CrossRef] [PubMed]
4. De Lourdes Moreno, M.; Pérez, D.; García, M.T.; Mellado, E. Halophilic bacteria as a source of novel hydrolytic enzymes. *Life* **2013**, *3*, 38–51. [CrossRef] [PubMed]
5. Amoozegar, M.A.; Siroosi, M.; Atashgahi, S.; Smidt, H.; Ventosa, A. Systematics of Haloarchaea and Biotechnological Potential of their Hydrolytic Enzymes. *Microbiology* **2017**, *163*, 623–645. [CrossRef] [PubMed]
6. Ozcan, B.; Ozcengiz, G.; Coleri, A.; Cokmus, C. Diversity of Halophilic Archaea From Six Hypersaline Environments in Turkey. *J. Microbiol. Biotechnol.* **2007**, *17*, 985–992. [PubMed]
7. Ghai, R.; Pašić, L.; Fernández, A.B.; Martin-Cuadrado, A.-B.; Mizuno, C.M.; McMahon, K.D.; Papke, R.T.; Stepanauskas, R.; Rodriguez-Brito, B.; Rohwer, F.; et al. New abundant microbial groups in aquatic hypersaline environments. *Sci. Rep.* **2011**, *1*, 135. [CrossRef]
8. Ghai, R.; Hernandez, C.M.; Picazo, A.; Mizuno, C.M.; Ininbergs, K.; Díez, B.; Valas, R.; DuPont, C.L.; McMahon, K.D.; Camacho, A. Metagenomes of Mediterranean coastal lagoons. *Sci. Rep.* **2012**, *2*, 490. [CrossRef] [PubMed]
9. Ventosa, A.; Fernández, A.B.; León, M.J.; Sánchez-Porro, C.; Rodriguez-Valera, F. The Santa Pola saltern as a model for studying the microbiota of hypersaline environments. *Extremophiles* **2014**, *18*, 811–24. [CrossRef] [PubMed]
10. Oren, A. Halophilic microbial communities and their environments. *Curr. Opin. Biotechnol.* **2015**, *33*, 119–124. [CrossRef]
11. Rodriguez-Valera, F.; Rodriguez-Brito, B.; Thingstad, T.F.; Rohwer, F.; Mira, A. Explaining microbial population genomics through phage predation. *Nat. Rev.* **2009**, *7*, 828–836.
12. Montalvo-Rodriguez, R.; Vreeland, R.H.; Aharon, O.; Martin, K.; Lopez-garriga, J.; Chester, W. *Halogemetricum borinquense* sp. nov., a novel halophilic archaeon from Puerto Rico. *Int. J. Syst. Bacteriol.* **1998**, *48*, 1305–1312. [CrossRef] [PubMed]
13. Montalvo-Rodrıguez, R.; Vreeland, R.H.; Lopez-garriga, J.; Oren, A.; Ventosa, A.; Kamekura, M.; Chester, W. *Haloterrigena thermotolerans* sp. nov., a halophilic archaeon from Puerto Rico. *Int. J. Syst. Evol. Microbiol.* **2000**, 1065–1071.
14. Soto-Ramirez, N.; Sanchez-Porro, C.; Rosas, S.; Gonzalez, W.; Quinones, M.; Ventosa, A.; Montalvo-Rodriguez, R. *Halomonas avicenniae* sp. nov., isolated from the salty leaves of the black mangrove *Avicennia germinans* in Puerto Rico. *Int. J. Syst. Evol. Microbiol.* **2007**, *57*, 900–905. [CrossRef]
15. Soto-Ramírez, N.; Sánchez-Porro, C.; Rosas-Padilla, S.; Almodóvar, K.; Jiménez, G.; Machado-Rodríguez, M.; Zapata, M.; Ventosa, A.; Montalvo-Rodríguez, R. *Halobacillus mangrovi* sp. nov., a moderately halophilic bacterium isolated from the black mangrove *Avicennia germinans*. *Int. J. Syst. Evol. Microbiol.* **2008**, *58*, 125–130. [CrossRef]
16. Sanchez-Porro, C.; de la Haba, R.R.; Soto-Ramirez, N.; Marquez, M.C.; Montalvo-Rodriguez, R.; Ventosa, A. Description of *Kushneria aurantia* gen. nov., sp. nov., a novel member of the family *Halomonadaceae*, and a proposal for reclassification of *Halomonas marisflavi* as *Kushneria marisflavi* comb. nov., of *Halomonas indalinina* as *Kushneria indalinina* comb. nov. and of *Halomonas avicenniae* as *Kushneria avicenniae* comb. nov. *Int. J. Syst. Evol. Microbiol.* **2009**, *59*, 397–405. [PubMed]
17. Sánchez-Nieves, R.; Facciotti, M.; Saavedra-Collado, S.; Dávila-Santiago, L.; Rodríguez-Carrero, R.; Montalvo-Rodríguez, R. Draft genome of *Haloarcula rubripromontorii* strain SL3, a novel halophilic archaeon isolated from the solar salterns of Cabo Rojo, Puerto Rico. *Genom. Data* **2016**, *7*, 287–289. [CrossRef]
18. Sánchez-Nieves, R.; Facciotti, M.T.; Saavedra-Collado, S.; Dávila-Santiago, L.; Rodríguez-Carrero, R.; Montalvo-Rodríguez, R. Draft genome sequence of *Halorubrum tropicale* strain V5, a novel halophilic archaeon isolated from the solar salterns of Cabo Rojo, Puerto Rico. *Genom. Data* **2016**, *7*, 284–286. [CrossRef] [PubMed]

19. Thomas, T.; Gilbert, J.; Meyer, F. Metagenomics—A guide from sampling to data analysis. *Microb. Inform. Exp.* **2012**, *2*, 3. [CrossRef] [PubMed]

20. Narasingarao, P.; Podell, S.; Ugalde, J.A.; Brochier-Armanet, C.; Emerson, J.B.; Brocks, J.J.; Heidelberg, K.B.; Banfield, J.F.; Allen, E.E. De novo metagenomic assembly reveals abundant novel major lineage of Archaea in hypersaline microbial communities. *ISME J.* **2012**, *6*, 81–93. [CrossRef] [PubMed]

21. Crits-Christoph, A.; Gelsinger, D.R.; Ma, B.; Wierzchos, J.; Ravel, J.; Davila, A.; Casero, M.C.; DiRuggiero, J. Functional interactions of archaea, bacteria and viruses in a hypersaline endolithic community. *Environ. Microbiol.* **2016**, *18*, 2064–2077. [CrossRef] [PubMed]

22. Rodríguez-García, C.M. Metagenomic Analysis of Prokaryotic Communities from Hypersaline Environments at Cabo Rojo, Puerto Rico through Pyrosequencing of 16S rRNA Genes. Master's Thesis, University of Puerto Rico, San Juan, Puerto Rico, 2016.

23. Martín-Cuadrado, A.-B.; López-García, P.; Alba, J.-C.; Moreira, D.; Monticelli, L.; Strittmatter, A.; Gottschalk, G.; Rodríguez-Valera, F. Metagenomics of the deep Mediterranean, a warm bathypelagic habitat. *PLoS ONE* **2007**, *2*, e914. [CrossRef]

24. Andrews, S. FastQC: A Quality Control Tool for High Throughput Sequence Data. 2010. Available online: http://www.bioinformatics.babraham.ac.uk/projects/fastqc (accessed on 1 April 2019).

25. Nurk, S.; Meleshko, D.; Korobeynikov, A.; Pevzner, P.A. metaSPAdes: A new versatile metagenomic assembler. *Genome Res.* **2017**, *27*, 824–834. [CrossRef]

26. Meyer, F.; Paarmann, D.; D'Souza, M.; Olson, R.; Glass, E.M.; Kubal, M.; Paczian, T.; Rodriguez, A.; Stevens, R.; Wilke, A.; et al. The metagenomics RAST server—A public resource for the automatic phylogenetic and functional analysis of metagenomes. *BMC Bioinform.* **2008**, *9*, 386. [CrossRef]

27. Kanehisa, M.; Goto, S. KEGG: Kyoto encyclopedia of genes and genomes. *Nucleic Acids Res.* **2000**, *28*, 27–30. [CrossRef]

28. Muller, J.; Szklarczyk, D.; Julien, P.; Letunic, I.; Roth, A.; Kuhn, M.; Powell, S.; von Mering, C.; Doerks, T.; Jensen, L.J.; et al. eggNOG v2.0: Extending the evolutionary genealogy of genes with enhanced non-supervised orthologous groups, species and functional annotations. *Nucleic Acids Res.* **2010**, *38*, D190–D195. [CrossRef]

29. Tatusov, R.L.; Fedorova, N.D.; Jackson, J.D.; Jacobs, A.R.; Kiryutin, B.; Koonin, E.V.; Krylov, D.M.; Mazumder, R.; Mekhedov, S.L.; Nikolskaya, A.N. The COG database: An updated version includes eukaryotes. *BMC Bioinform.* **2003**, *4*, 41. [CrossRef]

30. Li, H.; Durbin, R. Fast and accurate short read alignment with Burrows—Wheeler transform. *Bioinformatics* **2009**, *25*, 1754–1760. [CrossRef]

31. Kang, D.D.; Froula, J.; Egan, R.; Wang, Z. MetaBAT, an efficient tool for accurately reconstructing single genomes from complex microbial communities. *PeerJ* **2015**, *3*, e1165. [CrossRef]

32. Parks, D.H.; Imelfort, M.; Skennerton, C.T.; Hugenholtz, P.; Tyson, G.W. CheckM: Assessing the quality of microbial genomes recovered from isolates, single cells, and metagenomes. *Genome Res.* **2015**, *25*, 1043–1055. [CrossRef]

33. Rodriguez-R, L.M.; Gunturu, S.; Harvey, W.T.; Rosselí O-Mora, R.; Tiedje, J.M.; Cole, J.R.; Konstantinidis, K.T. The Microbial Genomes Atlas (MiGA) webserver: Taxonomic and gene diversity analysis of Archaea and Bacteria at the whole genome level. *Nucleic Acids Res.* **2018**, *46*, W282–W288. [CrossRef]

34. Letunic, I.; Bork, P. Interactive tree of life (iTOL) v3: An online tool for the display and annotation of phylogenetic and other trees. *Nucleic Acids Res.* **2016**, *44*, W242–W245. [CrossRef]

35. Overbeek, R.; Olson, R.; Pusch, G.D.; Olsen, G.J.; Davis, J.J.; Disz, T.; Edwards, R.A.; Gerdes, S.; Parrello, B.; Shukla, M.; et al. The SEED and the Rapid Annotation of microbial genomes using Subsystems Technology (RAST). *Nucleic Acids Res.* **2014**, *42*, D206–D214. [CrossRef]

36. Olson, N.D.; Treangen, T.J.; Hill, C.M.; Cepeda-espinoza, V.; Ghurye, J.; Koren, S.; Pop, M. Metagenomic assembly through the lens of validation: Recent advances in assessing and improving the quality of genomes assembled from metagenomes. *Brief. Bioinform.* **2017**, 1–11. [CrossRef] [PubMed]

37. Mikheenko, A.; Saveliev, V.; Gurevich, A. Genome analysis MetaQUAST: Evaluation of metagenome assemblies. *Bioinformatics* **2016**, *32*, 1088–1090. [CrossRef]

38. Rhodes, M.E.; Oren, A.; House, C.H. The Dynamics and Persistence of Dead Sea Microbial Populations as Shown by High Throughput Sequencing of Ribosomal RNA. *Appl. Environ. Microbiol.* **2012**, *8*, 2489–2492. [CrossRef]

39. Pasic, L.; Bartual, S.G.; Ulrih, N.P.; Grabnar, M.; Velikonja, B.H. Diversity of halophilic archaea in the crystallizers of an Adriatic solar saltern. *FEMS Microbiol. Ecol.* **2005**, *54*, 491–498. [PubMed]

40. Plominsky, A.M.; Henríquez-Castillo, C.; Delherbe, N.; Podell, S.; Ramirez-Flandes, S.; Ugalde, J.A.; Santibañez, J.F.; van den Engh, G.; Hanselmann, K.; Ulloa, O.; et al. Distinctive Archaeal Composition of an Artisanal Crystallizer Pond and Functional Insights Into Salt-Saturated Hypersaline Environment Adaptation. *Front. Microbiol.* **2018**, *9*, 1800. [CrossRef]

41. Podell, S.; Emerson, J.B.; Jones, C.M.; Ugalde, J.A.; Welch, S.; Heidelberg, K.B.; Banfield, J.F.; Allen, E.E. Seasonal fluctuations in ionic concentrations drive microbial succession in a hypersaline lake community. *ISME J.* **2014**, *8*, 979–90. [CrossRef]

42. Peter, H.; Hörtnagl, P.; Reche, I.; Sommaruga, R. Bacterial diversity and composition during rain events with and without Saharan dust influence reaching a high mountain lake in the Alps. *Environ. Microbiol. Rep.* **2014**, *6*, 618–624. [CrossRef]

43. Demergasso, C.; Escudero, L.; Casamayor, E.O.; Chong, G.; Balagué, V.; Pedrós-Alió, C. Novelty and spatio–temporal heterogeneity in the bacterial diversity of hypersaline Lake Tebenquiche (Salar de Atacama). *Extremophiles* **2008**, *12*, 491–504. [CrossRef]

44. Baricz, A.; Coman, C.; Andrei, A.Ş.; Muntean, V.; Keresztes, Z.G.; Păuşan, M.; Alexe, M.; Banciu, H.L. Spatial and temporal distribution of archaeal diversity in meromictic, hypersaline Ocnei Lake (Transylvanian Basin, Romania). *Extremophiles* **2014**, *18*, 399–413. [CrossRef] [PubMed]

45. Atanasova, N.S.; Demina, T.A.; Buivydas, A.; Bamford, D.H.; Oksanen, H.M. Archaeal viruses multiply: Temporal screening in a solar saltern. *Viruses* **2015**, *7*, 1902–1926. [CrossRef] [PubMed]

46. Almeida-Dalmet, S.; Sikaroodi, M.; Gillevet, P.M.; Litchfield, C.D.; Baxter, B.K. Temporal Study of the Microbial Diversity of the North Arm of Great Salt Lake, Utah, U.S. *Microorganisms* **2015**, *3*, 310–326. [CrossRef]

47. Javor, B.J. *Hypersaline Environments: Microbiology and Biogeochemistry*; Springer Science & Business Media: Berlin, Germany, 2012; ISBN 3642743706.

48. Joint, I.; Henriksen, P.; Garde, K.; Riemann, B. Primary production, nutrient assimilation and microzooplankton grazing along a hypersaline gradient. *FEMS Microbiol. Ecol.* **2002**, *39*, 245–257. [CrossRef] [PubMed]

49. Oren, A. The ecology of Dunaliella in high-salt environments. *J. Biol. Res.* **2014**, *21*, 1–8. [CrossRef] [PubMed]

50. Ríos, A.d.l.; Valea, S.; Ascaso, C.; Davila, A.F.; Kastovsky, J.; McKay, C.P.; Wierzchos, J. Comparative analysis of the microbial communities inhabiting halite evaporites of the Atacama Desert. *Int. Microbiol.* **2010**, *13*, 79–89.

51. Payler, S.J.; Biddle, J.F.; Sherwood Lollar, B.; Fox-Powell, M.G.; Edwards, T.; Ngwenya, B.T.; Paling, S.M.; Cockell, C.S. An Ionic Limit to Life in the Deep Subsurface. *Front. Microbiol.* **2019**, *10*, 426. [CrossRef] [PubMed]

52. Cabello, P.; Roldan, M.D.; Moreno-Vivian, C. Nitrate reduction and the nitrogen cycle in archaea. *Microbiology* **2004**, *150*, 3527–3546. [CrossRef]

53. Offre, P.; Spang, A.; Schleper, C. Archaea in Biogeochemical Cycles. *Annu. Rev. Microbiol.* **2013**, *16*, 437–457. [CrossRef] [PubMed]

54. Stahl, D.A.; Torre, R. De Physiology and Diversity of Ammonia-Oxidizing Archaea. *Annu. Rev. Microbiol.* **2012**, *66*, 83–101. [CrossRef]

55. Könneke, M.; Bernhard, A.E.; José, R.; Walker, C.B.; Waterbury, J.B.; Stahl, D.A. Isolation of an autotrophic ammonia-oxidizing marine archaeon. *Nature* **2005**, *437*, 543. [CrossRef] [PubMed]

56. Walker, C.B.; De La Torre, J.R.; Klotz, M.G.; Urakawa, H.; Pinel, N.; Arp, D.J.; Brochier-Armanet, C.; Chain, P.S.G.; Chan, P.P.; Gollabgir, A. Nitrosopumilus maritimus genome reveals unique mechanisms for nitrification and autotrophy in globally distributed marine crenarchaea. *Proc. Natl. Acad. Sci. USA* **2010**, *107*, 8818–8823. [CrossRef] [PubMed]

57. Oren, A. Bioenergetic aspects of halophilism. *Microbiol. Mol. Biol. Rev.* **1999**, *63*, 334–348. [PubMed]

58. Muyzer, G.; Stams, A.J.M. The ecology and biotechnology of sulphate-reducing bacteria. *Nat. Rev.* **2008**, *6*, 441–454. [CrossRef] [PubMed]

59. Barton, L.L.; Fauque, G.D. Biochemistry, physiology and biotechnology of sulfate-reducing bacteria. *Adv. Appl. Microbiol.* **2009**, *68*, 41–98. [PubMed]

60. Tourova, T.P.; Kovaleva, O.L.; Sorokin, D.Y.; Muyzer, G. Ribulose-1, 5-bisphosphate carboxylase/oxygenase genes as a functional marker for chemolithoautotrophic halophilic sulfur-oxidizing bacteria in hypersaline habitats. *Microbiology* **2010**, *156*, 2016–2025. [CrossRef]

61. Segerer, A.; Neuner, A.; Kristjansson, J.K.; Stetter, K.O. Acidianus infernus gen. nov., sp. nov., and Acidianus brierleyi comb. nov.: Facultatively aerobic, extremely acidophilic thermophilic sulfur-metabolizing archaebacteria. *Int. J. Syst. Evol. Microbiol.* **1986**, *36*, 559–564. [CrossRef]

62. Hafenbradl, D.; Keller, M.; Dirmeier, R.; Rachel, R.; Roßnagel, P.; Burggraf, S.; Huber, H.; Stetter, K.O. *Ferroglobus placidus* gen. nov., sp. nov., a novel hyperthermophilic archaeum that oxidizes Fe^{2+} at neutral pH under anoxic conditions. *Arch. Microbiol.* **1996**, *2*, 308–314. [CrossRef]

63. Rodriguez-R, L.M.; Konstantinidis, K.T. Bypassing Cultivation to Identify Bacterial Species Culture-independent genomic approaches identify credibly distinct clusters, avoid cultivation bias, and provide true insights into microbial species. *Microbe* **2014**, *9*, 111–118.

64. Arahal, D.R. *Whole-Genome Analyses: Average Nucleotide Identity*, 1st ed.; Elsevier Ltd.: Amsterdam, The Netherlands, 2014; Volume 41.

65. Goris, J.; Konstantinidis, K.T.; Klappenbach, J.A.; Coenye, T.; Vandamme, P.; Tiedje, J.M. DNA—DNA hybridization values and their relationship to whole-genome sequence similarities. *Int. J. Syst. Evol. Microbiol.* **2007**, *57*, 81–91. [CrossRef] [PubMed]

66. Konstantinidis, K.T.; Tiedje, J.M. Towards a Genome-Based Taxonomy for Prokaryotes. *J. Bacteriol.* **2005**, *187*, 6258–6264. [CrossRef] [PubMed]

67. Richter, M.; Rossello, R. Shifting the genomic gold standard for the prokaryotic species definition. *PNAS* **2009**, *106*, 19126–19131. [CrossRef]

68. Luo, C.; Rodriguez-R, L.M.; Konstantinidis, K.T. *A User's Guide to Quantitative and Comparative Analysis of Metagenomic Datasets*, 1st ed.; Elsevier Inc.: Amsterdam, The Netherlands, 2013; Volume 531, ISBN 9780124078635.

69. Oren, A.; Elevi, R.; Watanabe, S.; Ihara, K.; Corcelli, A. *Halomicrobium mukohataei* gen. nov., comb. nov., and emended description of *Halomicrobium mukohataei*. *Int. J. Syst. Evol. Microbiol.* **2002**, *52*, 1831–1835. [PubMed]

70. Nishimura, T.; Vertès, A.A.; Shinoda, Y. Anaerobic growth of *Corynebacterium glutamicum* using nitrate as a terminal electron acceptor. *Appl. Microbiol. Biotechnol.* **2007**, *75*, 889–897. [CrossRef] [PubMed]

71. Bolhuis, H.; Palm, P.; Wende, A.; Falb, M.; Rampp, M.; Rodriguez-valera, F.; Pfeiffer, F.; Oesterhelt, D. The genome of the square archaeon *Haloquadratum walsbyi*: Life at the limits of water activity. *BMC Genom.* **2006**, *7*, 1–12. [CrossRef]

 *genes*

Article

Halobacterium salinarum virus *ChaoS9*, a Novel Halovirus Related to PhiH1 and PhiCh1

Mike Dyall-Smith [1,2], **Peter Palm** [1], **Gerhard Wanner** [3], **Angela Witte** [4], **Dieter Oesterhelt** [1] and **Friedhelm Pfeiffer** [1,*]

[1]   Computational Biology Group, Max-Planck-Institute of Biochemistry, Am Klopferspitz 18, 82152 Martinsried, Germany; mike.dyallsmith@gmail.com (M.D.-S.); hp.palm@kabelmail.de (P.P.); oesterhe@biochem.mpg.de (D.O.)

[2]   Veterinary Biosciences, Faculty of Veterinary and Agricultural Sciences, University of Melbourne, Parkville, VIC 3052, Australia

[3]   AG Ultrastrukturforschung, Biozentrum der LMU, Großhadernerstrasse 2-4, 82152 Martinsried, Germany; wanner@lrz.uni-muenchen.de

[4]   Department of Microbiology, Immunobiology and Genetics, MFPL Laboratories, University of Vienna, Dr. Bohr-Gasse 9, 1030 Vienna, Austria; angela.witte@univie.ac.at

*   Correspondence: fpf@biochem.mpg.de; Tel.: +49-89-8578-2323

Received: 29 January 2019; Accepted: 25 February 2019; Published: 1 March 2019

Abstract: The unexpected lysis of a large culture of *Halobacterium salinarum* strain S9 was found to be caused by a novel myovirus, designated ChaoS9. Virus purification from the culture lysate revealed a homogeneous population of caudovirus-like particles. The viral genome is linear, dsDNA that is partially redundant and circularly permuted, has a unit length of 55,145 nt, a G + C% of 65.3, and has 85 predicted coding sequences (CDS) and one tRNA (Arg) gene. The left arm of the genome (0–28 kbp) encodes proteins similar in sequence to those from known caudoviruses and was most similar to myohaloviruses phiCh1 (host: *Natrialba magadii*) and phiH1 (host: *Hbt. salinarum*). It carries a tail-fiber gene module similar to the invertible modules present in phiH1 and phiCh1. However, while the tail genes of ChaoS9 were similar to those of phiCh1 and phiH1, the Mcp of ChaoS9 was most similar (36% aa identity) to that of *Haloarcula hispanica* tailed virus 1 (HHTV-1). Provirus elements related to ChaoS9 showed most similarity to tail/assembly proteins but varied in their similarity with head/assembly proteins. The right arm (29–55 kbp) of ChaoS9 encoded proteins involved in DNA replication (ParA, RepH, and Orc1) but the other proteins showed little similarity to those from phiH1, phiCh1, or provirus elements, and most of them could not be assigned a function. ChaoS9 is probably best classified within the genus *Myohalovirus*, as it shares many characteristics with phiH1 (and phiCh1), including many similar proteins. However, the head/assembly gene region appears to have undergone a recombination event, and the inferred proteins are different to those of phiH1 and phiCh1, including the major capsid protein. This makes the taxonomic classification of ChaoS9 more ambiguous. We also report a revised genome sequence and annotation of *Natrialba* virus phiCh1.

Keywords: halovirus; caudovirus; halobacteria; *Archaea*; *haloarchaea*; genome inversion; transposon

---

1. Introduction

Viruses infecting extremely halophilic archaea (haloarchaea) include a variety of morphotypes, such as caudoviruses (e.g., phiH1), round viruses (e.g., SH1), pleomorphic viruses (e.g., His2), and spindle-shaped viruses (e.g., His1) [1,2]. In hypersaline environments such as salt lakes and saltern crystallizer ponds, the vast majority of prokaryotes are usually haloarchaea (Class Halobacteria) with cell concentrations reaching up to 10^8 per mL, but the concentration of virus particles is often 10-fold higher [3], so that haloviruses are important regulators of host cell populations as well as major drivers of their evolution. Among the haloarchaeal caudoviruses, examples of myovirus-like and siphovirus-like viruses have been described, and both temperate and virulent (lytic) isolates have been reported [4–7]. With many haloarchaeal genome sequences now available, it is clear that proviruses are common and widespread, and can be present either as plasmids [8] or integrated into the host chromosome [9,10]. The high numbers of viruses in environmental samples, their stability, and the need for large quantities of salt for cultivation of haloarchaea, is a potential hazard for large scale culture of haloarchaea for biotechnological purposes, as virus contamination from medium components or the local environment could result in lysis of the cells.

For many decades, the biological function of bacteriorhodopsin was a focus of study in the Oesterhelt department of the Max Planck Institute (MPI) in Martinsried [11–13]. The bacteriorhodopsin producer strain *Halobacterium salinarum* strain S9 was derived from strain R1 (DSM 671). The R1 strain and derivative strains such as S9 have been widely used to produce commercial quantities of bacteriorhodopsin [14]. *Hbt. salinarum* S9 (originally strain R1S9) was first described in a 1979 review by Stoeckenius et al. [15] as a purple membrane overproducer strain derived by Lily Jan (unpublished) from strain R1 by nitrosoguanidine mutagenesis. It has been used in numerous studies, such as gene regulation and expression [16–19], and cell chemotaxis and phototaxis [20,21]. In some studies, this strain has also been labelled as *bat+* [18].

Commercial production of pure bacteriorhodopsin entails regular, large-scale cultivation, a practice known to increase the likelihood of virus contamination that can lead to sudden lysis of the microbial cells in a bioreactor [22]. Such an event occurred in 2007 in a 1 m^3 culture of *Hbt. salinarum* S9 being grown at the MPI laboratory. A similar, spontaneous lysis event had occurred in 1974 in the Oesterhelt group and a sample of culture fluid, which had been collected by Hartmut Michel, turned out to contain halovirus phiH [23]. Accordingly, it was initially assumed that phiH was also responsible for the 2007 event.

The aim of this study was to characterize the virus that caused the large-scale lysis of *Hbt. salinarum* S9 in 2007. This turned out to be a novel halovirus, related to phiH1 and phiCh1, which we named ChaoS9 (Chao: Caudovirus of haloarchaeal origin; S9: the affected strain). The virus morphology, proteins, and genome sequence were analyzed and compared with other described haloviruses and provirus elements in order to assess its novelty, evolution, and taxonomic position. During this study, we also resequenced Natrialba virus phiCh1.

2. Materials and Methods

2.1. Host Strain, Virus Isolation, Cultivation, and Purification

Hbt. salinarum S9 is a purple-membrane (Pum) constitutive strain [15]. It was grown aerobically in peptone/salts medium, as previously described [24]. Virus purification and DNA extraction followed the methods previously described for halovirus phiH [23]. Briefly, this involved filtration of the lysate through diatomaceous earth (DE), concentration of viral particles from the filtrate by PEG$_{6000}$ precipitation, and finally, banding twice on CsCl gradients. The plaque assay method followed that was described previously for phiH [23].

2.2. DNA Sequencing and Assembly of the ChaoS9 Genome

The ChaoS9 genome was sequenced by the whole-genome shotgun approach (7-fold coverage). Briefly, DNA was randomly sheared by sonication and fragments cloned into plasmid vectors and sequenced by the chain termination method using the BigDye system (Applied Biosystems, ABI, Foster City, CA, USA). Contig assembly used the Phred–Phrap–Consed package [25]. The remaining gaps were sequenced by targeted PCR amplification of viral DNA using custom primers (Supplementary Table S1), followed by sequencing of the amplimers using the BigDye system. Sequencing was performed at the MPI of Biochemistry core sequencing facility (Martinsried, Germany).

2.3. DNA Resequencing of Natrialba Virus phiCh1

The genome sequence of phiCh1 was determined using the Illumina HiSeq platform (Max-Planck Genome Centre, Cologne, Germany), as previously described for phiH1 [26]. This generated 309 Mbp of high-quality sequence data. Reads were mapped to the reference genome (GenBank:AF440695) using the "map to reference" option within the Geneious (version 10.2) environment, as described for phiH1 [26]. Average genome coverage was 3192-fold. Gene annotation used a combination of gene prediction with GeneMarkS-2 [27] and manual refinement using database searches (BLASTp/BLASTn). Refinement of the annotation took into consideration the original annotation of phiCh1, the annotation of *Nab. magadii* plasmid pNMAG03 [28] and the annotation of haloviruses phiH1 and ChaoS9, as well as the nr database at the NCBI webserver (https://blast.ncbi.nlm.nih.gov, accessed 10 December 2018).

2.4. Electron Microscopy of Virus

Samples of purified ChaoS9 virus were fixed iso-osmotically with 2.5% (*v/v*) glutaraldehyde. A drop of the sample was then placed on a carbon-coated copper grid, freshly treated by glow discharge to make it hydrophilic. After incubation for 2 min, the drop was quickly removed, and the grid was stained with a solution of 1% (*w/v*) uranyl acetate and 0.01% (*w/v*) glucose. Micrographs were taken with an EM 912 electron microscope (Zeiss, Oberkochen, Germany) equipped with an integrated OMEGA energy filter operated in the zero-loss mode. Head diameters were measured on micrographs from vertex-to-vertex, not including the axis in line with the tail.

2.5. Protein Analyses of Purified Virus

Samples of purified virus were added to Laemmli sample buffer (with 2-mercaptoethanol) [29] and heated at 95 °C for 5 min before loading on a precast NuPAGE 4%–12% Bis-Tris polyacrylamide gel (Invitrogen). The running buffer was NuPAGE MES buffer with 0.1% (*w/v*) sodium dodecyl sulfate (SDS). PageRulerTM prestained protein ladder—3 (Fermentas, #SM1819), containing proteins of 250, 130, 100, 70, 55, 35, 25, 15, 10 kDa, were loaded in parallel. The gel was stained with Coomassie blue and destained in 10% acetic acid.

2.6. Bioinformatics Analyses

Sequence alignments, editing, and phylogenetic tree reconstructions were performed within the Geneious (version 10.2) suite of programs (https://www.geneious.com/) [30]. For phylogenetic tree reconstructions, protein sequences were first aligned using CLUSTALW, and trees inferred using the Neighbor–Joining algorithm (within Geneious). Consensus trees were determined after 100 bootstrap repetitions. GeneMarkS-2 [27] was used for gene prediction. Protein and DNA sequence similarity searches used the programs BLASTp and BLASTn to search the nr databases at the NCBI webserver (https://blast.ncbi.nlm.nih.gov, accessed on 10 December 2018). The VIRFAM webserver (http://biodev.cea.fr/virfam/) [31] was used to classify ChaoS9. Halovirus genomes were compared by the dot plot method zPicture [32]. Average Nucleotide Identity (ANIb) values between viral genomes were calculated using EZbiocloud webserver [33], and a heatmap produced from these values using heatmapper [34]. Searches for matching CRISPR spacers were performed at the CRISPRs web server (http://crispr.i2bc.paris-saclay.fr/crispr/BLAST/CRISPRsBlast.php) [35] and at the IMG/VR server (https://img.jgi.doe.gov/cgi-bin/vr/main.cgi) [36], and also by direct searching of hypersaline metagenomes using the crass software [37], as previously described [26]. Identification of the *pac* site utilized the program PhageTerm [38] as implemented on the CPT Phage Galaxy (https://cpt.tamu.edu/galaxy-pub/). Correction of the molecular weight estimates of acidic proteins was based on the study of Guan et al. [39]. The following equations were used to convert the protein MW calculated from the inferred protein sequence into an apparent protein MW expected upon SDS-PAGE: $MW_{real} + Sh = MW_{app}$, where MW_{real} is the MW computed from the protein sequence, MW_{app} is the MW expected to be observed by SDS-PAGE, and Sh is the shift (or estimated error) computed by the formula

Sh = len $\times$ (276.5x − 31.33). Here, len is the length of the protein (in aa) and x represents the proportion of acidic amino acids (Glu, Asp).

Data Availability

The ChaoS9 genome sequence has been deposited at Genbank under the accession MK310226. The revised phiCh1 genome sequence has been deposited under accession MK450543 and the phiCh1 raw reads were submitted to the SRA and can be retrieved via BioProject PRJNA517034.

3. Results

3.1. Isolation of Halovirus ChaoS9

In 2007, a large-scale culture (1 m^3) of *Hbt. salinarum* S9 lysed spontaneously. Suspecting a virus infection, the lysate was processed by the method described for purifying halovirus phiH1 [23]. This involved filtration through diatomaceous earth, addition of PEG$_{6000}$ to precipitate virus particles, and the resulting pellets applied to CsCl gradients. Virus bands were observed on CsCl gradients, and negative-stain electron-microscopy of this fraction revealed a homogeneous population of tailed virus particles, some displaying contracted tails (Figure 1). The head diameter was 61 nm, uncontracted tails were 128 $\times$ 17 nm, and contracted tails had sheaths of 74 $\times$ 23 nm.

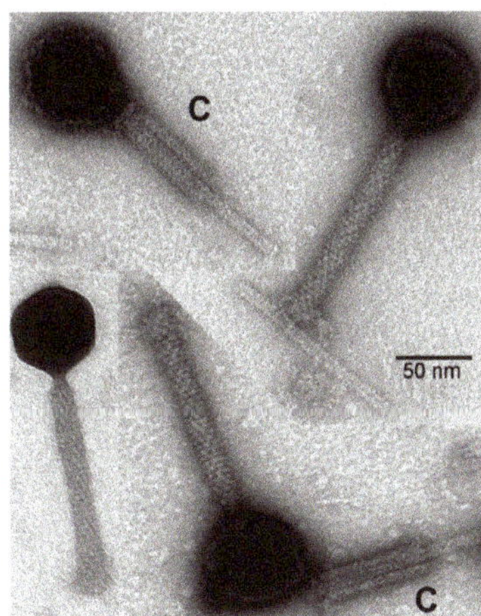

Figure 1. Morphology of ChaoS9 particles by negative-stain electron microscopy. Purified virus was fixed with 2.5% glutaraldehyde, negatively-stained with 1% uranyl acetate and examined under a Zeiss EM 912 electron microscope. Scale bar represents 50 nm. Examples of contracted tails (labeled with a C) are shown in the top left-hand corner and the bottom right-hand corner.

3.2. Virus Proteins

The proteins of purified virus were separated by SDS-PAGE and revealed four major protein bands (VP1, VP3, VP4, and VP7), three minor bands (VP2, VP5, and VP6), and several very faint bands (Figure 2).

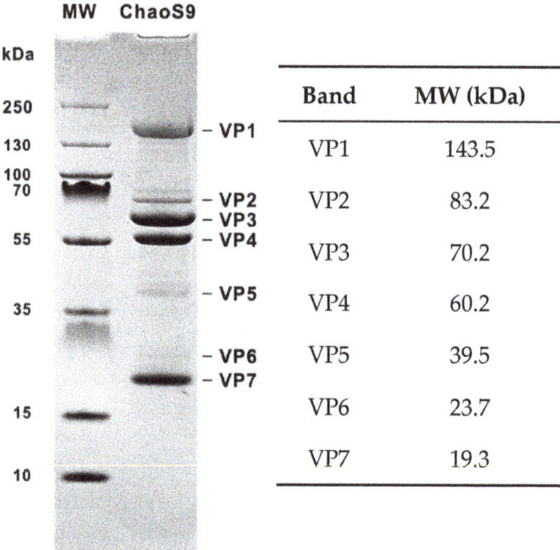

Band	MW (kDa)
VP1	143.5
VP2	83.2
VP3	70.2
VP4	60.2
VP5	39.5
VP6	23.7
VP7	19.3

Figure 2. Proteins of halovirus ChaoS9. Purified virus was heated in sample buffer and separated on a 4%–12% SDS-polyacrylamide gel, and the separated proteins stained with Coomassie blue. MW, molecular weight standards (PageRuler). The sizes of the protein standards are indicated at the left (in kDa). Virus proteins (VP1 to VP7) are numbered from largest to smallest. Molecular weight estimates of the major virus proteins are shown in the adjacent table.

3.3. ChaoS9 Genome and Sequence

Nucleic acids were extracted from virus preparations, treated with several restriction enzymes and the digests separated by agarose gel electrophoresis (Figure 3). Cleavage of the viral genome by these enzymes, and the different fragment patterns observed for each enzyme, indicated that the genome was dsDNA. The ChaoS9 genome was sequenced using the whole-genome shotgun approach (7-fold coverage; see Methods) and contig gaps were closed by PCR amplification of viral DNA using specific primers (Table S1). All sequences assembled to a single contig with a unit length of 55,145 nt and a G + C content of 65.3% (Table 1). Comparison of the observed restriction fragment patterns with in silico predictions based on linear and circular versions of the genome (Figure S1) not only showed a close correspondence, but also identified terminal fragments that were either underrepresented or not visible (white triangles) on gels, as well as bands predicted to occur only in longer than unit length genomes (blue triangles). These results were consistent with the viral genome being partially redundant and circularly permuted, as is typical for headful packaging.

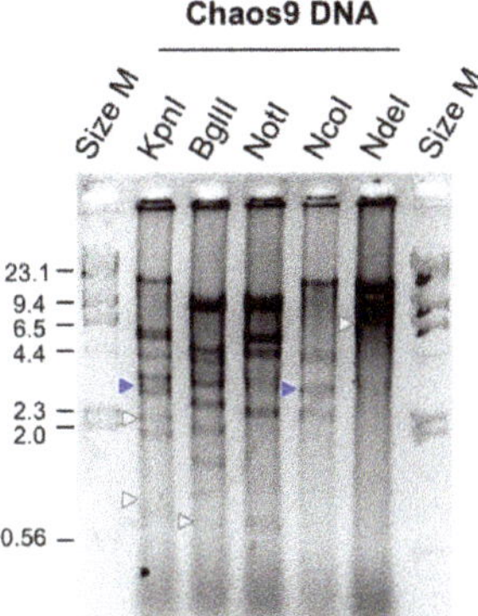

Figure 3. Restriction digests of ChaoS9 DNA. Enzymes used are indicated above each well. The outside wells (Size M) were loaded with DNA size markers (Lambda-HindIII), and the fragment sizes (in kbp) are shown at the left edge. White triangles; terminal fragments predicted from the DNA sequence that were either underrepresented or not visible on gels. Blue triangles; bands predicted from the DNA sequence to occur only in longer than unit length genomes. See also Supplementary Figure S1.

A dotplot comparison of this sequence with 17 other tailed haloviruses showed a specific and close relationship with phiCh1 and phiH1 (Figure 4a), and average nucleotide identity (ANIb) values between ChaoS9, phiCh1 and phiH1 were ≥74% (Figure 4b). In a recent study, the complete genome sequence of phiH1 was compared to the previously published sequence of phiCh1 [26], and they shared 63% (BLASTn) nucleotide identity. As part of the present study, we have resequenced phiCh1 (see below and Table S2 for details). Since phiH1 is a valid species of the genus *Myohalovirus* [40], phiCh1 should be placed in the same genus. While phiCh1 and phiH1 show strong similarity over most of their genomes, their similarity to ChaoS9 is largely confined to a central region covering from about 10–30 kbp. This can be seen in Figure 4a, but is more clearly evident in the annotated genome comparison depicted in Figure 5.

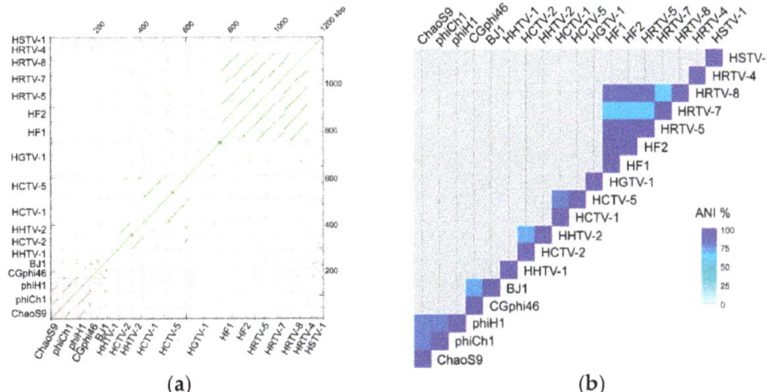

(a)   (b)

Figure 4. Comparison of the ChaoS9 genome with the genomes of 17 other published halovirus isolates. (a) Dot plot showing sequence similarity using the method of zPicture (Blastz) [32]. Names of viruses are shown along lower and left edges, and their lengths are indicated by the pink dashed lines. The scale, in kbp, is shown along the upper and right edges. The pink shaded box at lower left highlights the similarity between ChaoS9, phiCh1 and phiH1. (b) Average Nucleotide Identity (ANIb) heatmap. ANIb percentage values were calculated using EZbiocloud [33] and the heatmap produced using heatmapper [34]. Color scale indicates % ANIb.

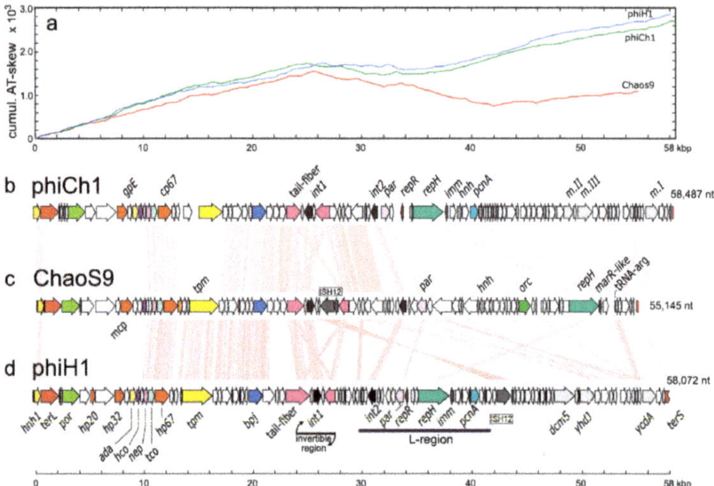

Figure 5. Genome map of ChaoS9 compared to phiCh1 and phiH1. (**a**) Cumulative AT-skew of all three viral genomes. (**b**) Genome map of phiCh1; (**c**) Genome map of ChaoS9; (**d**) Genome map of phiH1. Scale bar at bottom shows length, in kbp. Pink shading between genome maps indicates regions of similarity (tBLASTx, ≥30% amino acid identity). Functionally similar genes have been colored the same and are labeled nearby on one or more of the genomes (see Table 2 for details). The colors, gene labels (and encoded proteins) are: red, *terL* (large subunit terminase); light green, *por* (portal protein); brown, capsid protein genes, such as *mcp*, *gpE* and *hp32* (major capsid protein), hp20, hp67, cp67; yellow, *tpm* (tape measure protein); blue, *bpj* (baseplate J family protein); pink, tail fiber genes; light purple, *par* (plasmid partition); dark green, *orc1*, (replication protein Orc1); light blue, *repH* (plasmid replication protein); light grey, *dcm5*, *yhdJ*, *ycdA*, *m.II*, *m.III*, *m.I* (methyltransferases); dark grey, ISH12 transposase; black, integrases (*int1*, *int2*). Genes of unknown or uncertain function are uncolored (white). The L-region of phiH1 has been described by [41].

Table 1. Characteristics of ChaoS9 and related haloviruses phiH1 and phiCh1.

Virus [1]	Head Diameter (nm)	Tail Length × Width (nm)	Morphotype	Plaque Morphology	Unit Genome Length [2] (nt)	%G + C	Genome Ends in Virus [3]	Temperate (Genome Form)
ChaoS9	61	128 × 17	myovirus	turbid	55,145	65.3	ds, linear, TR, CP, >1 unit length	?
phiH1	64	170 × 18	myovirus	turbid	58,072	63.7	ds, linear, TR, CP, >1 unit length	Yes, provirus is a plasmid (circular, ds, 1 unit length)
phiCh1	70	130 × 20	myovirus	turbid	58,487	61.9	ds, linear, TR, CP, >1 unit length	Yes, provirus is a plasmid (circular, ds, 1 unit length)

[1] Data from this study for ChaoS9, [23] for phiH and [42] for phiCh1. [2] Data from this study (ChaoS9), Dyall-Smith et al. [26] (phiH1), and Witte et al., and this study [43] (phiCh1). [3] TR, CP; terminally redundant, circularly permuted.

Table 2. Annotated coding sequences (CDS) of the ChaoS9 genome (accession MK450543).

Start (nt)	Stop (nt)	Locus_Tag	Length (bp)	Direction	Gene	Product	Homologs [1]: phiCh1, pNMAG03	Homologs [2]: phiH1, [Other]
100	669	ChaoS9_005	570	+	-	HTH domain protein	PhiCh1_005, PhiCh1p02, ORF1, Nmag_4251	PhiH1_005
656	2302	ChaoS9_010	1647	+	*terL*	terminase large subunit TerL	-	[HALG_00007]
2316	3944	ChaoS9_015	1629	+	*por*	portal protein Por	-	[HGTV1_7]
3937	4083	ChaoS9_020	147	+	-	CxxC motif protein	-	ORPHAN
4086	5273	ChaoS9_025	1188	+	-	putative phage head assembly protein, SPP1_gp7 family	-	[C478_10461]
5384	7300	ChaoS9_030	1917	+	-	probable prohead protease protein	-	[HLASA_2034]
7303	7755	ChaoS9_035	453	+	-	uncharacterized protein	-	[HLASA_2033]
7801	8928	ChaoS9_040	1128	+	-	major capsid protein MCP	-	[3] [HLASA_2032; HHTV1_21]
8944	9363	ChaoS9_045	420	+	-	uncharacterized protein	PhiCh1_055, PhiCh1p13, ORF12, Nmag_4261	[6] PhiH1_050, [HLASA_2031]
9398	9784	ChaoS9_050	387	+	-	uncharacterized protein	-	[HLASA_2030]
9781	10191	ChaoS9_055	411	+	*hco*	head closure protein Hco, type 1	PhiCh1_065, PhiCh1p15, ORF14, Nmag_4263	[5] PhiH1_060, [WP_054519912]
10188	10421	ChaoS9_060	234	+	-	uncharacterized protein	-	ORPHAN
10414	10701	ChaoS9_065	288	+	-	uncharacterized protein	PhiCh1_070, PhiCh1p16, ORF15, Nmag_4264	[6] PhiH1_065, [HLASA_2029]
10703	11149	ChaoS9_070	447	+	*nep*	putative neck protein Nep, type 1	PhiCh1_075, PhiCh1p17, ORF16, Nmag_4265	PhiH1_070 [HLASA_2028]
11156	11746	ChaoS9_075	591	+	*tco*	tail completion protein Tco, type 1	PhiCh1_080, PhiCh1p18, ORF17, Nmag_4266	PhiH1_075 [HLASA_2027]
11767	13071	ChaoS9_080	1305	+	-	tail sheath protein	PhiCh1_085, PhiCh1p19, ORF18, Nmag_4267	PhiH1_080 [HLASA_2026]

Table 2. Cont.

Start (nt)	Stop (nt)	Locus_Tag	Length (bp)	Direction	Gene	Product	Homologs [1]: phiCh1, pNMAG03	Homologs [2]: phiH1, [Other]
13082	13480	ChaoS9_085	399	+	-	predicted tail tube protein	PhiCh1_090, PhiCh1p20, ORF19, Nmag_4268	PhiH1_085 [WP_054519907]
13492	13932	ChaoS9_090	441	+	-	uncharacterized protein	PhiCh1_095, PhiCh1p21, ORF20, Nmag_4269	PhiH1_090 [HLASA_2025]
13935	14144	ChaoS9_095	210	+	-	uncharacterized protein	-	[WP_054519905]
14147	16891	ChaoS9_100	2745	+	tpm	tape-measure tail protein Tpm	[7] PhiCh1_105, PhiCh1p23+PhiCh1p24, ORF22+ORF23, Nmag_4272	PhiH1_100 [HLASA_2024]
16895	17422	ChaoS9_105	528	+	-	uncharacterized protein	PhiCh1_110, PhiCh1p25, ORF24, Nmag_4273	PhiH1_105 [HLASA_2023]
17423	17767	ChaoS9_110	345	+	-	uncharacterized protein	PhiCh1_115, PhiCh1p26, ORF25, Nmag_4274	PhiH1_110 [HLASA_2022]
17771	18604	ChaoS9_115	834	+	-	uncharacterized protein	[7] PhiCh1_120, PhiCh1p27+PhiCh1p28, ORF26+ORF27, Nmag_4275	PhiH1_115 [HLASA_2021]
18644	18787	ChaoS9_120	144	+	-	CxxC motif protein	-	PhiH1_120
18784	19335	ChaoS9_125	552	+	-	uncharacterized protein	PhiCh1_125, PhiCh1p29, ORF28, Nmag_4276	PhiH1_125 [HLASA_2020]
19338	19700	ChaoS9_130	363	+	-	virus-related protein	-	PhiH1_135
19697	20062	ChaoS9_135	366	+	-	uncharacterized protein	PhiCh1_130, PhiCh1p30, ORF29, Nmag_4277	PhiH1_140 [HLASA_2019]
20069	21328	ChaoS9_140	1260	+	bpj	baseplate J family protein Bpj	PhiCh1_135, PhiCh1p31, ORF30, Nmag_4278	PhiH1_145 [HLASA_2018]
21321	21929	ChaoS9_145	609	+	-	uncharacterized protein	PhiCh1_140, PhiCh1p32, ORF31, Nmag_4279	PhiH1_150 [HLASA_2017]

Table 2. *Cont.*

Start (nt)	Stop (nt)	Locus_Tag	Length (bp)	Direction	Gene	Product	Homologs [1]: phiCh1, pNMAG03	Homologs [2]: phiH1, [Other]
21933	22517	ChaoS9_150	585	+	-	uncharacterized protein	PhiCh1_145, PhiCh1p33, ORF32, Nmag_4280	- [HLASA_2016]
22514	23122	ChaoS9_155	609	+	-	uncharacterized protein	PhiCh1_150, PhiCh1p34, ORF33, Nmag_4281	- [HLASA_2015]
23115	24698	ChaoS9_160	1584	+	-	repeat-containing tail fiber protein	[8] PhiCh1_155+ PhiCh1_175, PhiCh1p35+PhiCh1p37, ORF34+ORF36, Nmag_4282+Nmag_4286	PhiH1_165+ PhiH1_185
24702	24986	ChaoS9_165	285	+	-	uncharacterized protein	[8] PhiCh1_160+ PhiCh1_170, Nmag_4285+Nmag_4283	PhiH1_180+ PhiH1_170
25024	25698	ChaoS9_170	675	+	*int1*	tyrosine integrase/recombinase Int1	PhiCh1_165, PhiCh1p36, ORF35, Nmag_4284	PhiH1_175
25709	25996	ChaoS9_175	288	-	-	uncharacterized protein	[8] PhiCh1_160+ PhiCh1_170, Nmag_4285+Nmag_4283	PhiH1_180+ PhiH1_170
25999	26145	ChaoS9_180	147	-	-	repeat-containing tail fiber protein (C-term) (nonfunctional)[9]	* [9]	* [9]
26199	27455	ChaoS9_185	1257	-	*trpB*	IS1341-type transposase TnpB	-	PhiH1_340
27457	27849	ChaoS9_190	393	-	*tnpA*	IS200-type transposase TnpA	-	PhiH1_335
27906	28820	ChaoS9_195	915	-	-	repeat-containing tail fiber protein (N-term) (nonfunctional)	[8] PhiCh1_155+ PhiCh1_175, PhiCh1p35+PhiCh1p37, ORF34+ORF36, Nmag_4282+Nmag_4286	PhiH1_165+ PhiH1_185
28854	29579	ChaoS9_200	726	+	-	transmembrane domain protein	PhiCh1_180, PhiCh1p38, ORF37, Nmag_4287	-
29589	29861	ChaoS9_205	273	-	-	HTH domain protein	PhiCh1_185, PhiCh1p39, ORF38, Nmag_4288	-

Table 2. Cont.

Start (nt)	Stop (nt)	Locus_Tag	Length (bp)	Direction	Gene	Product	Homologs [1]: phiCh1, pNMAG03	Homologs [2]: phiH1, [Other]
29933	30241	ChaoS9_210	309	-	-	uncharacterized protein	PhiCh1_190, PhiCh1p40, ORF39, Nmag_4289	PhiH1_220
30238	30813	ChaoS9_215	576	-	-	glutamine amidotransferase domain protein, class-II	PhiCh1_195, PhiCh1p41, ORF40, Nmag_4290	-
30818	31891	ChaoS9_220	1074	-	-	uncharacterized protein	[7] PhiCh1_200, PhiCh1p42+PhiCh1p43, ORF41+ORF42, Nmag_4291	-
32030	32266	ChaoS9_225	237	+	-	uncharacterized protein	-	ORPHAN
32339	32584	ChaoS9_230	246	+	-	uncharacterized protein	-	PhiH1_225
32581	33009	ChaoS9_235	429	+	-	VapC family toxin	-	[BRC75_08225]
33098	33448	ChaoS9_240	351	+	-	uncharacterized protein	PhiCh1_230, Nmag_4297	PhiH1_250
33457	34059	ChaoS9_245	603	-	int2	tyrosine integrase/recombinase Int2	PhiCh1_215, PhiCh1p46, ORF45, Nmag_4294	PhiH1_240
34241	34432	ChaoS9_250	192	-	-	uncharacterized protein	-	ORPHAN
34507	35055	ChaoS9_255	549	-	-	uncharacterized protein	-	PhiH1_255 [C466_00612]
35048	35902	ChaoS9_260	855	-	-	Plasmid partition protein ParA	PhiCh1_220, PhiCh1p47, ORF46, Nmag_4295	PhiH1_265
35976	36440	ChaoS9_265	465	-	-	transmembrane domain protein	-	PhiH1_210
36460	38241	ChaoS9_270	1782	-	-	uncharacterized protein	-	[AV929_12240]
38243	38857	ChaoS9_275	615	-	-	uncharacterized protein	-	[CRI94_04435]
38850	39080	ChaoS9_280	231	-	-	CxxC motif protein	-	[HALLA_11930]
39073	39240	ChaoS9_285	168	-	-	transmembrane domain protein	-	ORPHAN
39233	40492	ChaoS9_290	1260	-	-	uncharacterized protein	-	[DM826_07215]
40485	40673	ChaoS9_295	189	-	-	CxxC motif protein	-	ORPHAN

Table 2. *Cont.*

Start (nt)	Stop (nt)	Locus_Tag	Length (bp)	Direction	Gene	Product	Homologs [1]: phiCh1, pNMAG03	Homologs [2]: phiH1, [Other]
40663	41181	ChaoS9_300	519	-	-	uncharacterized protein	-	[DJ71_18565]
41174	41611	ChaoS9_305	438	-	-	HNH-type endonuclease/MarR family transcription regulator	-	[4] [DJ70_12900; B4589_07635]
41613	41870	ChaoS9_310	258	-	-	HTH domain protein	-	[Natpe_3999]
41997	42602	ChaoS9_315	606	+	-	uncharacterized protein	-	[C480_10020]
42605	43051	ChaoS9_320	447	+	-	uncharacterized protein	-	[Natgr_3468]
43044	43250	ChaoS9_325	207	+	-	uncharacterized protein	-	ORPHAN
43250	43621	ChaoS9_330	372	+	-	uncharacterized protein	-	[OSG_eHP13_00215]
43621	44127	ChaoS9_335	507	+	-	uncharacterized protein	-	ORPHAN
44124	44255	ChaoS9_340	132	+	-	CxxC motif protein	-	[SAMN04488133_0114]
44342	45400	ChaoS9_345	1059	+	orc1	Orc1-type DNA replication protein	-	[HLASA_2006]
45482	45709	ChaoS9_350	228	-	-	uncharacterized protein	-	ORPHAN
45837	45992	ChaoS9_355	156	+	-	uncharacterized protein	-	[HALDL1_16575]
46369	46905	ChaoS9_360	537	+	-	uncharacterized protein	PhiCh1_295, PhiCh1p65, ORF64, Nmag_4216	[5] [C472_00499]
46902	47177	ChaoS9_365	276	+	-	uncharacterized protein	-	ORPHAN
47179	47979	ChaoS9_370	801	+	-	zinc-finger domain protein	-	[DJ84_18225]
47976	48068	ChaoS9_375	93	+	-	uncharacterized protein	-	ORPHAN
48065	48397	ChaoS9_380	333	+	-	uncharacterized protein	-	ORPHAN
48394	51690	ChaoS9_385	3297	+	repH	plasmid replication protein RepH	PhiCh1_245, PhiCh1p55, ORF54, Nmag_4299	PhiH1_285
51683	51925	ChaoS9_390	243	+	-	MarR family transcription regulator	-	[AV929_12115]
51932	52591	ChaoS9_395	660	+	-	CxxC motif protein	-	[C443_17983]
53208	53453	ChaoS9_400	246	+	-	transmembrane domain protein	PhiCh1_440, PhiCh1p93, ORF92, Nmag_4244	PhiH1_460

Table 2. *Cont.*

Start (nt)	Stop (nt)	Locus_Tag	Length (bp)	Direction	Gene	Product	Homologs [1]: phiCh1, pNMAG03	Homologs [2]: phiH1, [Other]
53446	53769	ChaoS9_405	324	+	-	transmembrane domain protein	PhiCh1_445, PhiCh1p94, ORF93, Nmag_4245	[10] PhiH1_465
53766	54455	ChaoS9_410	690	+	-	uncharacterized protein	-	[3] [halTADL_2427; HGTV1_34]
54460	54705	ChaoS9_415	246	+	-	DUF217 domain protein	PhiCh1_460, PhiCh1p97, ORF96, Nmag_4248	-
54734	54922	ChaoS9_420	189	+	-	CxxC motif protein	PhiCh1_465, PhiCh1p98, ORF97, Nmag_4249	PhiH1_480
54915>	<63	ChaoS9_425	294	+	*terS*	terminase small subunit TerS	PhiCh1_470, PhiCh1p01, ORF98, Nmag_4250	PhiH1_485

[1] PhiCh1/pNMAG03 homologs of ChaoS9 proteins. For phiCh1, three codes are given: the locus tag from the revised genome (PhiCh1_), the RefSeq PhiCh1p and the originally assigned ORF codes (ORF for open reading frame). For example, PhiCh1_005, PhiCh1p02, Orf1. Codes starting with PhiCh1_ represent the revised genome sequence and annotation (Genbank accession MK450543; this publication), ORF codes represent the original annotation of the phiCh1 genome [35] (Genbank accession AF440695.1), while codes beginning with PhiCh1p represent the RefSeq version of the annotation of the same genome sequence (GB accession NC_004084). The number shift between ORF and PhiCh1p is due to the *terS* gene, the N-terminal part being encoded at the end of the genome, and the C-terminal part at its beginning. This gene is ORF98 in the original annotation and PhiCh1p01 in the RefSeq annotation. Codes starting with Nmag_ represent the annotation of the *Nab. magadii* plasmid pNMAG03 [28] (accession CP001935.1), which is the provirus state of phiCh1. The point of ring opening in pNMAG03 was set between Nmag_4303 and Nmag_4211. The absence of ORF and PhiCh1p codes indicates missing gene calls in the original annotation of phiCh1. [2] phiH1 homologs, or else "other" homologs of ChaoS9 proteins. If a homolog exists in phiH1 then the code is provided; if a homolog is lacking from phiH1 but exists in phiCh1, this is indicated by a hyphen; if a homolog is lacking in both, phiH1 and phiCh1, then an existing "other" homolog is listed in square brackets; codes are either from UniProt (locus tags) or from NCBI nr (WP numbers). The term ORPHAN indicates a complete lack of homologs. [3] Multiple homologs, separated by semicolon, are listed when a homolog is found in a halovirus, but this is significantly more distant than the closest homolog. [4] Multiple homologs indicate a ChaoS9-specific gene fusion. [5] The combination of phiH1/phiCh1 and "other" homologs is used when the homologs from phiH1 or phiCh1 are especially distant. [6] The combination of phiH1/phiCh1 and "other" homologs is used for HLASA_ codes to illustrate a longer stretch of synteny to PVHS1 from *Halanaeroarchaeum sulfurireducens* (see later). [7] Multiple PhiCh1p/ORF codes indicate that the gene was split by a frameshift in the originally published genome sequence of that virus. [8] Multiple PhiCh1/Nmag codes indicate the existence of paralogs. [9] This ORF represents the C-terminal fragment of a pseudogene (indicated by the term nonfunctional) which has been targeted by ISH12. The asterisks (*) indicate that corresponding pseudogene fragments do not exist as independent ORFs in phiH1 or phiCh1. [10] PhiCh1_445 and PhiH1_465, like ChaoS9_405, have three predicted transmembrane domains and are suspected to function as a holin [43].

The *pac* site of the ChaoS9 genome was identified by alignment with phiH1 and phiCh1, for which *pac* sites have been previously reported [26]. Like the other two haloviruses, *pac* occurs within the *terS* gene, near the stop codon, at a well conserved GC-rich sequence motif. For convenience, base 1 of ChaoS9 was chosen so that it corresponds with the starting bases of phiH1 and phiCh1, which places the *pac* site terminal base at nt 46. A summary of the major features of ChaoS9 is given in Table 1, along with the characteristics of phiH1 and phiCh1.

Annotation of the ChaoS9 genome predicted 85 coding sequences (CDS) and one tRNA gene (Figure 5 and Table 2). Most CDS were closely spaced, with 31 overlapping at stop/start codons, and 30 separated by 0–15 nt. The majority of CDS were organized into groups having the same orientation, such as 0–25 kbp and 42–55 kbp, where all but one CDS are on the upper (forward) strand, and 33.5 to 41.9 kbp where all CDS are on the lower (reverse) strand. The most common stop codon was TGA (56; 65%), followed by TAA (19; 22%) and TAG (10; 12%), a pattern that is similar to the host species, *Hbt. salinarum*, that also prefers TGA stop codons (TGA, 49%; TAG, 28%; TAA, 23%) (see http://www.kazusa.or.jp/codon).

The ChaoS9 open reading frames (ORFs) were compared (BLASTp) to those from phiH1 and phiCh1 to identify homologs (E-values $\leq 10^{-10}$). In some cases, this threshold was relaxed because there was additional support from a conserved gene neighborhood; in other cases, a more stringent threshold was applied in case of casual matches caused by a strong compositional bias. The same process was used to identify homologs in haloarchaeal proviruses (see Section 3.6).

3.4. Resequencing the Genome of Halovirus PhiCh1

Given the close similarity of phiCh1 to ChaoS9, it was decided to check the phiCh1 genome sequence by high throughput sequencing using Illumina HiSeq (see methods). This revealed a number of differences to the existing sequence (Genbank: AF440695.1), which are listed in Table S2 along with the genes and proteins affected. Briefly, a total of 40 bases were affected by the revision; 13 point mutations, 9 one-base indels, and one 18 base indel. Overall, the sequence revision made phiCh1 more similar to phiH1 and to ChaoS9. The *pac* terminal base was determined to be nt 46 ($p = 1.23 \times 10^{-23}$), based on analysis of the Illumina reads using the program PhageTerm [38]. This position is consistent with previous studies [26,43].

3.5. Organisation of the ChaoS9 Genome

A gene map of ChaoS9 (Figure 5, panel c) is shown between the maps of phiCh1 and phiH1 (panels b and d). Pink shading between the maps indicates regions encoding similar proteins ($\geq$30% aa identity). Over the first 25 kbp, the three viruses share a similar gene synteny, while beyond 25 kbp, ChaoS9 differs considerably from the other two viruses in both gene composition and order. The same pattern is reflected by cumulative AT-skew plots (panel a), which show a similar, steady rise over the first 25 kbp for all three virus genomes, but after this, the plot for ChaoS9 diverges significantly from those of phiCh1 and phiH1. In general, the cumulative AT-skew plots appear to parallel the transcription directions of genes of the three viruses.

The left arm (0–28 kbp) of the ChaoS9 genome. All genes are in the forward direction (top strand) and form a long, functional module specifying proteins putatively involved in DNA packaging, virus structure, and assembly. They include genes encoding the large subunit terminase (TerL) and portal protein (Por), the major virus capsid protein (Mcp), tail sheath, tail-tube, and tape measure (Tmp) proteins, and the tail fiber protein. The gene for the latter protein is also part of an invertible region (see next section). In phiCh1 and phiH1, these genes are expressed during the late phase of lytic infection [44]. The major capsid protein and the tail sheath protein are likely to produce the most prominent bands on SDS-PAGE, which were VP3 and VP4 (Figure 2). The protein molecular weights of Mcp and of the tail sheath protein calculated from their amino acid sequences were 42 kDa (Mcp) and 46.1 kDa (tail sheath), but these values are considerably lower than the observed MWs of VP3 and VP4 (70.2 and 60.2 kDa, respectively). After applying the compensatory adjustment for acidic proteins reported by Guan et al. [39], the predicted gel sizes of the Mcp (50.4 kDa) and tail sheath protein (56.1 kDa) were still lower than VP3 and VP4. The MW of the tail tube protein (14.8 kDa), after applying the Guan et al. adjustment [39] was predicted to be 19.3 kDa on SDS gels, a value identical to that of VP7. The VP1 band (143.5 kDa) is much larger than any of the annotated virus structural proteins, and may represent a multimeric form.

Sequences of head-neck-tail module proteins can be used to classify caudoviruses [45], and this classified ChaoS9 within the Myoviridae (Type1, Cluster 6). While the gene composition and synteny were well conserved between ChaoS9 and the other viruses, the sequence similarity of genes and proteins revealed major differences between them, suggesting a long history of recombination. For example, the proteins encoded by genes *terL* to *hco* of ChaoS9 showed no significant similarity to the corresponding proteins from phiCh1 or phiH1, and this segment includes many of the most highly conserved genes used for virus classification, such as the terminase, portal, and major capsid proteins. The putative assignments of ChaoS9_015 as portal protein and ChaoS9_030 as prohead protease are based on VIRFAM predictions. From 10–25 kbp, the majority of the encoded proteins are related to tail assembly proteins, and most share sequence similarity with the corresponding phiH1 and phiCh1 proteins, with the obvious exceptions (see Figure 5) of the phiCh1 tape measure protein (Tpm) and two hypothetical proteins (PhiH_155, PhiH_160) of phiH1 that immediately precede the tail fiber protein (PhiH_165).

The invertible region (23–29 kbp). The inflection in AT-skew at around 25 kbp occurs at the end of the tail-fiber gene, which is embedded in a segment containing an integrase/recombinase and another tail-fiber related gene (probably a pseudogene) that is inversely oriented to the first one. The similarity of the two fiber genes (pink arrows in Figure 5) can be seen by the crossing of shaded lines (light pink shaded) in this region of Figure 5. Similar gene arrangements to this are found in phiH1 and phiCh1, where it has been shown that the central recombinase allows inversion of the nearby tail-fiber genes, so altering the sequence of the actively expressed copy [46]. The ChaoS9 invertible region contains an ISH12 transposon that is not present in the corresponding invertible regions of the other viruses. This transposon is identical to ISH12 from strain R1 and has targeted the inactive copy of the tail fiber protein (N-term part: ChaoS9_195, C-term part: ChaoS9_160). Curiously, the same transposon is also integrated into phiH1, but at a different genome position.

The right arm (29–55 kbp). This region is the most divergent compared to phiCh1 and phiH1, and contains genes putatively involved in replication (*parA*, *orc1* and *repH*), a tRNA-Arg gene, and numerous genes specifying proteins of unknown function. In both phiH1 and phiCh1, this region has been shown to control lysogeny and the provirus state, maintaining the viral genome as a circular, extrachromasomal dsDNA element [47–49]. Lytic phase gene expression in phiH1 has been shown to be repressed by RepR, a coliphage-like repressor [50–52], but a homologous gene similar to this was not detected in ChaoS9.

There are no DNA methylase genes in ChaoS9, while the other viruses each carry three (e.g., m.I, m.II, and m.III of phiCh1). The corresponding regions of phiH1 and phiCh1 are similar to each other, but contain relatively few genes with matching protein sequences to ChaoS9, and even in these cases the arrangement usually differs. For example, the ISH12 elements of ChaoS9 and phiH1 are 15 kbp apart, in opposite orientation, and in distinct modules. Also, while the RepH proteins show weak similarity to each other, the position of *repH* in ChaoS9 is about 15 kbp further right compared to the *repH* genes of phiCh1 and phiH1. Even the type of replication related genes differs, with ChaoS9 carrying a gene similar to Orc (*orc1*) that is not present in the other viruses, while phiH1 and phiCh1 carry a gene similar to PCNA (*pcnA*) that is not found in ChaoS9. In phiH1, the L-region has been shown to be able to replicate independently as a plasmid [41], and also contains an immunity gene (*imm*) near *repH* that protects L-plasmid containing host cells from lytic infection by phiH1 [41]. The same arrangement is found in phCh1. While a gene related to *imm* was not detected near *repH* of ChaoS9, a gene specifying a MarR-like repressor (ChaoS9_390) is present just downstream of *repH*.

ChaoS9 is the only virus of the three predicted to carry a tRNA gene, tRNA-Arg(TCG). In BLASTn searches, this sequence is unlike other tRNAs except for a conserved region of the right arm (nt 43-70), which matches many haloarchaeal and bacterial tRNAs. Curiously, the best matches are to cyanobacteria tRNAs, such as tRNA Gly (CCC) of *Synechococcus* sp. KORDI-100 (CP006269 nt 191738-191667), which gives a 28 nt perfect alignment. In haloarchaea, the best match found was 19 nt. Although the ChaoS9 tRNA appears to be complete, its function is less clear. Ostensibly, it specifies arginine (anticodon TCG), and the corresponding codon is the third most frequent Arg codon used in both the host species *Hbt. salinarum* [53] and ChaoS9, but there is no large difference in usage between the two (7% in *Halobacterium* and 11% in ChaoS9), and there are much rarer codons for this amino acid used by both host and virus. It could also represent an *att* sequence used for integration of the viral genome into tRNA genes of host strains (see later) or may have a regulatory role.

3.6. Related Provirus-Like Matches in Haloarchaea

Several haloarchaeal genomes carried provirus elements related to ChaoS9, and the gene maps of two examples (PVH3A1 and PVHS1) are depicted in Figure 6. The *attL* and *attR* sites of both proviruses indicate they were integrated into tRNA genes (tRNA-Met and tRNA-Cys), probably mediated by the integrases encoded at their right ends. Both appear to be intact and probably functional, as their left halves possess complete suites of virus structural and assembly genes. Their right halves predominantly carry genes for uncharacterized proteins unrelated to ChaoS9. A few genes in this half, and at the extreme left end (before the structural/assembly genes), could be assigned functions, such as the replication protein Orc, and DNA methylases (Mtase, Dam). Neither provirus carried a region corresponding to the invertible tail-fiber genes of ChaoS9.

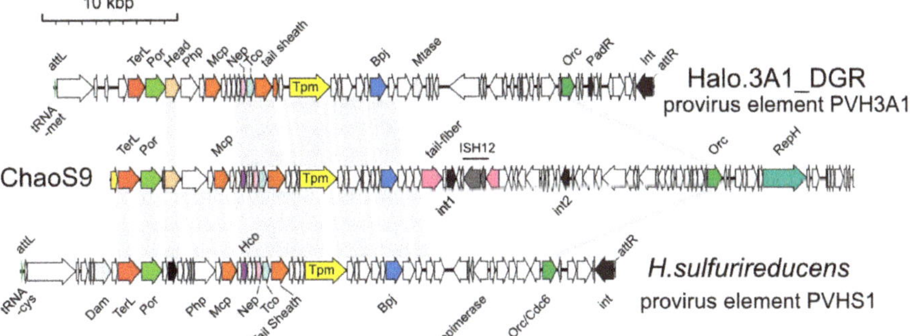

Figure 6. Provirus elements PVH3A1 and PVHS1 compared to ChaoS9. PVH3A1 from haloarchaeon strain 3A1_DGR (accession NZ_KK033114; nt 1475908-1520418), and PVHS1 from *Haa. sulfurireducens* strain M27-SA2 (accession NZ_CP011564; nt 1971038- 1927125), are compared to ChaoS9. Grey bands connecting genome maps represent similarity (tBLASTx, >30% identity) between the inferred proteins of ChaoS9 and each provirus. Positions of several annotated proteins and their genes are indicated by color and name; TerL, large subunit terminase (red); Por, portal protein (light green); Head, SPP1_gp7 family head assembly protein, (light brown); capsid proteins such as Mcp (major capsid protein) and Tail sheath protein (brown); Tpm, tape measure protein (yellow); Dam, adenine methyltransferase (light grey); Bpj, baseplate J family protein (blue); Orc1, replication protein Orc1 (green); RepH, plasmid replication protein (turquoise). Scale bar, in kbp, is shown at top.

The right arm (29–55 kbp) of ChaoS9 shows little similarity to either provirus except for Orc1. Across the left arm (0–28 kbp), the region covering virus structural and assembly proteins, ChaoS9 maintains good synteny and protein similarity to PVHS1, but with PVH3A1 there is a distinct break in similarity within the virus structural/assembly gene module. PVH3A1 neck and tail proteins are similar to ChaoS9 but the head/assembly proteins are unrelated, except for the head assembly protein (labeled Head; Figure 6, panel a). PhiH1, and to a lesser extent phiCh1, show a comparable break in similarity to ChaoS9 head and tail genes (Figure 5). All of these examples suggest that ancestral recombination events between ChaoS9-like viruses have occurred within genes located between the head morphogenesis genes and the tail morphogenesis genes, and that these often result in viable progeny.

The distinct right halves of the proviruses suggest that they have also undergone extensive recombination relative to ChaoS9, perhaps reflecting differences in virulence and/or provirus state.

3.7. A Diverse Family of Haloviruses

In order to examine evolutionary relationships between ChaoS9 and other viruses and proviruses, genes were sought that were both sufficiently conserved and present in all examples. As shown in Figures 5 and 6, relatively few genes matched these criteria. Even the large subunit terminase (TerL) and major capsid protein (Mcp) were poorly conserved, and these have been widely used in previous studies of bacteriophages for delineating virus taxa [4]. The Bpj (baseplate J family protein) and the tail sheath proteins were selected to infer phylogenetic relationships as they were relatively long, conserved in sequence and present in all examples. BLASTp searches (accessed 20 November 2018) with ChaoS9 Bpj retrieved over 50 high scoring matches (E value < 10^{-30}), with the closest matches all being from haloarchaea or haloviruses, while less similar matches included proteins from Bacteria and bacteriophages. An inferred phylogenetic tree based on alignment of Bpj (Figure 7a) shows that ChaoS9 is part of a robust clade (100% bootstrap confidence) which includes haloviruses phiCh1, phiH1, and provirus-like elements of four haloarchaea belonging to at least three different genera. A separate clade, also with high bootstrap confidence, contains four haloviruses (HF1, HF2, HRTV-8, and HSTV-2) and a provirus element of *Haloferax larsenii*.

BLASTp searches of the NCBI database with the ChaoS9 tail sheath protein (ChaoS9_080) retrieved only nine high scoring matches (E value < 10^{-35}), and these were all from haloarchaea or haloviruses present in the Bpj tree, and most closely related to ChaoS9. The inferred phylogenetic tree based on this protein (Figure 7b) reveals a topology similar to that of Bpj proteins.

The major capsid protein (Mcp) of ChaoS9 was used to search the NCBI database (BLASTp, nr database, accessed 22 November 2018) and retrieved only five matches (Figure 7c), which varied in similarity from 36 to 76% (aa identity). Three were from organisms previously identified as specifying ChaoS9-related proteins (*Haa. sulfurireducens*, *Saliphagus* sp. LR7 and *Hpt. malekzadehii*) and are present in Figures 6 and 7, while the other two sequences were from *Salinigranum rubrum* and the tailed halovirus HHTV-1 [5,54].

The DNA sequences of HHTV-1 and ChaoS9 share no significant similarity (Figure 4), but a BLASTp comparison of all ChaoS9 and HHTV-1 proteins found that Mcp was the only protein with significant similarity (36%) shared between these haloviruses. Such a pattern of similarity between tailed viruses that is strictly limited to the Mcp appears to be uncommon in the published literature. A less clear-cut but comparable example occurs between actinophages Jace and Tweety (accessions EF536069 and MH153804), which share similar Mcp (32%) and integrase (39%) protein sequences but weak or negligible similarity between all other proteins. Cases where head genes and tail genes derive from different virus lineages are slightly more common [55,56] (see Discussion).

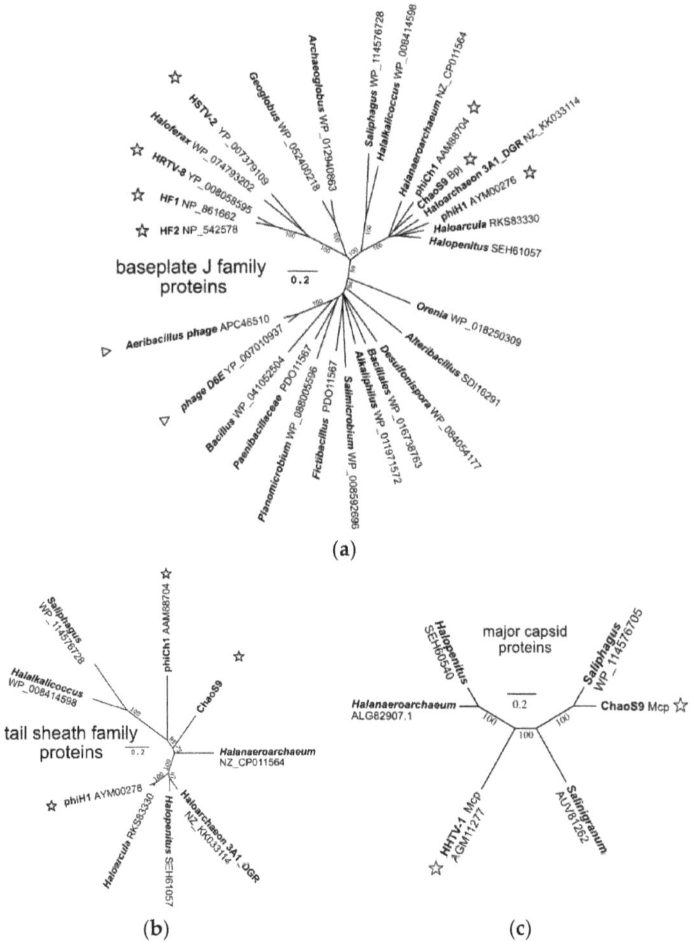

Figure 7. Phylogenetic tree reconstructions using (**a**) baseplate J family proteins, (**b**) major tail sheath proteins and (**c**), major capsid proteins. Halovirus proteins are marked by stars and bacteriophage proteins are marked by triangles. The consensus trees were produced using the Neighbor–Joining algorithm and 100 bootstrap replications. Scale bars represent the estimated number of amino acid replacements per position.

3.8. CRISPR Spacer Matches to ChaoS9

The ChaoS9 genome was used to search for matching CRISPR spacer sequences available from public databases (see methods). A total of 277 spacers were identified that closely matched ChaoS9, with the majority of spacers originating from Antarctic lake metagenomes (Deep Lake, Rauer Lake and Club Lake). After removing duplicates, the number of distinct spacers was reduced to 39, and in Table S3 the matching spacers have been ordered by their position along the ChaoS9 genome. The distribution of spacers is highly skewed, with most (34/39) targeting sequences within the right arm, and particularly the *repH* gene, for which there were 14 distinct spacer sequences. The direct repeats (DR) of the matching spacers were most similar or identical to those found in CRISPR arrays of sequenced haloarchaea, particularly antarctic isolates such as *Halorubrum lacusprofundi*, *Halobacterium* sp. DL1 and haloarchaeon DL3. For comparison, the phiH1 and phiCh1 genomes were also scanned for matching spacers in the same antarctic lake metagenomes (IMG/VR webserver;

(https://img.jgi.doe.gov/cgi-bin/vr/main.cgi; accessed 20 December 2018). This returned only 2 (phiCh1) and 3 (phiH1) significant matches (supplementary Table S4), of which all of the phiH1 matching spacers were to the tail fiber gene (PhiH1_165), as was one of the spacers matching phiCh1 (PhiCh1_155). The remaining spacer to phiCh1 matched a sequence within the prohead protease gene (gpB, PhiCh1_045). All six spacers were from Deep Lake and Rauer Lake metagenomes.

4. Discussion

This study focused on identifying the cause of a lysis event in a large, laboratory culture of *Hbt. salinarum* S9. A novel myovirus was recovered, ChaoS9, with morphological and molecular characteristics specifically resembling myoviruses phiH and phiCh1, but which differed significantly in sequence from both of these. An earlier lysis event in the same laboratory that affected a culture of *Hbt. salinarum* R1 was shown to have been caused by phiH, however, the 2007 event was not a recurrent infection. The sources of both infections are unknown, possibly raw salt, but these events highlight the need for preventative measures, even though the high salt conditions used for cultivation of haloarchaea are generally regarded as providing a strong barrier to contamination by non-halophilic microorganisms. However, tailed viruses (caudoviruses) such as ChaoS9 are not only the most common type of prokaryotic virus, but together with other bacterial and archaeal viruses, they represent the most abundant biological entity on Earth, estimated to be 10^{31} virions [57]. It is not surprising, then, that virus contamination and lysis events are a constant threat in large-scale commercial fermentations [22], and cause such significant losses that systematic preventative programs have been formulated, such as PACCP (phage analysis and critical control point) [58].

ChaoS9 was most similar in morphology, genome type, genome length, and sequence to the tailed haloviruses phiH1, and phiCh1 [26,43]. The dsDNA genome was terminally redundant and circularly permuted, consistent with headful packaging, as has been shown for the related viruses. The likely *pac* site was identified by sequence homology. Gene synteny of the virus morphogenesis genes of the left arm of the genome was similar to those of phiH1 and phiCh1, while the right arm comprised genes for DNA replication, plasmid partition, and a tRNA, as well as many genes specifying proteins of unknown function. The right arm corresponds to the lysogeny, replication, and accessory gene region of phiH1 and phiCh1, and probably serves the same general function in ChaoS9. A characteristic feature of phiH1 and phiCh1 is an invertible tail fiber gene module, which was also present in ChaoS9, except that it also included an ISH12 transposon.

Comparison of the ChaoS9 genome with those of phiH1 and phiCh1 revealed a distinctive pattern of similarity and difference, suggesting an evolutionary history involving large recombination events. While the tail gene region of ChaoS9 was similar to the other two viruses, the head/assembly genes of the left arm of the genome, as well as most of right arm (the replication/lysogeny region), were not. This means that the major capsid protein, terminase (large subunit) and portal proteins of ChaoS9 are all unrelated to those of phiH1 and phiCh1. The most parsimonious explanation is that a recombination event has replaced the head morphogenesis module while leaving the tail morphogenesis module intact. It appears to be a chimeric recombinant, with the point of similarity disjuncture occurring between the putative head closure (*hco*, ChaoS9_055) and neck protein (*nep*, ChaoS9_070) genes. Similar examples where head morphogenesis and tail morphogenesis modules appear to derive from different virus types are not uncommon in the literature, such as the *Gordonia* spp. phage Kita (Figure 4 of [55]), and *Xylella fastidiosa* phage Xfas53 [56], phi80 and Gifsy-2 [59]. The likelihood of such recombination events producing viable progeny is increased because of the way caudoviruses are assembled, with heads and tails produced as separate structures that are then joined together [60].

Provirus elements related to ChaoS9 were found among haloarchaeal genome sequences, integrated into their host chromosomes via tRNA genes, which supports the view that ChaoS9, like phiH1 and phiCh1, has a temperate lifestyle [61]. One provirus (PVHS1) showed good sequence similarity to ChaoS9 proteins across most of the virus morphogenesis genes, while a second provirus (PVH3A1) showed a general pattern of sequence similarity to ChaoS9 like that of phiH1 and phiCh1;

with mainly unrelated head morphogenesis genes (including dissimilar major capsid proteins), similar tail morphogenesis genes, and largely unrelated right arms (lysogeny/replication). The right arms of both proviruses differed significantly from the right arm of the ChaoS9 genome, but all carried a putative *orc1* gene, presumably for use in replication as a (provirus) plasmid if the circularized genome is unable to integrate into the host chromosome via *attP*. These two examples again point to frequent recombination events that exchange the head morphogenesis module while retaining the tail morphogenesis genes.

Curiously, the ChaoS9 major capsid protein is most similar to that of the haloarchaeal siphovirus HHTV-1, but this is the only protein between them that is similar, and they are otherwise unrelated. In studies of enterobacterial viruses, major capsid protein (MCP) gene exchange was estimated to be extremely rare between phage clusters or types [61]. How this occurred in the evolutionary history of these viruses will be intriguing to resolve, but presumably indicates widespread recombination between tailed viruses of haloarchaea.

A surprisingly large number of identical or near identical CRISPR spacers to ChaoS9 were detected, and even more surprising was their identification in the metagenomes from Antarctic salt lakes. The majority of target sites were located in the right arm of the genome, in the lysogeny/replication region. A few were to sites within the left arm of the genome, such as the tail fiber gene. In contrast to ChaoS9, the genomes of phiH1 and phiCh1 matched only a few spacers from the same Antarctic metagenomes, and all but one targeted the tail fiber gene. This suggests that these lakes harbor tailed viruses with a lysogeny/replication module, similar to ChaoS9. Their tail fiber genes also show significant sequence similarity to ChaoS9, phiH1, and phiCh1, but the other virus morphogenesis genes are distinct.

The precise relationship between ChaoS9, phiCh1, and phiH1 was explored in several ways. They share average nucleotide identity (ANIb) values of $\geq$74%, and phylogenetic tree reconstructions using baseplate J virus protein sequences clustered ChaoS9, phiH1, and phiCh1 into a well-supported clade, distinct from other tailed haloviruses. The close relationship is consistent with the other comparative data (morphology, genome size and packaging, gene synteny, and inferred protein sequences). *Halobacterium virus phiH* is the type species of the genus *Myohalovirus* and a previous study has shown that phiCh1 should be included in the same genus, as their genomes share 63% nucleotide identity, are mostly colinear, and their proteins show, on average, 70% amino acid identity [26]. PhiCh1 proteins affected by genome revision became more similar to those of phiH1. ChaoS9 shows many similarities and correspondences to phiH1 and phiCh1 but also considerable differences, including a distinct major capsid protein, terminase (large subunit) and portal protein, and is certainly a distinct species to phiH1 and phiCh1, but whether it should be classified outside of the *Myohalovirus* genus requires further consideration.

Supplementary Materials: The following are available online at http://www.mdpi.com/2073-4425/10/3/194/s1, Figure S1: Experimental and in silico restriction digests of ChaoS9 DNA. Table S1; DNA primers used for PCR or sequencing viral DNA. Table S2; Revisions of phiCh1 genes after re-sequencing the genome. Table S3; CRISPR spacers matching ChaoS9; Table S4, CRISPR spacers matching phiH1 and phiCh1.

Author Contributions: Conceptualization, visualization, investigation, M.D.-S.; investigation (electron microscopy) G.W.; investigation (genome sequencing) P.P.; data curation, F.P., M.D.-S.; funding acquisition, D.O.; project administration, F.P.; resources, A.W.; writing—original draft, F.P., M.D.-S.; writing—review & editing, A.W., D.O., F.P., M.D.-S.; validation, A.W.

Funding: This research received no external funding.

Acknowledgments: We wish to thank Cyril Boulegue, Beatrix Scheffer, and Snezan Marinkovic (core facility, MPI) for their contribution to genome sequencing. We thank Graham Hatfull for advice on bacteriophage diversity.

Conflicts of Interest: The authors declare no conflict of interest.

References

1. Dyall-Smith, M.; Tang, S.L.; Bath, C. Haloarchaeal viruses: How diverse are they? *Res. Microbiol.* **2003**, *154*, 309–313. [CrossRef]
2. Krupovic, M.; Cvirkaite-Krupovic, V.; Iranzo, J.; Prangishvili, D.; Koonin, E.V. Viruses of archaea: Structural, functional, environmental and evolutionary genomics. *Virus Res.* **2017**, *244*, 181–193. [CrossRef] [PubMed]
3. Oren, A.; Bratbak, G.; Heldal, M. Occurrence of virus-like particles in the Dead Sea. *Extremophiles* **1997**, *1*, 143–149. [CrossRef] [PubMed]
4. Krupovic, M.; Forterre, P.; Bamford, D.H. Comparative analysis of the mosaic genomes of tailed archaeal viruses and proviruses suggests common themes for virion architecture and assembly with tailed viruses of bacteria. *J. Mol. Biol.* **2010**, *397*, 144–160. [CrossRef] [PubMed]
5. Sencilo, A.; Jacobs-Sera, D.; Russell, D.A.; Ko, C.C.; Bowman, C.A.; Atanasova, N.S.; Osterlund, E.; Oksanen, H.M.; Bamford, D.H.; Hatfull, G.F.; et al. Snapshot of haloarchaeal tailed virus genomes. *RNA Biol.* **2013**, *10*, 803–816. [CrossRef] [PubMed]
6. Sencilo, A.; Roine, E. A glimpse of the genomic diversity of haloarchaeal tailed viruses. *Front. Microbiol.* **2014**, *5*, 84. [PubMed]
7. Atanasova, N.S.; Bamford, D.H.; Oksanen, H.M. Virus-host interplay in high salt environments. *Environ. Microbiol. Rep.* **2016**, *8*, 431–444. [CrossRef] [PubMed]
8. Zhang, Z.; Liu, Y.; Wang, S.; Yang, D.; Cheng, Y.; Hu, J.; Chen, J.; Mei, Y.; Shen, P.; Bamford, D.H.; et al. Temperate membrane-containing halophilic archaeal virus SNJ1 has a circular dsDNA genome identical to that of plasmid pHH205. *Virology* **2012**, *434*, 233–241. [CrossRef] [PubMed]
9. Liu, Y.; Wang, J.; Liu, Y.; Wang, Y.; Zhang, Z.; Oksanen, H.M.; Bamford, D.H.; Chen, X. Identification and characterization of SNJ2, the first temperate pleolipovirus integrating into the genome of the SNJ1-lysogenic archaeal strain. *Mol. Microbiol.* **2015**, *98*, 1002–1020. [CrossRef] [PubMed]
10. Demina, T.A.; Pietila, M.K.; Svirskaite, J.; Ravantti, J.J.; Atanasova, N.S.; Bamford, D.H.; Oksanen, H.M. Archaeal *Haloarcula californiae* Icosahedral virus 1 highlights conserved elements in icosahedral membrane-containing DNA viruses from extreme environments. *MBio* **2016**, *7*, e00699-16. [CrossRef] [PubMed]
11. Hartmann, R.; Oesterhelt, D. Bacteriorhodopsin-mediated photophosphorylation in *Halobacterium halobium*. *Eur. J. Biochem.* **1977**, *77*, 325–335. [CrossRef] [PubMed]
12. Michel, H.; Oesterhelt, D. Electrochemical proton gradient across the cell membrane of *Halobacterium halobium*: Effect of N,N'-dicyclohexylcarbodiimide, relation to intracellular adenosine triphosphate, adenosine diphosphate, and phosphate concentration, and influence of the potassium gradient. *Biochemistry* **1980**, *19*, 4607–4614. [PubMed]
13. Haupts, U.; Tittor, J.; Oesterhelt, D. Closing in on bacteriorhodopsin: Progress in understanding the molecule. *Annu. Rev. Biophys. Biomol. Struct.* **1999**, *28*, 367–399. [CrossRef] [PubMed]
14. Mirfeizollahi, A.; Yakhchali, B.; Deldar, A.A.; Karkhane, A.A. In silico and experimental improvement of bacteriorhodopsin production in *Halobacterium salinarum* R1 by increasing DNA-binding affinity of Bat through Q661R/Q665R substitutions in HTH motif. *Extremophiles* **2019**, *23*, 59–67. [CrossRef] [PubMed]
15. Stoeckenius, W.; Lozier, R.H.; Bogomolni, R.A. Bacteriorhodopsin and the purple membrane of halobacteria. *Biochim. Biophys. Acta (BBA)—Rev. Bioenerg.* **1979**, *505*, 215–278. [CrossRef]
16. Baliga, N.S.; DasSarma, S. Saturation mutagenesis of the TATA box and upstream activator sequence in the haloarchaeal *bop* gene promoter. *J. Bacteriol.* **1999**, *181*, 2513–2518. [PubMed]
17. Baliga, N.S.; Kennedy, S.P.; Ng, W.V.; Hood, L.; DasSarma, S. Genomic and genetic dissection of an archaeal regulon. *Proc. Natl. Acad. Sci. USA* **2001**, *98*, 2521–2525. [CrossRef] [PubMed]
18. Baliga, N.S.; Pan, M.; Goo, Y.A.; Yi, E.C.; Goodlett, D.R.; Dimitrov, K.; Shannon, P.; Aebersold, R.; Ng, W.V.; Hood, L. Coordinate regulation of energy transduction modules in *Halobacterium* sp. analyzed by a global systems approach. *Proc. Natl. Acad. Sci. USA* **2002**, *99*, 14913–14918. [CrossRef] [PubMed]
19. Yang, C.F.; Kim, J.M.; Molinari, E.; DasSarma, S. Genetic and topological analyses of the *bop* promoter of *Halobacterium halobium*: Stimulation by DNA supercoiling and non-B-DNA structure. *J. Bacteriol.* **1996**, *178*, 840–845. [CrossRef] [PubMed]
20. Rudolph, J.; Nordmann, B.; Storch, K.F.; Gruenberg, H.; Rodewald, K.; Oesterhelt, D. A family of halobacterial transducer proteins. *FEMS Microbiol. Lett.* **1996**, *139*, 161–168. [CrossRef] [PubMed]

21. Storch, K.F.; Rudolph, J.; Oesterhelt, D. Car: A cytoplasmic sensor responsible for arginine chemotaxis in the archaeon *Halobacterium salinarum*. *EMBO J.* **1999**, *18*, 1146–1158. [CrossRef] [PubMed]

22. Marco, M.B.; Moineau, S.; Quiberoni, A. Bacteriophages and dairy fermentations. *Bacteriophage* **2012**, *2*, 149–158. [CrossRef] [PubMed]

23. Schnabel, H.; Zillig, W.; Pfaffle, M.; Schnabel, R.; Michel, H.; Delius, H. *Halobacterium halobium* phage ΦH. *EMBO J.* **1982**, *1*, 87–92. [CrossRef] [PubMed]

24. Soppa, J.; Oesterhelt, D. Bacteriorhodopsin mutants of *Halobacterium* sp. GRB. I. The 5-bromo-2′-deoxyuridine selection as a method to isolate point mutants in halobacteria. *J. Biol. Chem.* **1989**, *264*, 13043–13048. [PubMed]

25. Gordon, D. Viewing and editing assembled sequences using Consed. *Curr. Protoc. Bioinform.* **2003**, *2*, 11.2.1–11.2.43.

26. Dyall-Smith, M.; Pfeifer, F.; Witte, A.; Oesterhelt, D.; Pfeiffer, F. Complete genome sequence of the model halovirus phiH1 (ΦH1). *Genes* **2018**, *9*, 493. [CrossRef] [PubMed]

27. Lomsadze, A.; Gemayel, K.; Tang, S.; Borodovsky, M. Modeling leaderless transcription and atypical genes results in more accurate gene prediction in prokaryotes. *Genome Res.* **2018**, *28*, 1079–1089. [CrossRef] [PubMed]

28. Siddaramappa, S.; Challacombe, J.F.; Decastro, R.E.; Pfeiffer, F.; Sastre, D.E.; Gimenez, M.I.; Paggi, R.A.; Detter, J.C.; Davenport, K.W.; Goodwin, L.A.; et al. A comparative genomics perspective on the genetic content of the alkaliphilic haloarchaeon *Natrialba magadii* ATCC 43099^T. *BMC Genom.* **2012**, *13*, 165. [CrossRef] [PubMed]

29. Laemmli, U.K. Cleavage of structural proteins during the assembly of the head of the bacteriophage T4. *Nature* **1970**, *227*, 680–685. [CrossRef] [PubMed]

30. Kearse, M.; Moir, R.; Wilson, A.; Stones-Havas, S.; Cheung, M.; Sturrock, S.; Buxton, S.; Cooper, A.; Markowitz, S.; Duran, C.; et al. Geneious Basic: An integrated and extendable desktop software platform for the organization and analysis of sequence data. *Bioinformatics* **2012**, *28*, 1647–1649. [CrossRef] [PubMed]

31. Lopes, A.; Tavares, P.; Petit, M.A.; Guerois, R.; Zinn-Justin, S. Automated classification of tailed bacteriophages according to their neck organization. *BMC Genom.* **2014**, *15*, 1027. [CrossRef] [PubMed]

32. Ovcharenko, I.; Loots, G.G.; Hardison, R.C.; Miller, W.; Stubbs, L. zPicture: Dynamic alignment and visualization tool for analyzing conservation profiles. *Genome Res.* **2004**, *14*, 472–477. [CrossRef] [PubMed]

33. Yoon, S.H.; Ha, S.M.; Lim, J.; Kwon, S.; Chun, J. A large-scale evaluation of algorithms to calculate average nucleotide identity. *Antonie Van Leeuwenhoek* **2017**, *110*, 1281–1286. [CrossRef] [PubMed]

34. Babicki, S.; Arndt, D.; Marcu, A.; Liang, Y.; Grant, J.R.; Maciejewski, A.; Wishart, D.S. Heatmapper: Web-enabled heat mapping for all. *Nucleic Acids Res.* **2016**, *44*, W147–W153. [CrossRef] [PubMed]

35. Grissa, I.; Vergnaud, G.; Pourcel, C. CRISPRcompar: A website to compare clustered regularly interspaced short palindromic repeats. *Nucleic Acids Res.* **2008**, *36*, W145–W148. [CrossRef] [PubMed]

36. Paez-Espino, D.; Chen, I.A.; Palaniappan, K.; Ratner, A.; Chu, K.; Szeto, E.; Pillay, M.; Huang, J.; Markowitz, V.M.; Nielsen, T.; et al. IMG/VR: A database of cultured and uncultured DNA Viruses and retroviruses. *Nucleic Acids Res.* **2017**, *45*, D457–D465. [CrossRef] [PubMed]

37. Skennerton, C.T.; Imelfort, M.; Tyson, G.W. Crass: Identification and reconstruction of CRISPR from unassembled metagenomic data. *Nucleic Acids Res.* **2013**, *41*, e105. [CrossRef] [PubMed]

38. Garneau, J.R.; Depardieu, F.; Fortier, L.C.; Bikard, D.; Monot, M. PhageTerm: A tool for fast and accurate determination of phage termini and packaging mechanism using next-generation sequencing data. *Sci. Rep.* **2017**, *7*, 8292. [CrossRef] [PubMed]

39. Guan, Y.; Zhu, Q.; Huang, D.; Zhao, S.; Jan Lo, L.; Peng, J. An equation to estimate the difference between theoretically predicted and SDS PAGE-displayed molecular weights for an acidic peptide. *Sci. Rep.* **2015**, *5*, 13370. [CrossRef] [PubMed]

40. ICTV Report. ICTV Online (10th) Report on Virus Taxonomy. Available online: https://talk.ictvonline.org/taxonomy/p/taxonomy-history?taxnode_id=20170459 (accessed on 19 March 2018).

41. Schnabel, H. An immune strain of *Halobacterium halobium* carries the invertible L segment of phage ΦH as a plasmid. *Proc. Natl. Acad. Sci. USA* **1984**, *81*, 1017–1020. [CrossRef] [PubMed]

42. Witte, A.; Baranyi, U.; Klein, R.; Sulzner, M.; Luo, C.; Wanner, G.; Krüger, D.H.; Lubitz, W. Characterization of *Natronobacterium magadii* phage φCh1, a unique archaeal phage containing DNA and RNA. *Mol. Microbiol.* **1997**, *23*, 603–616. [CrossRef] [PubMed]

43. Klein, R.; Baranyi, U.; Rössler, N.; Greineder, B.; Scholz, H.; Witte, A. *Natrialba magadii* virus φCh1: First complete nucleotide sequence and functional organization of a virus infecting a haloalkaliphilic archaeon. *Mol. Microbiol.* **2002**, *45*, 851–863. [CrossRef] [PubMed]

44. Stolt, P.; Zillig, W. Gene regulation in halophage ΦH; more than promoters. *Syst. Appl. Microbiol.* **1993**, *16*, 591–596. [CrossRef]

45. VIRFAM. Remote Homology Detection of Viral Protein Families. Available online: http://biodev.cea.fr/virfam/ (accessed on 8 October 2018).

46. Klein, R.; Rossler, N.; Iro, M.; Scholz, H.; Witte, A. Haloarchaeal myovirus phiCh1 harbours a phase variation system for the production of protein variants with distinct cell surface adhesion specificities. *Mol. Microbiol.* **2012**, *83*, 137–150. [CrossRef] [PubMed]

47. Schnabel, H.; Zillig, W. Circular structure of the genome of phage ΦH in a lysogenic *Halobacterium halobium*. *Mol. Gen. Genet.* **1984**, *193*, 422–426. [CrossRef]

48. Iro, M.; Klein, R.; Galos, B.; Baranyi, U.; Rossler, N.; Witte, A. The lysogenic region of virus phiCh1: Identification of a repressor-operator system and determination of its activity in halophilic Archaea. *Extremophiles* **2007**, *11*, 383–396. [CrossRef] [PubMed]

49. Gropp, F.; Grampp, B.; Stolt, P.; Palm, P.; Zillig, W. The immunity-conferring plasmid pφHL from the *Halobacterium salinarium* phage φH: Nucleotide sequence and transcription. *Virology* **1992**, *190*, 45–54. [CrossRef]

50. Ken, R.; Hackett, N.R. *Halobacterium halobium* strains lysogenic for phage phiH contain a protein resembling coliphage repressors. *J. Bacteriol.* **1991**, *173*, 955–960. [CrossRef] [PubMed]

51. Stolt, P.; Zillig, W. In vivo studies on the effects of immunity genes on early lytic transcription in the *Halobacterium salinarium* phage φH. *Mol. Gen. Genet.* **1992**, *235*, 197–204. [CrossRef] [PubMed]

52. Stolt, P.; Zillig, W. Transcription of the halophage ΦH repressor gene is abolished by transcription from an inversely oriented lytic promoter. *FEBS Lett.* **1994**, *344*, 125–128. [CrossRef]

53. Pfeiffer, F.; Schuster, S.C.; Broicher, A.; Falb, M.; Palm, P.; Rodewald, K.; Ruepp, A.; Soppa, J.; Tittor, J.; Oesterhelt, D. Evolution in the laboratory: The genome of *Halobacterium salinarum* strain R1 compared to that of strain NRC-1. *Genomics* **2008**, *91*, 335–346. [CrossRef] [PubMed]

54. Svirskaite, J.; Oksanen, H.M.; Daugelavicius, R.; Bamford, D.H. Monitoring physiological changes in haloarchaeal cell during virus release. *Viruses* **2016**, *8*, 59. [CrossRef] [PubMed]

55. Pope, W.H.; Mavrich, T.N.; Garlena, R.A.; Guerrero-Bustamante, C.A.; Jacobs-Sera, D.; Montgomery, M.T.; Russell, D.A.; Warner, M.H.; Science Education Alliance-Phage Hunters Advancing, G.; Evolutionary, S.; et al. Bacteriophages of *Gordonia* spp. display a spectrum of diversity and genetic relationships. *MBio* **2017**, *8*, e01069-17. [CrossRef] [PubMed]

56. Summer, E.J.; Enderle, C.J.; Ahern, S.J.; Gill, J.J.; Torres, C.P.; Appel, D.N.; Black, M.C.; Young, R.; Gonzalez, C.F. Genomic and biological analysis of phage Xfas53 and related prophages of *Xylella fastidiosa*. *J. Bacteriol.* **2010**, *192*, 179–190. [CrossRef] [PubMed]

57. Hendrix, R.W.; Smith, M.C.; Burns, R.N.; Ford, M.E.; Hatfull, G.F. Evolutionary relationships among diverse bacteriophages and prophages: All the world's a phage. *Proc. Natl. Acad. Sci. USA* **1999**, *96*, 2192–2197. [CrossRef] [PubMed]

58. Samson, J.E.; Moineau, S. Bacteriophages in food fermentations: New frontiers in a continuous arms race. *Annu. Rev. Food Sci. Technol.* **2013**, *4*, 347–368. [CrossRef] [PubMed]

59. Grose, J.H.; Casjens, S.R. Understanding the enormous diversity of bacteriophages: The tailed phages that infect the bacterial family Enterobacteriaceae. *Virology* **2014**, *468–470*, 421–443. [CrossRef] [PubMed]

60. Fokine, A.; Rossmann, M.G. Common evolutionary origin of procapsid proteases, phage tail tubes, and tubes of bacterial type VI secretion systems. *Structure* **2016**, *24*, 1928–1935. [CrossRef] [PubMed]

61. Casjens, S.R.; Grose, J.H. Contributions of P2- and P22-like prophages to understanding the enormous diversity and abundance of tailed bacteriophages. *Virology* **2016**, *496*, 255–276. [CrossRef] [PubMed]

 *genes*

Article

Complete Genome Sequence of the Model Halovirus PhiH1 (ΦH1)

Mike Dyall-Smith [1,2], **Felicitas Pfeifer** [3], **Angela Witte** [4], **Dieter Oesterhelt** [1] **and Friedhelm Pfeiffer** [1,*]

[1] Computational Biology Group, Max-Planck-Institute of Biochemistry, Am Klopferspitz 18, 82152 Martinsried, Germany; mike.dyallsmith@gmail.com (M.D.-S.); oesterhe@biochem.mpg.de (D.O.)

[2] Veterinary Biosciences, Faculty of Veterinary and Agricultural Sciences, University of Melbourne, Parkville, VIC 3052, Australia

[3] Department of Biology; Microbiology and Archaea, TU Darmstadt, Schnittspahnstrasse 10, 64287 Darmstadt, Germany; pfeifer@bio.tu-darmstadt.de

[4] Department of Microbiology, Immunobiology and Genetics, MFPL Laboratories, University of Vienna, Dr. Bohr-Gasse 9, Vienna 1030, Austria; angela.witte@univie.ac.at

* Correspondence: fpf@biochem.mpg.de; Tel.: +49-89-8578-2323

Received: 11 September 2018; Accepted: 8 October 2018; Published: 12 October 2018

Abstract: The halophilic myohalovirus *Halobacterium virus phiH* (ΦH) was first described in 1982 and was isolated from a spontaneously lysed culture of *Halobacterium salinarum* strain R1. Until 1994, it was used extensively as a model to study the molecular genetics of haloarchaea, but only parts of the viral genome were sequenced during this period. Using Sanger sequencing combined with high-coverage Illumina sequencing, the full genome sequence of the major variant (phiH1) of this halovirus has been determined. The dsDNA genome is 58,072 bp in length and carries 97 protein-coding genes. We have integrated this information with the previously described transcription mapping data. PhiH could be classified into Myoviridae Type1, Cluster 4 based on capsid assembly and structural proteins (VIRFAM). The closest relative was *Natrialba* virus phiCh1 (φCh1), which shared 63% nucleotide identity and displayed a high level of gene synteny. This close relationship was supported by phylogenetic tree reconstructions. The complete sequence of this historically important virus will allow its inclusion in studies of comparative genomics and virus diversity.

Keywords: halovirus; virus; halophage; *Halobacterium salinarum*; Archaea; haloarchaea; halobacteria; genome inversion

1. Introduction

The temperate myovirus *Halobacterium virus phiH* (ΦH) infects the extremely halophilic archaeon *Halobacterium salinarum* strain R1 (DSM 671) and was isolated after the spontaneous lysis of a culture of its host [1]. Purified virions require 3.5 M NaCl for stability, have an isometric head of 64 nm diameter and a long, contractile tail (170 × 18 nm) with short tail fibres [1,2]. Virus preparations contain 3 major and 10 minor proteins [3]. The virus genome is linear dsDNA with a G+C content of 64%, contains a *pac* site, is about 3% terminally redundant and partially circularly permuted, and estimated to be 59 kb in length [4,5]. In the provirus state, the genome is extrachromosomal, covalently closed and circular, and 57 kb in length [4]. While always classified within the *Myoviridae*, the genus name has changed over the years from phiH-like viruses to *Phihlikevirus*, and most recently to *Myohalovirus* [6,7]. The species name itself has changed from *Halobacterium phage phiH* to *Halobacterium virus phiH* [6,7] but for convenience we will refer to it from here onwards simply as phiH, and the analysed variant as phiH1 or halovirus phiH1.

The original lysate of phiH was found to consist of a mixture of several distinct variants that appeared to have arisen from the activity of insertion sequences. The predominant variant, phiH1, was plaque-purified and a restriction map determined [5]. This was used for further study [3]. PhiH1 became a key model in the study of gene expression and regulation in haloarchaea and was instrumental in the development of genetic tools and methods in these extremophiles. Examples include the polyethylene glycol (PEG)-mediated transfection method [8], the pUBP1 cloning/expression vector [9], the identification of archaeal promoters, mapping transcription start and stop sites [10] and the analysis of gene regulation via repression [11,12]. The presence and function of antisense RNA in haloarchaea was first described in this virus [13]. An 11 kb invertible segment of the virus genome, called the L-region, was found to be flanked on one or both sides by the insertion sequence ISH1.8, and could also circularize to form a 12 kb plasmid (including one copy of ISH1.8), with subsequent loss of the remaining phage DNA [14]. A strain carrying this plasmid was immune to infection [14].

Unfortunately, work on phiH stopped in 1994 [15,16] but a related virus, *Natrialba* virus phiCh1 (φCh1), was described a few years later [17] and continues to be studied in the Witte laboratory [18,19]. PhiCh1 infects a haloalkaliphilic archaeon, *Natrialba magadii*, and the genomes of both host and virus are fully sequenced [18,20]. The provirus state of phiCh1 corresponds to plasmid pNMAG03 carried by *Nab. magadii*. A full comparison between phiCh1 and phiH1 was prevented as only parts of the phiH genome were ever determined. This deficit also prevented the inclusion of phiH in broad-ranging studies of virus diversity, taxonomy, and evolution. The aim of this study was to complete the phiH1 genome sequence and provide a thorough annotation. This will not only provide a better understanding of the results from previous studies on this virus but also allow complete genomic comparisons with a wealth of other datasets, including other sequenced viruses, haloarchaeal proviruses, metaviromic/metagenomic and environmental RNA sequences.

2. Materials and Methods

2.1. Virus DNA and Sequencing Methods

Purified phiH1 DNA [1] was originally provided to F. Pfeifer by Hans-Peter Klenk while both were working in the department of W. Zillig [21]. The DNA was stored frozen at −80 °C until use. Sequencing was performed in two stages. For the first stage, all available sequences of phiH1 were downloaded from National Center for Biotechnology Information (NCBI) [22] and imported into the Phred–Phrap–Consed package [23]. Overlapping sequences were assembled and primers designed to gather additional sequences using Sanger technology. This consisted either of primer-walking directly on virus DNA, or on polymerase chain reaction (PCR) amplimers, or PCR-sequencing across gaps. The resulting sequence reads were progressively assembled into contigs, base calls inspected manually and corrected where needed, and new primers designed for further rounds of sequencing until all gaps were closed. Except for overlaps, this approach left most of the previously published sequences unchecked.

In the second stage, short-read Illumina HiSeq sequencing of phiH1 DNA was performed (Max-Planck Genome Centre, Cologne, Germany). This returned 243 Mb of high quality sequence data (coverage = 4200-fold). De-novo assembly did not produce a single contig, due to short read-lengths and the presence of repeat sequences within the viral genome, but reads could be confidently mapped to the genome sequence obtained in the first stage (*Map to Reference* option; Geneious mapper method) in order to improve the sequence reliability.

2.2. CRISPR Spacer Searches

The crass v0.3.12 software [24] was used to extract CRISPR spacer sequences from genomic/metagenomic data available at the NCBI SRA database (accessed 27 July 2018) [25], as described previously [26]. These included all available genomes of members of the class Halobacteria, and metagenomes of hypersaline environments. CRISPR direct repeats (DR) identified by crass were used to search the CRISPRfinder database (accessed 25 July 2018) [27] for haloarchaea with matching or closely matching DR.

2.3. Bioinformatic Methods

Gene annotation used a combination of gene prediction with GeneMarkS-2 [28] and manual refinement using database searches (BLASTp/BLASTn; nr databases) at the NCBI webserver [29]. Repeats were identified by BLASTn, dot-plot comparison in Yass [30], and with tools within the Geneious software suite [31]. Circos plots were performed via the circoletto webserver [32]. Plots are coloured by the 'score/max' ratio of tBLASTx bitscores (real score/maximal score). Colours are: blue ≤ 0.25, green ≤ 0.50, orange ≤ 0.75, red > 0.75. Sequence mapping, alignments, editing and phylogenetic tree reconstructions were performed with Geneious software version 10.2 [31]. For phylogenetic tree reconstructions, protein sequences were first aligned using CLUSTALW, and trees inferred using the Neighbor-Joining algorithm (within Geneious). Consensus trees were determined after 100 bootstrap repetitions. Protein structural modelling used the I-Tasser webserver [33]. Identification of the *pac* site utilised the program PhageTerm [34] as implemented on the CPT Phage Galaxy [35]. The VIRFAM webserver [36] uses proteins of the phage head-neck-tail module to cluster phages into related groups, and was used to classify phiH1.

2.4. Data Availability

The phiH1 genome sequence has been deposited at Genbank under the accession MK002701. Raw reads were submitted to the SRA archive under accession SRP159490.

3. Results and Discussion

3.1. Sequence and Annotation of PhiH1

The previously sequenced regions of the phiH1 genome represented about 50% of the complete sequence (Figure 1, red lines). Using virus DNA as template, the gaps between these sequences were PCR amplified and Sanger sequenced. However, the quality of the previously sequenced regions was of uncertain reliability. High-coverage Illumina sequencing (ca. 4200-fold) was then used to enhance sequence confidence. Sequence revisions were only found to be required in previously deposited sequences but not to Sanger sequencing results of the first stage of the project. While the virus DNA found in capsid particles is linear, the head-full packaging process produces a population of molecules that are terminally redundant and partially circularly permuted [1]. The complete genome sequence determined in the current study is represented as the provirus form; a circular sequence of 58,072 bp. This value is close to the published size of 57 kb, estimated from restriction fragment sizes [4,37]. The G+C content of the genome was 63.7%, almost identical to the published value of 64% [3] but slightly lower than that of the host chromosome (68.0%) [38].

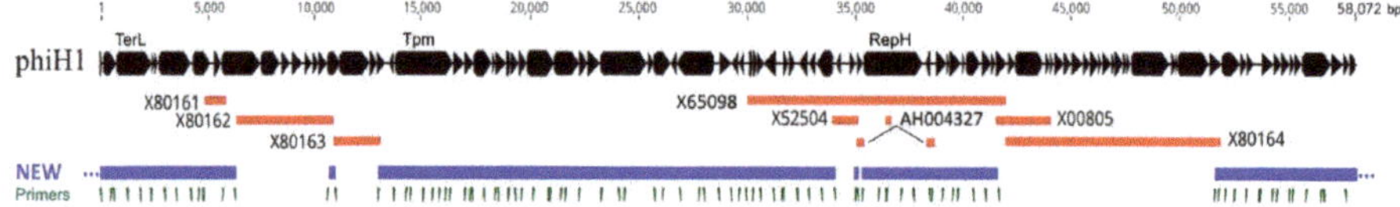

Figure 1. Diagram of the phiH1 genome with lines below showing regions previously sequenced (red) along with their database accessions. The blue lines (NEW) indicate regions sequenced in the present study by Sanger sequencing. Tick marks (dark green) below the blue lines show the positions of oligonucleotide primers used for PCR and primer-walking. Dots at the right and left contig ends indicate sequence continuity between them. Scale bar at top shows position in bp.

The original restriction map of phiH1 DNA, as determined by [5], corresponded closely with the *in silico* map inferred from the phiH1 genome sequence (Figure S1). The *pac* site located at the left end of the restriction map matched closely to the corresponding *pac* sequence of phiCh1. While the *pac* site of phiCh1 had been localized by restriction mapping [18], it had not been precisely mapped. For consistency, the start point of phiH1 was set to the corresponding start of phiCh1 even though this splits the *terS* gene. Using this numbering, the program PhageTerm [34] was used to analyse the mapping of Illumina reads to the phiH1 genome, and this located the *pac* site terminal base at nt 46, with high probability ($p = 2.5 \times 10^{-238}$). This is within the *terS* coding sequence (CDS) close to the stop codon and within a GC-rich region that is strongly conserved between phiH1 and phiCh1.

Annotation of the phiH1 genome resulted in 97 CDS (Table 1), most of which were encoded on the plus strand (86/97, Figure 2 panel b), and were frequently closely spaced, with 45 overlapping at start/stop codons and 23 separated by $\leq$8 nt. Many genes were in functional groupings typical of bacteriophages (Figure 2 panel b). The left end of the genome encodes DNA packaging proteins (e.g., terminase, portal protein), then virus assembly and structural proteins (e.g., major capsid protein, tape-measure protein, tail proteins). The three main proteins of purified virus were originally labelled by their estimated sizes on sodium dodecylsufate (SDS)-polyacrylamide gels (22, 53 and 80 kDa) [1], which were later revised to 27, 46 and 80 kDa [3] but in 1994, Stolt et al. [39] determined their N-terminal amino acid sequences and used this information to map the proteins (HP20, HP32 and HP67) to their genes (*hp20, hp32, hp67*) and sequence them. The inferred molecular weights (MWs) of proteins HP20 and HP67 were noted by these authors to be much smaller than previous estimates. In the present study, the locations of these genes on the full genome sequence have been resolved, an error in the *hp20* (accession X80161) coding sequence was corrected, and the MWs of the inferred proteins calculated (11.6, 35.4 and 45.5 kDa). For consistency we have retained the original gene names (Table 1).

The next genomic region is a replication/regulatory module (the L-region) that encodes RepR (repressor), a ParA-family protein (partition) and RepH (replication). There is also a VapC-like protein that together with the small overlapping upstream CDS may form a toxin-antitoxin pair that could be involved in plasmid maintenance [40]. The right end of the genome carries many genes with unknown function but includes genes specifying DNA methylases and cell lysis proteins. The taxonomic position of phiH1 was assessed using the VIRFAM webserver [36], which classifies bacteriophages and archaeal viruses based on the order and similarity of capsid assembly/structural proteins. Consistent with previous studies [6], phiH1 was classified by this system as a member of the Myoviridae (Type1, Cluster 4).

Table 1. Annotated coding sequences (CDS) of halovirus phiH1.

Start (nt)	Stop (nt)	Locus_tag	Length (bp)	Direction	Gene	Product	*Homologs* [1]: phiCh1, ORF pNMAG03 [Other]
115	717	PhiH1_005	603	+	-	uncharacterized protein	PhiCh1p02, ORF1 Nmag_4251
710	2371	PhiH1_010	1662	+	*terL*	terminase large subunit TerL	PhiCh1p03, ORF2 Nmag_4252
2377	2505	PhiH1_015	129	+	-	uncharacterized protein	Nmag_4253
2498	2689	PhiH1_020	192	+	-	uncharacterized protein	PhiCh1p05, ORF4 Nmag_4255
2686	4242	PhiH1_025	1557	+	*por*	portal protein Por	PhiCh1p07, ORF6 Nmag_4257
4246	5187	PhiH1_030	942	+	-	head morphogenesis protein	PhiCh1p08, ORF7 Nmag_4258
5261	5587	PhiH1_035	327	+	*hp20*	capsid protein HP20	[AJF28118.1]
5667	7466	PhiH1_040	1800	+	-	prohead protease	[4] PhiCh1p09, ORF8 [4] PhiCh1p10, ORF9 Nmag_4259
7506	8468	PhiH1_045	963	+	*hp32*	major capsid protein HP32	PhiCh1p12, ORF11 Nmag_4260
8481	8933	PhiH1_050	453	+	-	uncharacterized protein	PhiCh1p13, ORF12 Nmag_4261
8940	9542	PhiH1_055	603	+	*ada*	head-tail adaptor protein Ada	PhiCh1p14, ORF13 Nmag_4262
9539	9919	PhiH1_060	381	+	*hco*	head closure protein type 1 Hco	PhiCh1p15, ORF14 Nmag_4263
9921	10,202	PhiH1_065	282	+	-	uncharacterized protein	PhiCh1p16, ORF15 Nmag_4264
10,202	10,636	PhiH1_070	435	+	*nep*	probable neck protein type 1 Nep	PhiCh1p17, ORF16 Nmag_4265

Table 1. *Cont.*

Start (nt)	Stop (nt)	Locus_tag	Length (bp)	Direction	Gene	Product	Homologs [1]: phiCh1, ORF pNMAG03 [Other]
10,643	11,239	PhiH1_075	597	+	*tco*	tail completion protein type 1 Tco	PhiCh1p18, ORF17 Nmag_4266
11,259	12,557	PhiH1_080	1299	+	*hp67*	tail sheath protein HP67	PhiCh1p19, ORF18 Nmag_4267
12,607	13,002	PhiH1_085	396	+	-	probable structural protein	PhiCh1p20, ORF19 Nmag_4268
13,006	13,407	PhiH1_090	402	+	-	uncharacterized protein	PhiCh1p21, ORF20 Nmag_4269
13,572	13,745	PhiH1_095	174	−	-	DUF4177 domain protein	[SEH60446.1]
13,792	16,581	PhiH1_100	2790	+	*tpm*	tape-measure tail protein Tpm	[4] PhiCh1p23, ORF22 [4] PhiCh1p24, ORF23 Nmag_4272
16,583	17,104	PhiH1_105	522	+	-	uncharacterized protein	PhiCh1p25, ORF24 Nmag_4273
17,108	17,446	PhiH1_110	339	+	-	uncharacterized protein	PhiCh1p26, ORF25 Nmag_4274
17,450	18,298	PhiH1_115	849	+	-	uncharacterized protein	PhiCh1p27, ORF26 Nmag_4275
18,306	18,446	PhiH1_120	141	+	-	CxxC motif protein	[SEH61109.1]
18,443	18,988	PhiH1_125	546	+	-	uncharacterized protein	PhiCh1p29, ORF28 Nmag_4276
18,988	19,146	PhiH1_130	159	+	-	uncharacterized protein	-
19,143	19,508	PhiH1_135	366	+	-	virus-related protein	[AGM10900.1]
19,505	19,867	PhiH1_140	363	+	-	uncharacterized protein	PhiCh1p30, ORF29 Nmag_4277
19,874	21,148	PhiH1_145	1275	+	*bpj*	baseplate J family protein Bpj	PhiCh1p31, ORF30 Nmag_4278

Table 1. *Cont.*

Start (nt)	Stop (nt)	Locus_tag	Length (bp)	Direction	Gene	Product	*Homologs* [1]: phiCh1, ORF pNMAG03 [Other]
21,135	22,277	PhiH1_150	1143	+	-	uncharacterized protein	PhiCh1p32, ORF31 Nmag_4279
22,295	22,678	PhiH1_155	384	+	-	virus-related protein	[AFH21897.1]
22,683	23,249	PhiH1_160	567	+	-	virus-related protein	[AFH21653.1]
23,252	25,504	PhiH1_165	2253	+	-	repeat-containing tail fibre protein	PhiCh1p37, ORF36 Nmag_4282 PhiCh1p35, ORF34 Nmag_4286
25,506	25,787	PhiH1_170	282	+	-	uncharacterized protein	Nmag_4285
25,825	26,499	PhiH1_175	675	+	*int1*	tyrosine integrase/recombinase Int1	PhiCh1p36, ORF35 Nmag_4284
26,490	26,792	PhiH1_180	303	−	-	uncharacterized protein	Nmag_4283
26,798	27,766	PhiH1_185	969	−	-	repeat-containing tail fibre protein [2]	PhiCh1p37, ORF36 Nmag_4282 PhiCh1p35, ORF34 Nmag_4286
27,803	28,150	PhiH1_190	348	+	-	YncB-like endonuclease	[AGM11801.1]
28,153	28,386	PhiH1_195	234	+	-	virus-related protein	[AGC34510.1]
28,379	28,675	PhiH1_200	297	+	-	uncharacterized protein	[EMA49173.1]
28,682	28,783	PhiH1_205	102	+	-	uncharacterized protein	-
28,788	29,357	PhiH1_210	570	+	-	transmembrane domain protein	-
29,394	29,642	PhiH1_215	249	−	-	uncharacterized protein	-
29,651	29,941	PhiH1_220	291	−	-	uncharacterized protein	PhiCh1p40, ORF39 Nmag_4289
30,104	30,244	PhiH1_225	144	+	-	uncharacterized protein	-

Table 1. *Cont.*

Start (nt)	Stop (nt)	Locus_tag	Length (bp)	Direction	Gene	Product	*Homologs* [1]: phiCh1, ORF pNMAG03 [Other]
30,250	30,414	PhiH1_230	165	+	-	uncharacterized protein	PhiCh1p44, ORF43 Nmag_4292
30,411	30,806	PhiH1_235	396	+	-	VapC family toxin	PhiCh1p45, ORF44 Nmag_4293
30,803	31,465	PhiH1_240	663	−	*int2*	tyrosine integrase/recombinase Int2	PhiCh1p46, ORF45 Nmag_4294
31,680	31,934	PhiH1_245	255	+	-	uncharacterized protein	-
31,939	32,271	PhiH1_250	333	+	-	uncharacterized protein	Nmag_4297
32,420	32,857	PhiH1_255	438	−	-	HNH-type endonuclease	PhiCh1p48, ORF47 Nmag_4296
32,854	33,255	PhiH1_260	402	−	-	uncharacterized protein	[ELY96531.1]
33,248	34,024	PhiH1_265	777	−	-	parA domain protein	PhiCh1p47, ORF46 Nmag_4295
34,161	34,430	PhiH1_270	270	−	*repR*	repressor protein RepR	[5] PhiCh1p49, ORF48 [5] Nmag_4298 [ELZ06324.1]
34,730	35,071	PhiH1_275	342	+	-	uncharacterized protein	-
35,068	35,424	PhiH1_280	357	+	-	uncharacterized protein	PhiCh1p50, ORF49
35,381	38,167	PhiH1_285	2787	+	*repH*	plasmid replication protein RepH	[4] PhiCh1p54, ORF53 [4] PhiCh1p55, ORF54 Nmag_4299
38,262	38,489	PhiH1_290	228	−	*imm*	probable immunity protein Imm	PhiCh1p56, ORF55 Nmag_4300
38,733	39,263	PhiH1_295	531	+	-	transcriptional regulator, PadR-like family	PhiCh1p57, ORF56 Nmag_4301
39,260	39,385	PhiH1_300	126	+	-	CxxC motif protein	-
39,382	39,978	PhiH1_305	597	+	-	uncharacterized protein	PhiCh1p59, ORF58 Nmag_4303

Table 1. *Cont.*

Start (nt)	Stop (nt)	Locus_tag	Length (bp)	Direction	Gene	Product	*Homologs* [1]: phiCh1, ORF pNMAG03 [Other]
39,975	40,133	PhiH1_310	159	+	-	uncharacterized protein	-
40,153	40,902	PhiH1_315	750	+	*pcnA*	DNA polymerase sliding clamp PcnA	PhiCh1p60, ORF59 Nmag_4211
40,908	41,339	PhiH1_320	432	+	-	uncharacterized protein	PhiCh1p61, ORF60 Nmag_4212
41,339	41,554	PhiH1_325	216	+	-	uncharacterized protein	PhiCh1p62, ORF61 Nmag_4213
41,547	42,041	PhiH1_330	495	+	-	uncharacterized protein	-
42,098	42,490	PhiH1_335	393	+	*tnpA*	IS200-type transposase TnpA	[CAP12925.1]
42,492	43,748	PhiH1_340	1257	+	*tnpB*	IS1341-type transposase TnpB	[CAP12926.1]
43,808	44,014	PhiH1_345	207	+	-	uncharacterized protein	-
44,007	44,234	PhiH1_350	228	+	-	uncharacterized protein	PhiCh1p66, ORF65 Nmag_4217
44,231	44,656	PhiH1_355	426	+	-	CxxC motif protein	PhiCh1p68, ORF67 Nmag_4219
44,646	45,026	PhiH1_360	381	+	-	uncharacterized protein	PhiCh1p69, ORF68 Nmag_4220
45,023	45,646	PhiH1_365	624	+	-	HNH-type endonuclease	[KYG11427.1]
45,639	45,926	PhiH1_370	288	+	-	uncharacterized protein	PhiCh1p71, ORF70 Nmag_4222
45,919	46,350	PhiH1_375	432	+	-	DUF4326 domain protein	PhiCh1p72, ORF71 Nmag_4223
46,343	46,441	PhiH1_380	99	+	-	uncharacterized protein	-
46,438	46,884	PhiH1_385	447	+	-	CxxC motif protein	PhiCh1p74, ORF73 Nmag_4225
46,865	47,038	PhiH1_390	174	+	-	uncharacterized protein	[5] PhiCh1p73, ORF72 [5] Nmag_4224

Table 1. *Cont.*

Start (nt)	Stop (nt)	Locus_tag	Length (bp)	Direction	Gene	Product	Homologs [1]: phiCh1, ORF pNMAG03 [Other]
47,031	47,447	PhiH1_395	417	+	-	uncharacterized protein	-
47,440	47,739	PhiH1_400	300	+	-	NTPase protein	[PLX87675.1]
47,732	49,618	PhiH1_405	1887	+	*dcm5*	C-5 cytosine-specific DNA methylase Dcm5	[5] PhiCh1p81, ORF80 [PCR88664.1]
49,611	49,931	PhiH1_410	321	+	-	uncharacterized protein	PhiCh1p82, ORF81 Nmag_4234
49,918	50,037	PhiH1_415	120	+	-	CxxC motif protein	-
50,091	51,452	PhiH1_420	1362	+	*yhdJ*	DNA methylase N-4/N-6 domain protein YhdJ	PhiCh1p83, ORF82 Nmag_4235
51,449	52,024	PhiH1_425	576	+	-	uncharacterized protein	PhiCh1p84, ORF83 Nmag_4236
52,021	52,791	PhiH1_430	771	+	-	uncharacterized protein	PhiCh1p85, ORF84 Nmag_4237
52,784	53,152	PhiH1_435	369	+	-	uncharacterized protein	PhiCh1p88, ORF87 Nmag_4240
53,145	53,504	PhiH1_440	360	+	-	uncharacterized protein	PhiCh1p89, ORF88 Nmag_4241
53,788	54,369	PhiH1_445	582	+	-	CxxC motif protein	PhiCh1p90, ORF89 Nmag_4242
54,403	54,771	PhiH1_450	369	+	-	uncharacterized protein	PhiCh1p91, ORF90 Nmag_4243
54,794	55,147	PhiH1_455	354	+	-	uncharacterized protein	-
55,144	55,401	PhiH1_460	258	+	-	transmembrane domain protein	PhiCh1p93, ORF92 Nmag_4244
55,394	55,729	PhiH1_465	336	+	-	transmembrane domain protein [3]	PhiCh1p94, ORF93 Nmag_4245
55,794	57,053	PhiH1_470	1260	+	*ycdA*	DNA methylase N-4/N-6 domain protein YcdA	PhiCh1p95, ORF94 Nmag_4246

Table 1. *Cont.*

Start (nt)	Stop (nt)	Locus_tag	Length (bp)	Direction	Gene	Product	*Homologs* [1]: phiCh1, ORF pNMAG03 [Other]
57,046	57,564	PhiH1_475	519	+	-	uncharacterized protein	PhiCh1p96, ORF95 Nmag_4247
57,621	57,830	PhiH1_480	210	+	-	CxxC motif protein	PhiCh1p98, ORF97 Nmag_4249
57,827>	<63	PhiH1_485	309	+	*terS*	terminase small subunit TerS	PhiCh1p01, ORF98 Nmag_4250

[1] PhiCh1/pNMAG03 homologs of phiH1 proteins show BLASTp E-values $< 10^{-20}$. For phiCh1 proteins, both the PhiCh1p and originally assigned ORF codes (ORF for open reading frame) are shown (e.g., PhiCh1p02, ORF1). Codes starting with ORF represent the original annotation of the phiCh1 genome [17] (GB accession AF440695.1); and codes starting with PhiCh1p represent the RefSeq version of the annotation of the same genome sequence (GB accession NC_004084). The number shift is due to the *terS* gene, the N-terminal part being encoded at the end of the genome, and the C-terminal part at its beginning. This ORF is complete in the provirus state due to circularization and in the linear virus state due to terminal redundancy. This gene is *ORF98* in the original annotation and PhiCh1p01 in the RefSeq annotation. Codes starting with Nmag_ represent the annotation of the *Natrialba magadii* plasmid pNMAG03 [20] (accession CP001935.1). The point of ring opening in pNMAG03 was set between Nmag_4303 and Nmag_4211. Codes in square brackets represent NCBI accessions referring to homologous proteins (BLASTp E-values $\leq 10^{-11}$), which are from other sources. [2] Gene PhiH1_185 is encoded on an invertible segment. In the current sequence version, it is inactivated because it is uncoupled from a start codon. By genome inversion, it becomes activated while its partner gene PhiH1_165 becomes inactivated. Overall, this results in tail fibre protein switching. [3] This protein (PhiH1_465) has three predicted transmembrane domains and has been suspected to function as a holin [18]. [4] In these cases, the phiCh1 gene is split into two CDS but is continuous in phiH1. [5] These proteins are more distantly related (show less than 39% sequence identity or fall above BLASTp E-values of 10^{-20}). In these cases, a similar genetic context supports their stated relationship.

A GC-profile plot [41] of the phiH1 genome shows a major low point inflection within the L-region (Figure 2, panel a), indicating a potential replication origin. The L-region is ~12 kb in length, can replicate as a plasmid in *Halobacterium* [14], and carries genes encoding a replication protein (RepH), and a DNA-binding repressor (RepR). It can also provide cells with immunity to infection by phiH1 virus. The transcription program of phiH1 during lytic growth (panels c, d and e) has been summarized from previous studies, and shows temporal changes (early, middle and late transcripts). The broad directions of transcription reflect the closely spaced and similarly directed gene clusters as well as the correspondence with functional gene groupings (panel b). The lowest two panels (d, e) summarize the results of hybridizing labelled transcripts from infected cells to Southern blots of restricted phiH1 DNA [42], so mapping transcripts to fragments of the virus genome. Panel c shows a summary of the virus-specific transcripts that were sized by agarose gel electrophoresis and had 5′ start sites mapped. While transcription across the L-region has been examined in more detail compared to the rest of the genome, there remains much that is incomplete or uncertain. For example, the 3′ end of the late transcript labelled T_{LL}, which is depicted ending in a dotted line and question mark (at ~21 kb), has not been determined. This transcript could potentially extend for another 5.5 kb. Counter-transcripts are commonly produced by prokaryotes and their viruses and play important roles in gene regulation. Their presence and activity in phiH1 gene expression has been studied and was one of the first reports of antisense RNA in Archaea [13]. However, this interesting topic remains to be fully explored.

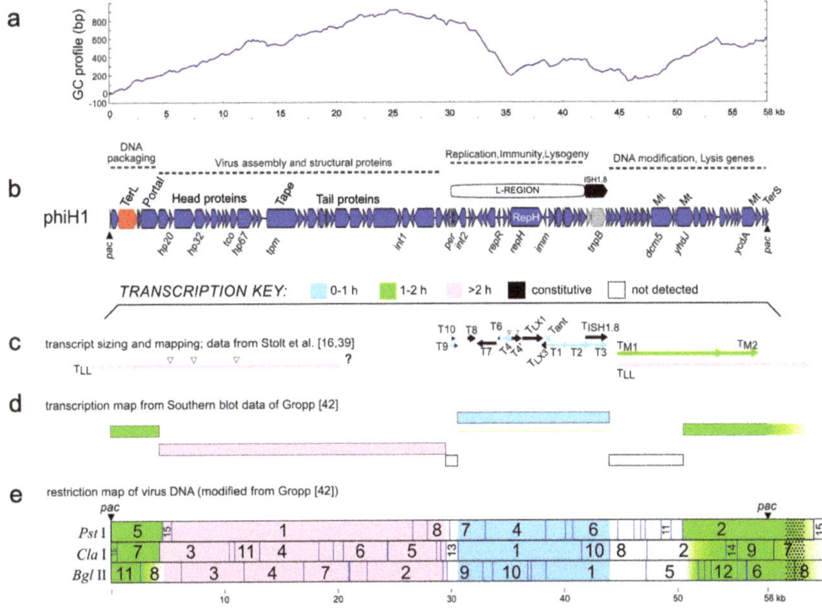

Figure 2. PhiH1 GC-profile, genetic map, and corresponding transcription program (adapted from [16,42]). (**a**) GC-profile of the phiH1 genome. (**b**) Genetic map of the phiH1 genome, showing coding sequences as red, blue or grey arrows. Dotted lines above indicate gene clusters involved in particular functions. Some CDS are labelled above the map, e.g., TerL, terminase large subunit; Portal, portal protein; Tape, tape-measure protein; RepH, replicase (label within CDS arrow); Mt, DNA methylases; TerS, terminase small subunit. Some genes are shown below the map, such as *hp32*, encoding the major capsid protein HP32. Panels c, d and e summarise transcription data from previously published studies, and above them is a colour key that indicates the time of appearance of early (0–1 h, blue), middle (1–2 h, green) and late (>2 h, pink) transcripts. (**c**) Precise mapping of viral transcripts, including start and termination sites [16,39]. (**d**) Summary transcription program of lytic

infection based on hybridisation of labelled infected-cell transcripts to restriction fragments of virus DNA [42]. Thin coloured lines indicate whether continuing transcription persists over time. (**e**) The transcription map data of [42] are shown, projected onto the in silico restriction map of phiH1, as determined from the complete genome sequence (this study). Enzymes are indicated at the left. Numbers on the restriction map refer to those of the original publication of [42] (see also Figure S1). Coloured shading follows that of panels c and d. Dotted pattern shown beyond the right-hand *pac* site indicates terminal redundancy of virus DNA. Scale bars (in kb) are shown below panels a and e.

Corrections to the previously sequenced regions resulted in significant changes to several coding sequences. For example, the *tnpB* gene of transposon ISH1.8 (nt 41,906–43,789) was thought to be inactive as it was split into three CDS by multiple mutations [43]. The high-quality Illumina sequence data show, however, that the gene is intact and that the previously reported transposon ISH1.8 (X00805) is actually an exact copy of transposon ISH12 from the host *Hbt. salinarum* strain R1 [44]. The element plays a key role in the mobilisation of the L-region of the genome to form the 12 kb plasmid, pΦHL [14,43]. Another case is the Dcm5 cytosine methylase, which was also reported as being split [39]. The revised sequence shows that the gene codes for a single, probably functional protein (PhiH1_405, nt 47,732–49,618) and not for the two parts (dcm5a, dcm5b) as previously reported. Although phiH1 carries three potentially active DNA methylase genes (*dcm5*, *yhdJ* and *ycdA*), the presence of modified bases in phiH1 DNA was not detected in the chromatographic (high-pressure liquid chromatography) profiles of deoxyribonucleosides released by enzymatic hydrolysis [45]. In that study, the genomes of phiH and another, unrelated virus (phiN) were analysed, and while unmodified dC was detected in phiH, the phiN genome contained only methylated dC (Figure 3 in [45]). The related halovirus phiCh1 carries homologs of two of the phiH1 methylases, and one of them, N6-adenine methylase (ORF94/M·φCh1-I, corresponding to YcdA of phiH1) has been shown to methylate DNA at GATC motifs [46] but the proportion of sites found to be modified in virus DNA by M·φCh1-I varies from 5% to 50%, depending upon the infection conditions. Modifying only some of the available sites is presumably advantageous to avoid host restriction, as distinct enzymes may target either unmethylated or methylated sites.

3.2. Matches to CRISPR Spacers

The phiH1 genome was used to search for matching CRISPR spacers among metagenomic datasets of hypersaline environments downloaded from the NCBI sequence read archive (SRA; see methods). Only four spacers showing close to moderate similarity to phiH1 were detected (Table 2). These spacers match to virus genes encoding structural and non-structural proteins, and the DRs of these spacers show that they are carried by haloarchaea. The datasets include metagenomes from the USA and Iran, as well as an isolate from the Andaman Islands, India. The results suggest that phiH1-like viruses are geographically widespread.

Table 2. CRISPR spacers matching phiH1.

No.	CRISPR Spacer Matches to phiH1 [1]		Translation [2]
	DR: GCTTCAACCCCACGAGGGTTCGTCTGAAAC (Haloarcula/Halorhabdus/Natronomonas)		
1	phiH1	18677 CACGCCCAGGAGCCGCTCGCGAAGCAGATCCAGCAG 18712	HAQEPLAKQIQQ phiH1 (tail module)
	PRJNA337743	36 CACGCCAAGGAGCCACTGGAGAAGCAGATCCAGCAG 1	HAKEPLEKQIQQ Alviso ponds, USA
		****** ******* ** * ***************	**:*** *****
	DR: GCTTCGACCCCACAAGGGTCCGTCTGAAAC (Natrinema/Haloarcula)		
2	phiH1	25264 CCTCACGAACGCCAGCAGCCTCACCGTCTACGA 25296	LTNASSLTVY phiH1 putative tail fiber
	PRJEB18068	1 CCTCACGAACGCCAGCAACCTCACTGTCTACGA 33	LTNASNLTVY Lake Meyghan, Iran
		***************** ****** ********	*****.****
	DR: GCTTCGACCCCACAAGGGTCCGTCTGAAAC (Natrinema/Haloarcula)		
3	phiH1	40543 GTCCTCGAAGGGCGGCACATCGACCGCGGCATCACCGC 40580	VLEGRHIDRGIT phiH1 PcnA
	JMIP02000025.1	38 GTCCTCGAAGGCCGGCGCATCACCCGCGGCGTCACCGC 1	VLEGRRITRGVT Halostagnicola sp. A56
		*********** **** **** ******* *******	*****:* **:*
	DR: GCTTCAACCCCACGAGGGTTCGTCTGAAAC (Haloarcula/Halorhabdus/Natronomonas)		
4	phiH1	50186 CATGGTGATGACCTCGCCTCCGTACTTCGGGTTGA 50220	MVMTSPPYFGL phiH1 methylase YhdJ
	PRJNA337743	35 CATGGTGATGACCTCGCCTCCGTACTTCGGGCTGC 1	MVMTSPPYFGL Alviso ponds, USA
		******************************** **	***********

[1] The matching spacer sequences were found in the following NCBI bioprojects using the crass program: PRJNA337743, (SRA SRR4030040; Alviso Ponds, San Francisco, CA, USA; metagenome); PRJNA245787 (*Halostagnicola* sp. A56 26 genome; Andaman Islands, India); PRJEB18068 (Lake Meyghan, Iran; metagenome). Aligned sequences show nt positions for phiH1, and asterisks indicated identical bases. DR: direct repeat (with haloarchaea containing most closely matching DR shown in brackets). [2] Symbols under alignment (*:.) indicate identical, similar and weakly similar residues, respectively (based on Gonnet PAM 250 matrix).

3.3. Relatives and Phylogeny of PhiH1

The only close matches to the phiH1 genome in the GenBank database were phiCh1 and the corresponding *Nab. magadii* plasmid pNMAG03 (BLASTn, accessed 20 July 2018). A dot-plot comparison of phiH with phiCh1 (Figure S2) revealed a largely colinear relationship (green line) and an overall nucleotide similarity of 63%. The plot also highlights several indels (line gaps) and two regions showing inversions (red lines). Inversion 1 (nt 24,227–27,767) corresponds closely in sequence and arrangement to the invertible region described in phiCh1 (ORF34-36) that has a central XerD type integrase/recombinase gene flanked by inverted repeats, and facilitates switching between two related tail fibre genes, each containing numerous short repeats [18]. The phiH1 orthologous integrase is PhiH1_175. In the current sequence version, PhiH1_165 is active while PhiH1_185 is uncoupled from a start codon and thus is inactivated. Upon inversion of the genome segment, PhiH1_185 is activated while PhiH1_165 is inactivated. Overall, this results in tail fibre protein switching, which may affect receptor binding specificity and host range of phiH1. The similarity in the tail fibre protein repeats is high enough to be detectable at the DNA level, which results in the X-shaped pattern for this region in the dot-plot. Inversion 2 (nt 31,932–34,126) occurs within the phiH1 L-segment, encompasses four CDS including a ParA-domain protein, and is nearby a different integrase/recombinase gene (Int2, PhiH1_240). Protein searches (BLASTp) of the phiH1 genome returned matches to phiCh1, a limited number of haloarchaeal genomes (often 5–10, which may flag proviral regions) and the haloarchaeal caudoviruses BJ1 [47] and CGphi46 (NC_021537), both of which infect *Halorubrum* spp. Pairwise alignments (BLASTp) between all phiH1 and matching phiCh1 proteins gave an average protein sequence identity of 70% (range 39–95%; with a few exceptions, see footnote 5, Table 1). Figure 3 is a graphical comparison of phiH1 proteins (tBLASTx) with those of phiCh1, BJ1, CGphi46 and, as an outlier, HSTV-1. The *Haloarcula* caudovirus HSTV-1 [48] shows very low similarity to phiH1. The figure summarizes the close similarity of phiH1 and phiCh1 proteins. BJ1 and CGphi46 show far fewer matching regions, mainly to proteins encoded near the left end of the phiH1 genome, a region specifying portal and capsid proteins. The three significant matches to HSTV-1 were to a methyltransferase (HSTV1_52), a hypothetical protein (HSTV1_53), and a DNA polymerase sliding clamp protein (HSTV1_40).

While several caudovirus proteins have been used to infer virus phylogenies, the major capsid protein (MCP) is often used because of its functional constraints maintaining a conserved structure [49]. Figure 4 shows a tree reconstruction using an alignment of phiH1 MCP (HP32, 35.4 kDa) and related sequences. Haloarchaeal proteins are seen to branch together (pink shading) and within this cluster the phiH1 and phiCh1 MCPs form a distinct and closely branching clade. These two proteins share 82% amino acid identity. The MCPs of CGphi46 and BJ1 branch at distant locations from each other and from phiH1 MCP. Structures of close homologs of phiH1 HP32 have not yet been determined. However, the major capsid proteins of bacterial caudoviruses and eukaryotic herpesviruses share a common folding structure, the archetype of which is the phage HK97 MCP (gp5) [50]. Consistent with this, modelling of the phiH1 MCP (I-Tasser) returned bacteriophage HK97 gp5 (PDB 2fs3A) as the closest matching structure (Template Modeling (TM)-score = 0.848, Root-Mean-Square Deviation (RMSD) = 1.17). Based on structure prediction and homology modelling, the HK97-fold may also be present in the MCP of phiCh1 [49]. The structure of the MCP of the haloarchaeal podovirus, HSTV-1, has recently been shown to be of the HK97 type [48].

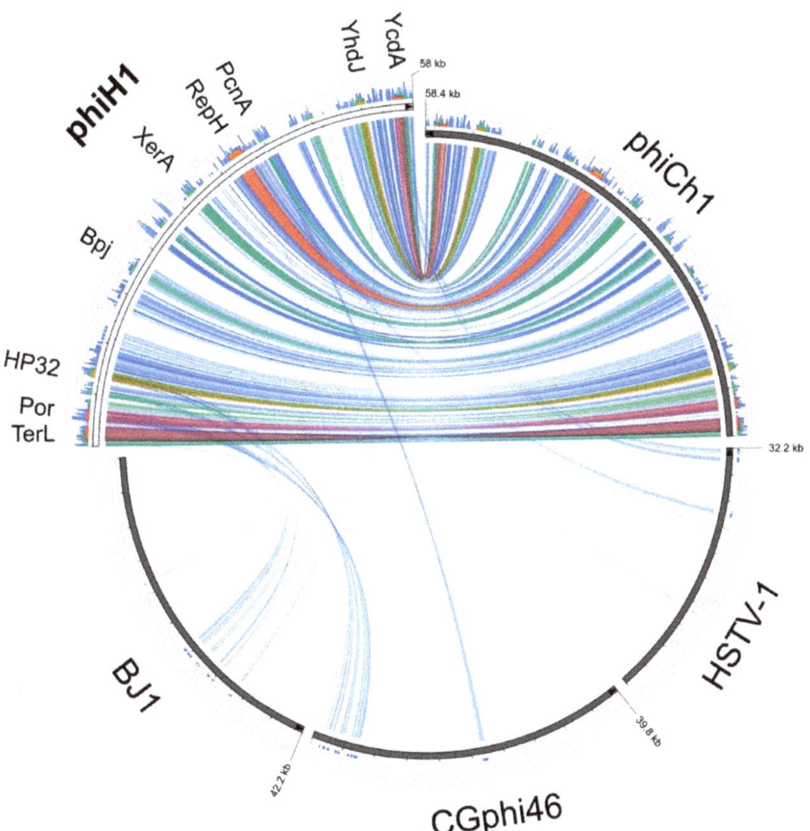

Figure 3. Circos plot of amino acid similarity (tBLASTx) between phiH1 and the haloviruses phiCh1, BJ1, CGphi46 and HSTV-1. The threshold for connecting lines was E-value $\leq 10^{-40}$, with line colours reflecting the ratio of actual tBLASTx bitscore to the maximal score (using 'score/max' ratio colouring with blue ≤ 0.25, green ≤ 0.50, orange ≤ 0.75, red > 0.75). The outer histogram counts how many times each colour has hit the specific part of the sequence and uses an equivalent colouring scheme. The distance between successive tick marks shown along each virus genome represents 0.1 of the full genome length. Protein names shown along the phiH1 genome indicate the positions of the corresponding genes.

PhiH1 and phiCh1 display a close sequence similarity across most of their genomes yet infect physiologically and biochemically different haloarchaeal hosts. *Hbt. salinarum* is a widely distributed neutrophilic heterotroph with glycolipid-containing membranes, and has often been isolated from spoilage of salted products while *Nab. magadii* is a haloalkaliphile (optimum pH 9.5) that lacks glycolipids [51] and is restricted in its distribution to highly alkaline salt lakes [52]. Looking more widely, the presence of phiH1 MCP homologs in diverse genera of haloarchaea and two haloviruses (Figure 4) indicates that the *Myohalovirus* genus and related viruses are a highly successful group, the reasons for which are worthy of more detailed study, particularly when large-scale cultivation of *Halobacterium* becomes more common [53]. PhiH1 has been well studied in the past, and the completion of its genome sequence now allows it to be included in much of the sequence-based studies used today, including comparative virology, detection of proviruses in archaeal genomes, virus evolution and the microbial ecology of hypersaline environments.

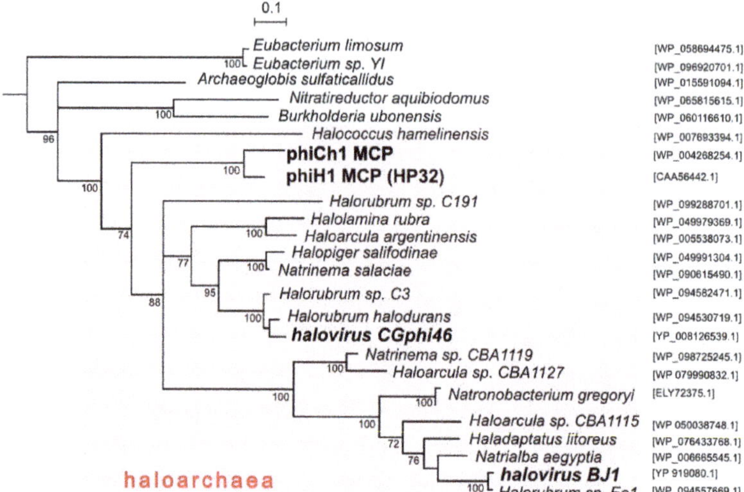

Figure 4. Phylogenetic tree reconstruction (NJ method) of major capsid proteins (MCP) of phiH1, other haloviruses and related proteins of haloarchaea. Species names of haloarchaeal species are shown, with accession numbers given at the right side. Bootstrap confidence values (100 repetitions) are shown at branch points. The pink shading highlights taxa belonging to the class Halobacteria. Scale bar (expected changes per site) is shown at top. The outgroup (not shown) consisted of distantly related MCP sequences of *Bacillus* spp. (WP_001060157.1, WP_098773561.1, WP_001064748.1 and WP_000178926.1).

Supplementary Materials: The following are available online at http://www.mdpi.com/2073-4425/9/10/493/s1, Figure S1: Original phiH1 restriction map compared to in silico map from genome sequence. Figure S2: Dot plot sequence comparison of phiH1 and phiCh1 genomes.

Author Contributions: Conceptualization, visualization, investigation, M.D.-S.; data curation, Fr.P., M.D.-S.; funding acquisition, D.O., Fr.P.; project administration, Fr.P.; resources, A.W., Fe.P.; writing—original draft, Fr.P., M.D.-S.; writing—review & editing, A.W., D.O., Fe.P., Fr.P., M.D.-S.; validation, A.W.

Funding: This research was funded by the Max Planck Society, Germany, to Dieter Oesterhelt, emeritus Director of the Department of Membrane Biochemistry, Max Planck Institute, Martinsried, Germany.

Conflicts of Interest: The authors declare no conflict of interest.

References

1. Schnabel, H.; Zillig, W.; Pfaffle, M.; Schnabel, R.; Michel, H.; Delius, H. *Halobacterium halobium* phage ΦH. *EMBO J.* **1982**, *1*, 87–92. [CrossRef] [PubMed]

2. Zillig, W.; Gropp, F.; Henschen, A.; Neumann, H.; Palm, P.; Reiter, W.D.; Rettenberger, M.; Schnabel, H.; Yeats, S. Archaebacteria virus host systems. *Syst. Appl. Microbiol.* **1986**, *7*, 58–66. [CrossRef]

3. Zillig, W.; Reiter, W.-D.; Palm, P.; Gropp, F.; Neumann, H.; Rettenberger, M. Viruses of Archaebacteria. In *The Bacteriophages*; Calendar, R., Ed.; Plenum Publishing Corpn: New York, NY, USA, 1988.

4. Schnabel, H.; Zillig, W. Circular structure of the genome of phage ΦH in a lysogenic *Halobacterium halobium*. *Mol. Gen. Genet.* **1984**, *193*, 422–426. [CrossRef]

5. Schnabel, H.; Schramm, E.; Schnabel, R.; Zillig, W. Structural variability in the genome of phage ΦH of *Halobacterium halobium*. *Mol. Gen. Genet.* **1982**, *188*, 370–377. [CrossRef]

6. ICTV Report, C. ICTV Online (10th) Report on Virus Taxonomy. Available online: https://talk.ictvonline.org/taxonomy/p/taxonomy-history?taxnode_id=20170459 (accessed on 19 March 2018).

7. Krupovic, M.; Dutilh, B.E.; Adriaenssens, E.M.; Wittmann, J.; Vogensen, F.K.; Sullivan, M.B.; Rumnieks, J.; Prangishvili, D.; Lavigne, R.; Kropinski, A.M.; et al. Taxonomy of prokaryotic viruses: Update from the ICTV bacterial and archaeal viruses subcommittee. *Arch. Virol.* **2016**, *161*, 1095–1099. [CrossRef] [PubMed]

8. Cline, S.W.; Doolittle, W.F. Efficient transfection of the archaebacterium *Halobacterium halobium*. *J. Bacteriol.* **1987**, *169*, 1341–1344. [CrossRef] [PubMed]

9. Blaseio, U.; Pfeifer, F. Transformation of *Halobacterium halobium*: Development of vectors and investigation of gas vesicle synthesis. *Proc. Natl. Acad. Sci. USA* **1990**, *87*, 6772–6776. [CrossRef] [PubMed]

10. Stolt, P.; Zillig, W. In vivo studies on the effects of immunity genes on early lytic transcription in the *Halobacterium salinarium* phage φH. *Mol. Gen. Genet.* **1992**, *235*, 197–204. [CrossRef] [PubMed]

11. Ken, R.; Hackett, N.R. *Halobacterium halobium* strains lysogenic for phage phiH contain a protein resembling coliphage repressors. *J. Bacteriol.* **1991**, *173*, 955–960. [CrossRef] [PubMed]

12. Stolt, P.; Zillig, W. In vivo and in vitro analysis of transcription of the L region from the *Halobacterium salinarium* phage φH: Definition of a repressor-enhancing gene. *Virology* **1993**, *195*, 649–658. [CrossRef] [PubMed]

13. Stolt, P.; Zillig, W. Antisense RNA mediates transcriptional processing in an archaebacterium, indicating a novel kind of RNase activity. *Mol. Microbiol.* **1993**, *7*, 875–882. [CrossRef] [PubMed]

14. Schnabel, H. An immune strain of *Halobacterium halobium* carries the invertible L segment of phage ΦH as a plasmid. *Proc. Natl. Acad. Sci. USA* **1984**, *81*, 1017–1020. [CrossRef] [PubMed]

15. Stolt, P.; Zillig, W. Transcription of the halophage ΦH repressor gene is abolished by transcription from an inversely oriented lytic promoter. *FEBS Lett.* **1994**, *344*, 125–128. [CrossRef]

16. Stolt, P.; Zillig, W. Gene regulation in halophage ΦH; more than promoters. *Syst. Appl. Microbiol.* **1993**, *16*, 591–596. [CrossRef]

17. Witte, A.; Baranyi, U.; Klein, R.; Sulzner, M.; Luo, C.; Wanner, G.; Krüger, D.H.; Lubitz, W. Characterization of *Natronobacterium magadii* phage φCh1, a unique archaeal phage containing DNA and RNA. *Mol. Microbiol.* **1997**, *23*, 603–616. [CrossRef] [PubMed]

18. Klein, R.; Baranyi, U.; Rössler, N.; Greineder, B.; Scholz, H.; Witte, A. *Natrialba magadii* virus φCh1: First complete nucleotide sequence and functional organization of a virus infecting a haloalkaliphilic archaeon. *Mol. Microbiol.* **2002**, *45*, 851–863. [CrossRef] [PubMed]

19. Selb, R.; Derntl, C.; Klein, R.; Alte, B.; Hofbauer, C.; Kaufmann, M.; Beraha, J.; Schoner, L.; Witte, A. The viral gene ORF79 encodes a repressor regulating induction of the lytic life cycle in the haloalkaliphilic virus phiCh1. *J. Virol.* **2017**, *91*. [CrossRef] [PubMed]

20. Siddaramappa, S.; Challacombe, J.F.; Decastro, R.E.; Pfeiffer, F.; Sastre, D.E.; Gimenez, M.I.; Paggi, R.A.; Detter, J.C.; Davenport, K.W.; Goodwin, L.A.; et al. A comparative genomics perspective on the genetic content of the alkaliphilic haloarchaeon *Natrialba magadii* ATCC 43099^T. *BMC Genom.* **2012**, *13*, 165. [CrossRef] [PubMed]

21. Zillig, W.; Palm, P.; Reiter, W.D.; Gropp, F.; Puhler, G.; Klenk, H.P. Comparative evaluation of gene expression in Archaebacteria. *Eur. J. Biochem.* **1988**, *173*, 473–482. [CrossRef] [PubMed]

22. National Center for Biotechnology Information. Available online: https://www.ncbi.nlm.nih.gov/ (accessed on 11 October 2018).

23. Gordon, D. Viewing and editing assembled sequences using Consed. *Curr. Protoc. Bioinform.* **2003**, *2*. [CrossRef]

24. Skennerton, C.T.; Imelfort, M.; Tyson, G.W. Crass: Identification and reconstruction of CRISPR from unassembled metagenomic data. *Nucleic Acids Res.* **2013**, *41*, e105. [CrossRef] [PubMed]

25. Leinonen, R.; Sugawara, H.; Shumway, M. The Sequence Read Archive. *Nucleic Acids Res.* **2010**, *39*, D19–D21. [CrossRef] [PubMed]

26. Dyall-Smith, M.; Pfeiffer, F. The PL6-family plasmids of *Haloquadratum* are virus-related. *Front. Microbiol.* **2018**, *9*, 1070. [CrossRef] [PubMed]

27. CRISPRs Web Server. Available online: http://crispr.i2bc.paris-saclay.fr/ (accessed on 11 October 2018).

28. Lomsadze, A.; Gemayel, K.; Tang, S.; Borodovsky, M. Modeling leaderless transcription and atypical genes results in more accurate gene prediction in prokaryotes. *Genome Res.* **2018**, *28*, 1079–1089. [CrossRef] [PubMed]

29. National Center for Biotechnology Information BLAST. Available online: https://blast.ncbi.nlm.nih.gov/Blast.cgi (accessed on 11 October 2018).

30. YASS Genomic Similarity Search Tool. Available online: http://bioinfo.lifl.fr/yass/index.php (accessed on 11 October 2018).

31. Geneious. Available online: https://www.geneious.com/geneious/ (accessed on 11 October 2018).

32. Circoletto. Available online: http://tools.bat.infspire.org/circoletto/ (accessed on 11 October 2018).

33. I-Tasser, Protein Structure & Function Predictions. Available online: https://zhanglab.ccmb.med.umich.edu/I-TASSER (accessed on 11 October 2018).

34. Garneau, J.R.; Depardieu, F.; Fortier, L.C.; Bikard, D.; Monot, M. PhageTerm: A tool for fast and accurate determination of phage termini and packaging mechanism using next-generation sequencing data. *Sci. Rep.* **2017**, *7*, 8292. [CrossRef] [PubMed]

35. CPT Phage Galaxy. Available online: https://cpt.tamu.edu/galaxy-pub/ (accessed on 11 October 2018).

36. VIRFAM, Remote Homology Detection of Viral Protein Families. Available online: http://biodev.cea.fr/virfam/ (accessed on 11 October 2018).

37. Schnabel, H.; Schnabel, R.; Yeats, S.; Tu, J.; Gierl, A.; Neumann, H.; Zillig, W. Genome organization and transcription in Archaebacteria. *Folia Biol. (Praha)* **1984**, *30*, 2–6. [PubMed]

38. Pfeiffer, F.; Schuster, S.C.; Broicher, A.; Falb, M.; Palm, P.; Rodewald, K.; Ruepp, A.; Soppa, J.; Tittor, J.; Oesterhelt, D. Evolution in the laboratory: The genome of *Halobacterium salinarum* strain R1 compared to that of strain NRC-1. *Genomics* **2008**, *91*, 335–346. [CrossRef] [PubMed]

39. Stolt, P.; Grampp, B.; Zillig, W. Genes for DNA cytosine methyltransferases and structural proteins, expressed during lytic growth by the phage ΦH of the archaebacterium *Halobacterium salinarium*. *Biol. Chem. Hoppe Seyler* **1994**, *375*, 747–757. [CrossRef] [PubMed]

40. Jin, G.; Pavelka, M.S., Jr.; Butler, J.S. Structure-function analysis of VapB4 antitoxin identifies critical features of a minimal VapC4 toxin-binding module. *J. Bacteriol.* **2015**, *197*, 1197–1207. [CrossRef] [PubMed]

41. Gao, F.; Zhang, C.T. GC-Profile: A web-based tool for visualizing and analyzing the variation of GC content in genomic sequences. *Nucleic Acids Res.* **2006**, *34*, W686–W691. [CrossRef] [PubMed]

42. Gropp, F. Genexpression im Archaebakterium *Halobacterium halobium*: Der Phage ΦH und die DNA-abhängige RNA-Polymerase. Ph.D. Thesis, Ludwig-Maximilians-Universitaet Muenchen, Munich, Germany, 26 July 1989.

43. Gropp, F.; Grampp, B.; Stolt, P.; Palm, P.; Zillig, W. The immunity-conferring plasmid pΦHL from the *Halobacterium salinarium* phage ΦH: Nucleotide sequence and transcription. *Virology* **1992**, *190*, 45–54. [CrossRef]

44. ISfinder. Available online: https://isfinder.biotoul.fr/ (accessed on 11 October 2018).

45. Vogelsang-Wenke, H.; Oesterhelt, D. Isolation of a halobacterial phage with a fully cytosine-methylated genome. *MGG Mol. Gen. Genet.* **1988**, *211*, 407–414. [CrossRef]

46. Baranyi, U.; Klein, R.; Lubitz, W.; Kruger, D.H.; Witte, A. The archaeal halophilic virus-encoded Dam-like methyltransferase M. ΦCh1-I methylates adenine residues and complements *dam* mutants in the low salt environment of *Escherichia coli*. *Mol. Microbiol.* **2000**, *35*, 1168–1179. [CrossRef] [PubMed]

47. Pagaling, E.; Haigh, R.D.; Grant, W.D.; Cowan, D.A.; Jones, B.E.; Ma, Y.; Ventosa, A.; Heaphy, S. Sequence analysis of an archaeal virus isolated from a hypersaline lake in Inner Mongolia, China. *BMC Genom.* **2007**, *8*, 410. [CrossRef] [PubMed]

48. Pietilä, M.K.; Laurinmäki, P.; Russell, D.A.; Ko, C.C.; Jacobs-Sera, D.; Hendrix, R.W.; Bamford, D.H.; Butcher, S.J. Structure of the archaeal head-tailed virus HSTV-1 completes the HK97 fold story. *Proc. Natl. Acad. Sci. USA* **2013**, *110*, 10604–10609. [CrossRef] [PubMed]

49. Krupovic, M.; Forterre, P.; Bamford, D.H. Comparative analysis of the mosaic genomes of tailed archaeal viruses and proviruses suggests common themes for virion architecture and assembly with tailed viruses of bacteria. *J. Mol. Biol.* **2010**, *397*, 144–160. [CrossRef] [PubMed]

50. Baker, M.L.; Jiang, W.; Rixon, F.J.; Chiu, W. Common ancestry of herpesviruses and tailed DNA bacteriophages. *J. Virol.* **2005**, *79*, 14967–14970. [CrossRef] [PubMed]

51. Kamekura, M.; Dyall-Smith, M. Taxonomy of the family *Halobacteriaceae* and the description of two new genera *Halorubrobacterium* and *Natrialba*. *J. Gen. Appl. Microbiol.* **1995**, *41*, 333–350. [CrossRef]
52. Tindall, B.J.; Ross, H.N.M.; Grant, W.D. *Natronobacterium* gen. nov. and *Natronococcus* gen. nov. Two new genera of haloalkaliphilic archaebacteria. *Syst. Appl. Microbiol.* **1984**, *5*, 41–57. [CrossRef]
53. Kalenov, S.V.; Baurina, M.M.; Skladnev, D.A.; Kuznetsov, A.Y. High-effective cultivation of *Halobacterium salinarum* providing with bacteriorhodopsin production under controlled stress. *J. Biotechnol.* **2016**, *233*, 211–218. [CrossRef] [PubMed]

 *genes*

Article

Back to the Salt Mines: Genome and Transcriptome Comparisons of the Halophilic Fungus *Aspergillus salisburgensis* and Its Halotolerant Relative *Aspergillus sclerotialis*

Hakim Tafer, Caroline Poyntner *, Ksenija Lopandic, Katja Sterflinger and Guadalupe Piñar

VIBT EQ Extremophile Center, Department of Biotechnology, University of Natural Resources and Life Sciences, Muthgasse 18, 1190 Vienna, Austria; hakim.tafer@boku.ac.at (H.T.); ksenija.lopandic@boku.ac.at (K.L.); katja.sterflinger@boku.ac.at (K.S.); guadalupe.pinar@boku.ac.at (G.P.)
* Correspondence: caroline.poyntner@boku.ac.at

Received: 22 April 2019; Accepted: 15 May 2019; Published: 20 May 2019

Abstract: Salt mines are among the most extreme environments as they combine darkness, low nutrient availability, and hypersaline conditions. Based on comparative genomics and transcriptomics, we describe in this work the adaptive strategies of the true halophilic fungus *Aspergillus salisburgensis*, found in a salt mine in Austria, and compare this strain to the ex-type halotolerant fungal strain *Aspergillus sclerotialis*. On a genomic level, *A. salisburgensis* exhibits a reduced genome size compared to *A. sclerotialis*, as well as a contraction of genes involved in transport processes. The proteome of *A. sclerotialis* exhibits an increased proportion of alanine, glycine, and proline compared to the proteome of non-halophilic species. Transcriptome analyses of both strains growing at 5% and 20% NaCl show that *A. salisburgensis* regulates three-times fewer genes than *A. sclerotialis* in order to adapt to the higher salt concentration. In *A. sclerotialis*, the increased osmotic stress impacted processes related to translation, transcription, transport, and energy. In contrast, membrane-related and lignolytic proteins were significantly affected in *A. salisburgensis*.

Keywords: halophiles; halotolerant; fungi; transcriptomics; genomics

1. Introduction

Saline and hypersaline environments like salt marshes, saline soil, and salt water, as well as the Dead Sea harbor a diverse community of fungi [1]. Most of these fungi do not require salt for growth and have optimum growth in the absence of salt. Nevertheless, they are halotolerant and can withstand a salt concentration up to 30% [2].

Real halophilic microorganisms are adapted to conditions of high salinity and require a certain concentration of NaCl for their optimum growth. Halophiles are phylogenetically quite diverse and can be found in the domains of Archaea, Bacteria, and Eukarya. Until almost a decade ago, it was a general belief in mycology and in food microbiology that fungi growing on substrates with low water activity (a_w) have a general xerophilic phenotype [3] that is determined by the water potential of the medium, rather than by the chemical nature of the solute [4,5]. Therefore, fungi were considered xerophilic if they grew well at an a_w of 0.85, corresponding to 17% NaCl or 50% glucose added to their growth medium. Xerophilic Aspergilli, together with Penicillia, dominate the actively-growing mycoflora on dried food and also on materials stored in museums and archives. *Aspergillus vitricola* was first isolated from glass surfaces [6] and occurs frequently in house dust. Furthermore, *Aspergillus glabripes* occurs on books and archive material, as well as in house dust [7]. Xerophilic halotolerant Aspergilli were also shown to be common invaders of pipe organs in churches [8].

The term "halophile" for fungi was introduced in 1975 for those few xerophilic food-borne species that exhibit quite superior growth on media with NaCl as the controlling solute. Fungi have subsequently been described in moderately-saline environments, such as salt marshes, saline soil, and sea water, but were considered to be unable to grow in highly saline waters [2]. In 2000, fungi were isolated for the first time from the brine of solar salterns [2,9]. However, halophilic fungi are uncommon [2]. Currently, only *Wallemia ichthyophaga*, *Wallemia muriae*, *Aspergillus baarnensis*, *Aspergillus salinarium*, and *Aspergillus ruber* [10] are classified as halophiles. Recently, a fungus isolated from a historical wooden staircase in a salt mine in Austria was described as a new species, *Aspergillus salisburgiensis* [11,12], and was added to the list of halophilic fungi.

Cells living in natural saline systems, where high salt amounts cause high osmotic pressure, must maintain lower water potential than their surroundings in order to survive and proliferate. The ability to survive osmotic stress requires several adaptations. *Wallemia ichthyophaga* withstands high salt concentration by increasing the intracellular concentration of polyols, using high-affinity K^+ transporters and increasing the thickness of its cell-wall by a factor of three [13]. Under high salt concentration, *Aspergillus ruber* increases the number of ion transporters, exhibits a higher proportion of acidic amino acids, restructures its cell wall, and uses glycerol as a compatible solute [10]. Similar adaptation strategies are found in halotolerant species, where the production of hydrophilic compounds, such as amino acids, sugar alcohols, and soluble sugars, have been reported [14–18].

The mechanisms behind the adaptive capacities of halotolerant and halophilic fungi have only begun to be understood in recent years, with the first study of the transcriptome of *Wallemia ichthyophaga* dating from 2013 [13]. Since then, an additional halophile, *Aspergillus ruber*, has been published [10]. In order to extend the knowledge in the field, we present the results of the comparative sequencing of the genomes of two species, *A. salisburgensis*, a halophilic fungus from a salt mine, and *A. sclerotialis*, as a representative of halotolerant fungi. Additionally, the two species were exposed to two concentrations of salt (5% and 20%), and the cellular response was studied on the transcriptome level. The genetic content, as well as the transcriptome, were compared to other halophilic, halotolerant, and to other well-studied, less salt-tolerant fungi (named as control) in order to better understand which mechanisms were involved in the resistance against osmotic stress.

2. Materials and Methods

2.1. Strains, Media, and Cultivation

The strain *Aspergillus salisburgensis* (EXF-10247/MA6005) was isolated in a previous study from a wooden staircase, built in the year 1108 B.C. and discovered in 2003 in a salt mine in the Austrian region "Salzkammergut", Upper Austria [12]. The strain *A. sclerotialis* (Strain No. MA 5985, CBS 366.77) was supplied by the Fungal Biodiversity Centre (CBS), Utrecht, Netherlands. The two examined strains were grown on salt glucose media for the whole genome sequencing (5% glucose, 0.1% peptone, 0.1% malt extract, agar, and either 5% or 20% NaCl) and salt yeast media for the transcriptome sequencing (0.4% glucose, 0.4% yeast extract, 1.0% malt extract, 1.2% agar, and either 5% or 20 % NaCl). The two strains were incubated at their optimal temperature (*A. salisburgensis* at room temperature, *A. sclerotialis* at 37 °C) for 35 days.

2.2. Whole Genome Sequencing

Genomic DNA extraction of pure fungal strains was performed as previously described [19]. Approximately one cm^2 of fungal colonies, consisting of both hyphal and conidiogenous structures, were collected and placed in a 1.5-mL bead beater tube with 0.2-g glass beads (0.75–1 mm, Carl Roth GmbH Co., KG, Karlsruhe, Germany) and 500 µL of CTAB-buffer (1.2 g Tris-HCl, 8.2 g NaCl, 0.81 g EDTA × 2 H_2O, 2.0 g CTAB, and 0.2 g β-mercaptoethanol, pH 8.0) and processed twice in the Fast Prep FP120 Ribolyzer (Thermo Savant, Holbrook, AZ, USA) for 40 s at a speed of 4 m/s. Between these ribolyzing steps, samples were incubated at 65 °C for 10 min at 300 rpm. Further DNA extraction was performed with phenol/chloroform/isoamyl alcohol. Genome sequencing was carried out using the ION Proton Technology (Ion AmpliSeq Library Preparation kit, Template OT2 200 kit, Ion PI Sequencing 200 kit, Ion PI chip kit V2, Life Technologies, Carlsbad, CA, USA) following the instructions of the manufacturers.

2.3. Total RNA Extraction

After 35 days of cultivation, colonies were collected as described in Section 2.2, and RNA was extracted from the biomass using the FastRNA Pro RED Kit (MP Biomedicals, Santa Ana, CA, USA). For each condition and strain, three replicates of total RNA were extracted following the manufacture's protocol. The quality and quantity was measured using the Agilent 2100 Bioanalyzer (Agilent Technologies, Santa Clara, CA, USA) and a Qubit 2.0 (Life Technologies, Carlsbad, CA, USA).

2.4. RNA Library Preparation

The total RNA library was prepared as described before [20]. mRNA was isolated from the total RNA using the Dynabeads mRNA DIRECT Micro Kit (Ambion, Life Technologies, Carlsbad, CA, USA). Subsequently, the RNA library for sequencing was constructed using the Ion Total RNA-Seq kit v2 (Life Technologies, Carlsbad, CA, USA). Quality was checked using the Agilent 2100 Bioanalyzer instrument (Agilent Technologies, Santa Clara, CA, USA). The final library was size selected to 290 bp using Pippin Prep (Sage Science, Beverly, MA, USA). The sequencing was done using the Ion Torrent Chef instrument, Ion Torrent Proton instrument, and the HiQ sequencing kit (Life Technologies, Carlsbad, CA, USA).

2.5. Genome Annotation and Assembly

A. salisburgensis and *A. sclerotialis* were assembled with Newbler. Genome completeness was assessed with BUSCO [21]. The final protein annotation was obtained by passing the ab initio protein annotation generated by BUSCO with the help of Augustus [22] to BRAKER2 [23,24] together with the RNA sequencing data. The functional annotation was done with InterProScan 5.25.64 [25] and EggNog [26] with default settings. Homologies with proteins in the Transporter Classification Database (TCDB) [27], the peptidase database (MEROPS) [28], and the Carbohydrate-Active Enzymes Database (CAZY) [29] were assessed with Blastp [30] (E-value <0.001).

2.6. Comparative Genomics

Protein orthology was assessed with ProteinOrtho 5.16b [31], where BLAST was replaced with DIAMOND [32]. The species phylogenomic tree was generated with iqtree [33–35] based on the MAFFT-alignment of one-to-one homologues. The TimeTree was inferred using the Reltime [36,37] method and ordinary least squares estimates of branch length. The TimeTree was computed using 8 calibration constraints retrieved from TimeTree.org. All position containing gaps and missing data were eliminated. There was a total of 37,607 positions in the final dataset. Evolutionary analyses were conducted in MEGA X [38].

Protein family expansion and contraction were computed with CAFE [39] and the previously-obtained TimeTree. The significance of protein families over- and under-representation was computed with the Fisher exact test.

Amino acids in the one-to-one conserved proteins and proteins containing a signal peptide, as annotated by SignalP [40], were tested for distribution bias with Wilcoxon tests in R [41].

2.7. Differential Expression

Genes' differential expression was obtained with kallisto [42], tximport [43] and DeSeq2 [44]. All figures were made in R with ggplot2 [45], ggtree [46], and deeptime.

3. Results and Discussion

3.1. Highly Complete Genome and Gene Set

The genome assemblies of *A. salisburgensis* and *A. sclerotialis* used for this study were sequenced on the Ion Proton sequencing platform (Sections 2.2–2.4). The assembled genome of *A. salisburgensis* had a size of 21.9 Mb and contained a total of 8895 protein-coding loci for a total of 1351 contigs (>500 bp). *A. sclerotialis* had a 27% larger genome (27.9 Mb) with 27% more coding-genes (11307), compared to *A. salisburgensis*, and 1704 contigs (Supplementary Table S1). Among the 4406 highly-conserved genes in Eurohiomycetes [21], 96% were found in *A. sclerotialis* and 94% were present in *A. salisburgensis*. It was reported that compact genomes with a reduced number of protein-coding genes might be a characteristic of obligate halophiles, such as *Wallemia ichthyophaga* [15]. This trend could not be confirmed in Aspergilli where the genome of the halophile *Aspergillus ruber* [10] has a similar size and gene content compared to *A. sclerotialis*.

3.2. Gene Conservation

The gene conservation pattern between *A. salisburgensis*, *A. sclerotialis*, two halophiles (Table 1, marked as HH), six halotolerant (marked as H), and three control strains exhibiting no salt resistance (marked as C) were studied with ProteinOrtho [31].

Table 1. List of genomes compared in this study. For each species, the name, the range of salt tolerance, the type of tolerance, and reference are listed. H: halotolerant, HH: halophile, C: control.

Species	Salt Concentration	Type	Reference
A. salisburgensis	5–30%,optimal at 20%	HH	This publication
Aspergillus ruber	>10%, optimal at 18%	HH	[10]
Wallemia ichthyophaga	>8%, optimal 18%	HH	[13]
A. sclerotialis	0–20%, optimal at 10%	H	This publication
Hortaea werneckii	0–32%, optimal 3–9%	H	[15]
Penicillium chrysogenum	0–18%, optimal at 10%	H	[14]
Candida parapsilosis	0–12%, optimal 0 %	H	[16]
Debaryomyces fabryi	0–16%, optimal 0 %	H	[17]
Debaryomyces hansenii	0–24%, optimal 0 %	H	[17]
Wallemia mellicola	0–27%, optimal 0 %	H	[47]
Saccharomyces cerevisiae	0–8%, optimal 0 %	C	[48]
Schizosaccharomyces pombe	0–5%, optimal 0 %	C	[48]
Ustilago maydis	0–7%, optimal 0 %	C	[18,49]

ProteinOrtho identified 1169 groups of orthologs present in all 13 species, 1091 genes specific to *A. salisburgensis*, and 2362 genes specific to *A. sclerotialis*. The genes specific to *A. salisburgensis* were enriched in functional terms related to protein refolding (GO:0042046, GO:0061077, *HSP20, HSP70, HSP90, CLP, DNAJ/HSP40*), DNA modification (GO:0015074, GO:0005727, GO:0046821, IPR01584, IPR025668, PF13683, GO:0005727, GO:0046821, TCDB 3.A.7), tripartite ATP-independent periplasmic (TRAP), and ATP-binding cassette transporter (IPR010656, IPR000515, TCDB 3.A.1, 2.A.56), as well as chitin degradation (CAZY:GH23) (Figure 1). TRAP transporters were shown to mediate the uptake of compatible solutes in halophilic Proteobacteria such as *Halomonas elongata* [50], while the proteins related to refolding are known to be involved in stress-response [51].

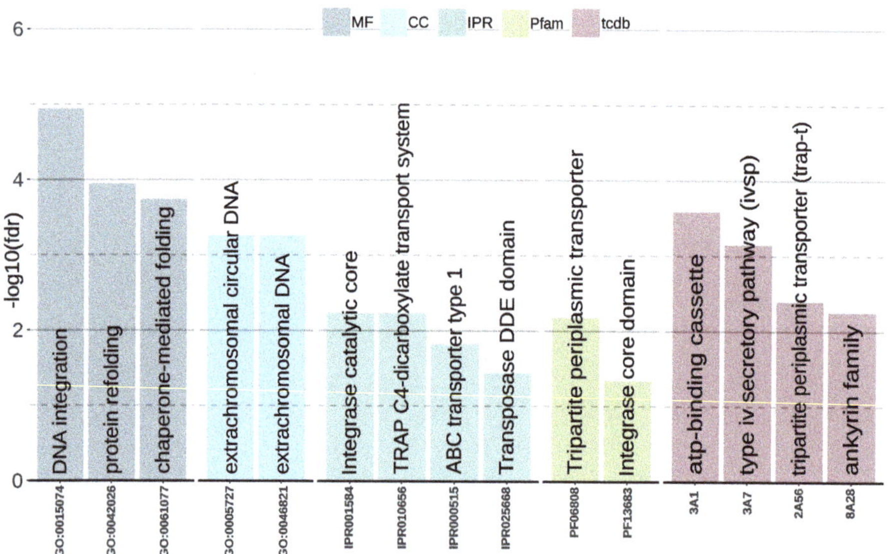

Figure 1. Graphical representation of the $-log_{10}(fdr)$ of the enriched categories in CAZY (Carbohydrate-Active Enzyme Database), Gene Ontology (GO), InterProScan (IPR), Protein Family (PFAM) and Transporter Classification Database (TCDB) in the set of *A. salisburgensis*-specific genes. Functional categories related to chitin degradation, protein folding, transport, and DNA integration are overrepresented.

In *A. sclerotialis*, the set of species-specific genes are functionally enriched in terms related to protein folding, heat shock protein 70, and helix-turn-helix (HTH)-binding domain protein (Supplementary Figure S1 and Supplementary Table S1). The helix-turn-helix binding-motif is found in many proteins involved in gene expression regulation [52].

3.3. Gene Family Evolution

A TimeTree of life for the 13 fungi (Table 1) was constructed from the multiple sequence alignments of the proteins shared by all strains and from nine time anchors extracted from the TimeTree database [53] (Figure 2). The results were consistent with the currently-accepted taxonomy [53]. *A. salisburgensis* and *A. sclerotialis* separated approximately 41 MYA (+/−40 MYA), 20 MY before the split between *Debaryomyces fabryi* and *Debaryomyces hansenii* and 40 MY after the split between *A. salisburgensis*, *A. sclerotialis*, and *Aspergillus ruber*. Interestingly, *A. salisburgensis* and *A. sclerotialis* exhibited the largest evolutionary rates, followed by the two *Debaryomyces* species.

This tree was used to estimate the numbers of rapidly-evolving annotation elements with CAFE [39]. The functional annotations used were derived for the carbohydrate enzymes (CAZY) [29], general enzymes from the EggNog annotation [26], InterProScan protein domains (IPR) [54], KEGG pathways [55], proteolytic enzymes (MEROPS) [28], PFAM protein domains [56], and transporters from TCDB [27].

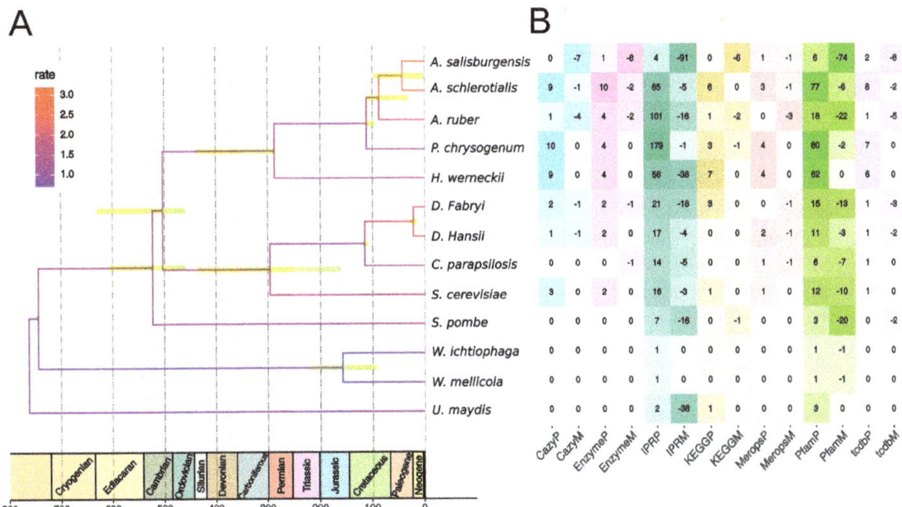

Figure 2. Phylogenomics analyses. (**A**) TimeTree generated from the merged alignment of the one-to-one homologues of the 13 species. The green bars represent the 95% confidence interval. Geologic periods are color-coded and indicated at the bottom of the tree. Evolutionary rate is color encoded. (**B**) CAFE analysis of the expansion and contraction of the annotation elements of KEGG, CAZY, TCDB, Enzyme, and MEROPSfor the 13 species. The suffixes Mand Pafter the annotation categories stand for contraction and expansion, respectively.

The internal node with the largest numbers of rapidly-evolving annotation families was the one leading to the common ancestor of *Hortaea* and *Aspergillus* (*leotiomyceta*) (Supplementary Table S1). CAFE reported 106 IPR and 105 PFAM rapidly-evolving families. The most significantly expanding families were related to fungal-specific transcription factors (*p*-value 4.75×10^{-19} and cytochrome P450 (*p*-value 1.127×10^{-16}). Cytochrome P450 was also shown to improve salt tolerance in plants [57] and is upregulated under salt stress in the fungus *Piriformosa indica*, a plant endophyte that confers salinity stress tolerance in rice plants [58]. An additional significantly-expanding family was DUF3468 (DUF, domain of unknown function, *p*-value 1.59×10^{-15}), a protein domain involved in transcription activation of genes related to asexual conidiation and sexual differentiation in *Aspergillus nidulans* and *Aspergillus flavus* [59].

The branch leading to *A. salisburgensis* exhibited the largest contraction of gene families from all species analyzed: 91 IPR, 74 PFAM, 7 CAZY, 8 TCDB, 1 MEROPS, and 8 Enzyme families contracted significantly, which is in line with the reduction in genome size of 27% compared to *A. sclerotialis*. Fungal transcription factors (PF11951) exhibited the largest loss in *A. salisburgensis* (-17). Protein families transporting amino acid and oligopeptide (2A3, -11; IPR002293, -9; 2A18, -5; PF03169, -3), fatty acid (4C1, -7), and iron (2A108, -5) were contracting in *A. salisburgensis*. Interestingly, the amino acid polyamine organocation superfamily (APC) contracted also in *Aspergillus ruber* (2A3, -8). APC is one of the largest families of secondary active transporters, and it is found in all living organisms [60]. Another family depleted in both *A. salisburgensis* and *Aspergillus ruber* is aminoglycoside phosphotransferase, a bacterial antibiotic resistance protein (IPR002575) with a reduction of 15 and 13 members, respectively.

A few annotation families expanded in *A. salisburgensis*. These families were related to the type IV secretory pathway, which exports proteins or DNA-protein complexes out of the cell [61] (3A7 +9), the TRAP-C4 transporter (IPR010656 +6, PF06808 +6), the fungal cellulose binding domain (PF00734, +5), the integrase catalytic core (IPR001584), cutinase (PF01083), sodium:bile-acid symporter/arsenical resistance protein ACR3 (PF01758 +2), and secretory lipase (PF03583, +2). The increase in cellulose degrading ability due to the fungal cellulose binding domain [62] and cutinase [63] is probably related to the environment, a wooden staircase, from which *A. salisburgensis* was isolated. In yeasts, proteins belonging to the PF01758 family confer arsenic resistance by extrusion of sodium arsenate and sodium arsenite [64].

Gene Family Enrichment

The significance of size variations of functional annotation families was studied by comparing species or a group of fungal species (Table 1) with the help of chi-squared tests. The most significant bias was seen for the major facilitator superfamily (MFS) transporters, which showed a peculiar pattern of enrichment. *A. salisburgensis* had significantly more MFS (2A1, IPR020846) transporters than *Wallemia ichthyophaga*, but significantly less than *A. sclerotialis* and *Aspergillus ruber* (Figure 3 and Supplementary Table S1). Further, there were significantly more MFS transporters in halotolerant fungi than in the halophiles (fdr = 0.018). Still, compared to the group of control fungi, both the halophilic (fdr = $3.76e^{-15}$) and the halotolerant (fdr = $7.19e^{-37}$) fungi were enriched in MFS.

Genes involved in protein translation, such as ribosomes (GO:0003735, GO:0005840), RNA binding (GO:0003723), and reverse transcriptase (GO:013103, GO:0015074), were depleted in halophiles and halotolerant compared to the control group. In contrast, GO terms involved in transcription, such as RNA polymerase II (GO:0000981), transcription (GO:0006531, GO:0006535), and transcription factor (IPR007219, IPR001138), were enriched in both H and HH compared to C. The terms related to cytochrome P450 E-class group I (GO:0016491, GO:0055114, IPR001128, IPR023753, IPR002401, IPR017972) were enriched in both groups of salt-resistant fungi compared to the control group.

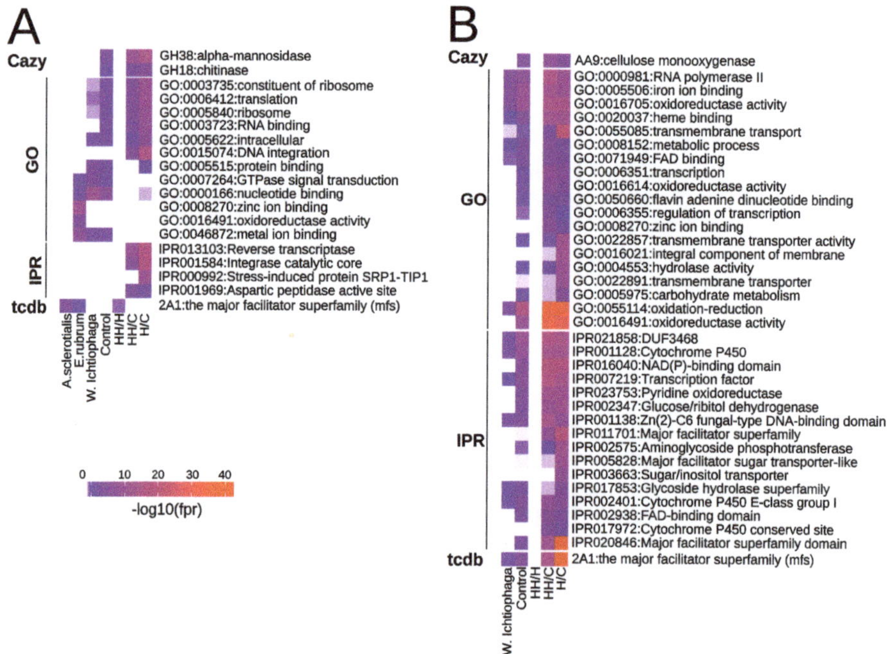

Figure 3. Heat map of the $-log(fdr)$ of the significantly-enriched or -depleted functional annotations in *A. salisburgensis* compared to *A. sclerotialis*, *Aspergillus ruber*, and *Wallemia ichthyophaga*. Depletion/enrichment found in HH vs. H, HH vs. C, and H vs. C is also shown. Only entries with an enrichment/depletion with an fdr < $1e^{-5}$ are shown. (**A**) Depletion. (**B**) Enrichment.

3.4. Amino Acid Composition

The protein amino-acid (AA) composition of the C, H, and HH groups for the set of orthologous genes and for the set of exported proteins, i.e., containing a signal-P annotation, was compared (Figure 4 and Supplementary Table S1). With respect to the control group, the conserved proteins in halophilic fungi were significantly enriched in glycine, proline, and arginine and depleted in isoleucine and lysine (Figure 4A). The same bias pattern for those 5 AA was previously reported in the extremely halophilic bacteria *Salinibacter ruber* and *Halomonas elongata* and in the Archaea *Halobacterium salinarum* and *Haloarcula marismortui*, compared to *Escherichia coli* [9]. A low occurrence of lysine and increase of arginine in halophilic proteins is a signature of halophilic microorganisms [65,66]. Psychrophilic organisms, which are similar to halophiles, face reduced water availability and exhibit an increased glycine protein content [67].

Among the set of exported proteins, the acidic AA aspartate and glutamate were overrepresented in halophiles compared to the control group. This is a general characteristic found in many osmotolerant bacteria [65,68]. Similarly, in the three studied HH species, the conserved proteins were enriched in glycine and proline. Finally, the serine depletion was previously reported in halophilic Archaea [66].

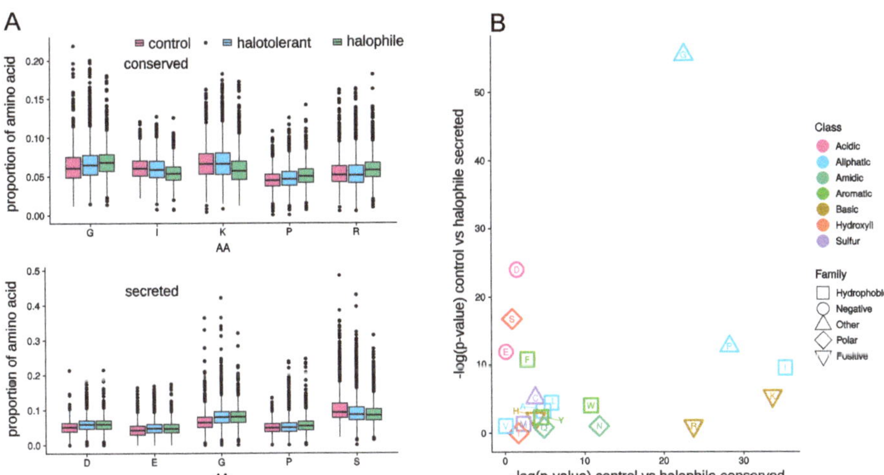

Figure 4. Bias in the amino-acid (AA) distribution between the sets of control, halotolerant, and halophilic fungi. (**A**) Boxplot representing the distribution of AA for the 5 AA with the most significant bias between the control and halophilic group for the set of conserved (top) and secreted proteins (bottom). (**B**) Scatter plot of the $-log(p\text{-value})$ computed with the Wilcoxon test for the AA distribution bias for the set of conserved proteins and the set of secreted proteins.

3.5. Differential Expression between 5% and 20% NaCl

The differential expression of genes between the low and high salt concentrations was studied. Upregulated genes are defined as the set of transcripts that exhibit an increase in expression at 20% salt concentration, while downregulated genes exhibit a reduced expression under high salinity.

3.5.1. Aspergillus Sclerotialis

A. sclerotialis differentially regulated 2097 genes (1121 up- and 976 downregulated fdr < 0.05, $|log_2FC| > 1$) between both salt concentrations. The most strongly upregulated gene (x3536) belonged to the major facilitator superfamily (MFS) transporter, which are membrane transport proteins that facilitate movement of small solutes across cell membranes in response to osmotic gradients [69].

Among the 96 genes upregulated by a factor larger than 50 in the halotolerant fungus, 14 genes were related to transmembrane transport (Supplementary Table S1). Cerato-ulmin hydrophobin, a parasitic fitness factor of the agents of Dutch elm disease [70], was upregulated by a factor of 2641. The regulation of this hydrophobin might indicate a role in salt resistance, as previously reported in *Wallemia ichthyophaga* [15]. Given the fact that the halotolerant strain is a dog pathogen, this hydrophobin might be involved in the pathogenicity of the strain [71].

The C-terminal dimerization domain found in transposases of elements belonging to the activator superfamily was increased by a factor 280 under high salt conditions, indicating that DNA modification might take place. A similar feature was seen in sunflower exposed to salt and drought stress [72]. Transcriptional regulators are also found among the most upregulated genes [73]. A gene involved in protein neddylation [74] had its expression increased by a factor of 1571. Inositol monophosphatase, which is upregulated by a factor of 571, is involved in the phosphatidylinositol signaling pathway and was shown to increase Na^+ resistance in yeast [75]. An F-box domain-containing protein was upregulated by a factor of 286 upon salt-induced stress, which is in line with previous reports in *Schizosaccharomyces pombe* [76]. Interestingly, two genes involved in the synthesis of antibiotics, Acyl-CoA 6-aminopenicillanic-acid-acyltransferase and pristinamycin IIA synthase subunit A, were upregulated by a factor of 369 and 338, respectively.

Among the genes downregulated by a factor larger than 50, eight were related to transporters, and one was related to SAM-dependent methyltransferase. Chorismatase-degrading enzymes were downregulated, in agreement with the salinity stress experiment in *Carthamus tinctorius* [77], where chorismatase synthase was upregulated upon salinity stress, indicating that chorismate plays a role in salt resistance. Two cupredoxin were downregulated by a factor of 247. A reverse transcriptase and a integrase were downregulated 95- and 129-times, respectively, while one Zn_2-Cys6 transcription factor was downregulated by a factor of 124.

The functional enrichment analysis for the downregulated genes indicates that transcriptional factors, MFS transporters, and urease activity (Enzyme 3.5.1.5 fdr = 0.0083, GO:0009039 fdr = $3.15e^{-4}$) were negatively impacted by the high salt concentration (Supplementary Table S1 and Figure 5A). It was previously shown that a depletion of urease improves salt resistance in *Arabidopsis thaliana* [78].

For the set of upregulated genes (Figure 5B), an enrichment in translation, glycine metabolism, oxidation-reduction process, mitochondrial ATP synthesis, and microtubule was seen. Osmotic stress was previously shown to increase the free glycine concentration in wheat transiently [79]. The five upregulated genes related to microtubules were two tubulins (jg1358, jg9339), one kinesin (jg11911), one dynein (jg2827), and one cytoskeleton-associated protein (jg228). These genes are involved in the transportation of vacuoles and organelles along the microtubules [80].

3.5.2. A. salisburgensis

In *A. salisburgensis*, 305 and 306 gene were up- and down-regulated, respectively. The most upregulated gene (x9013) was a peptidase S8 subtilisin protein with putative keratinolytic activity. Although the biological functions of most fungal subtilases are not yet described, some attempts led to the assumption that subtilisins (S8) could be related to a saprotrophic lifestyle [81]. More recently, a study showed striking correlations for subtilisins' (S8) expansion in pathogenic and soil-/dung-inhabiting fungi [82]. Subtilases have further been reported to be involved in drought and salt resistance mechanisms in plants [83]. An example is the *Arabidopsis thaliana* subtilase ATSBT6.1, which is associated with the unfolded protein response on salt stress through the cleavage of an ER-resident type II membrane protein (BZIP28). The cleavage of this protein is essential for the activation of genes associated with the salt stress response [84].

Among the proteins with at least 50-fold upregulation, eight proteins were glycoside hydrolases, three were involved in oxido-reduction processes, four exhibited a cupin-like domain, and two were involved in post-translational modification (Supplementary Table S1). Among the set of genes downregulated at least 50-fold, four ribitol dehydrogenases, seven transporters, two cyclic amino-acid related enzymes, two transcription factors, and two alcohol dehydrogenase were found.

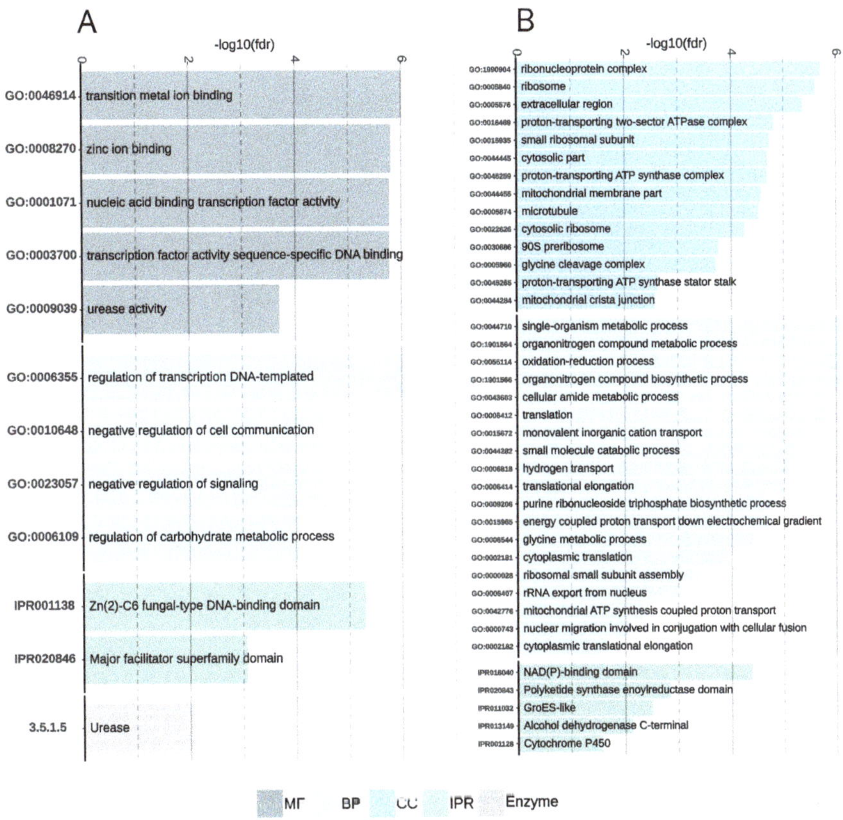

Figure 5. Bar plots showing overrepresented functional annotations in the set of regulated genes in *A. sclerotialis*. (**A**) Enriched functional terms in the set of downregulated genes. (**B**) Enriched functional terms in the set of upregulated genes.

Functional enrichment analyses of the downregulated genes indicated that the expression of membrane-located proteins, especially serine-/threonine-rich proteins, were downregulated (Figure 6A and Supplementary Table S1). Similarly, the gene coding for glutathione S-transferase, which was previously shown to play a negative role in salt stress tolerance in *A. thaliana* [85], was downregulated under high salinity condition. Chitin synthase genes were downregulated, which is in line with proteomic results in yeast [86].

The set of upregulated genes covered a broader spectrum of functions than that of the downregulated genes. Genes located at the cell periphery and extracellular region were overrepresented in the set of upregulated genes (Figure 6B and Supplementary Table S1). The over-representation of functional terms related to cell wall production like the D-alanine–D-alanine ligase and mannose-6-phosphate isomerase might indicate that cell wall damages are happening under high salinity [87]. Enrichment of superoxide-dismutase (K04564), which is an enzyme involved in the catalysis of superoxide, is a known marker of oxidative and salt stress [88]. Finally, the over-representation of MFS transporters is a paramount osmotic stress response, as these transporters are transporting small molecules in response to chemiosmotic ion gradients [69].

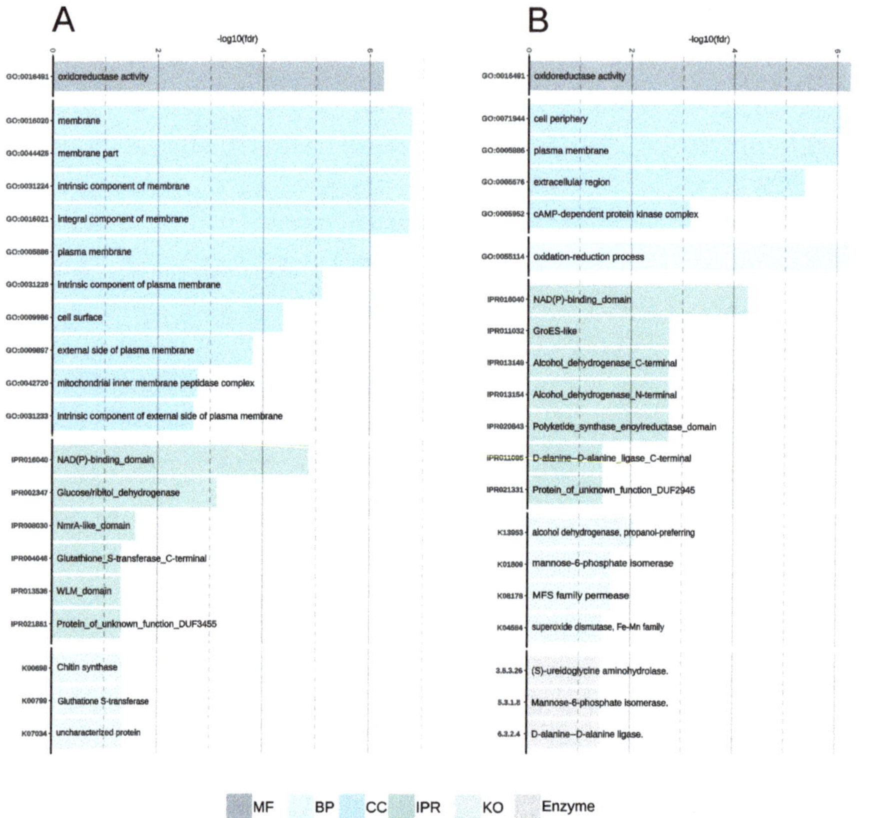

Figure 6. Bar plots showing overrepresented functional annotations in the set of regulated genes in *A. salisburgensis*. (**A**) Enriched functional terms in the set of downregulated genes. (**B**) Enriched functional terms in the set of upregulated genes.

3.5.3. Comparative Transcriptomics

Difference and similarities in expression of genes found in both studied fungal strains were assessed. A total of 84 genes was one-to-one homologous in both strains and was significantly regulated. Genes upregulated in the halophilic and halotolerant species were related to polyketide synthase, chloroperoxidase, haloacid dehalogenase, fructose-biphosphate aldolase, copper transport, and pectinolytic enzyme (Supplementary Table S1). Chloroperoxidases were shown to be potential chlorinators of lignin in plant materials and to contribute to lignin degradation [89]. This is similar to the use of chlorine for wood pulp delignification [90]. In the fungi *Fusarium fujikuroi* and *Neurospora crassa*, genes belonging to the haloacid dehalogenase family were shown to be involved in the osmotic stress and glycerol metabolism [91]. Conserved genes downregulated in both species are related to amino acid transport, trehalose-phosphatase, thiolase, and the p53 transcription factor.

Genes solely upregulated in *A. sclerotialis* were related to oxido-reductive processes, zinc transport, and polyketide synthase. In contrast, genes upregulated in the halophilic fungus and downregulated in the halotolerant fungus were related to the cation/H$^+$ exchanger, threonine/serine exporter, voltage-dependent channel, intradiol cleavage, the putative sensor protein SUR7/RIM9, which is a pH-response regulator protein, and the isochorismatase, a gene involved in salicylic acid metabolism. The upregulation of the last gene only in the halophilic strain may indicate an adaptive advantage to survive permanently in a hypersaline environment. Salicylic acid (SA) is a natural phenolic compound known to control many physiological and biochemical functions in plants such as growth, development, and responses to abiotic stresses [92]. Some studies have shown that SA plays a role in the response to salinity in plants and that its exogenous application improves tolerance to salt stress in several species [93].

3.6. Comparison to Fungal Osmoadaptation and Osmoregulation

The literature on osmoadaptation and osmoregulation in fungi was compared to the genome content and transcription patterns of *A. sclerotialis* and *A. salisburgensis*.

3.6.1. Osmosensing

All members of both branches of the high osmolarity signaling pathway were found in *A. salisburgensis* and *A. sclerotialis* [94,95]. Histidine kinase 7A/B [96] from *Hortaea werneckii*, the mitogen activated protein kinase (KSS1) [97] from *Saccharomyces cerevisiae*, both involved in signal transduction, and the cell wall integrity sensor (MID2) [98] were found in the halotolerant, but not in the halophile genome. Interestingly, the irritation of the cell wall and membrane by shrinking triggers induction of the HOG signaling pathway (e.g., in *S. cerevisiae* [99,100]). Inversely, *A. salisburgensis* has a homologue of the tyrosine-protein phosphatase 3 (PTP3), which is involved in the repression of the *HOG1* gene [101] (Supplementary Table S1).

Protein kinase A (PKA1, TPK1), an enzyme generally involved in the regulation of the glycogen, sugar, and lipid metabolism that negatively regulates the transcription factors (MSN2/MSN4) [102], was upregulated six-fold in *A. salisburgensis*, but downregulated by a factor of three in *A. sclerotialis*. *TOR1*, another gene that negatively regulates the MSN2/MSN4 transcription factors [102], was also downregulated by a factor five in *A. sclerotialis*, but not in *A. salisburgensis*. The PBS2 MAP kinase kinase, which is involved in the HOG pathway, and GSP2, a histidine kinase involved in the nuclear import of HOG1 [103], were upregulated 5.5- and 4.8-fold at a high salt concentration in *A. sclerotialis*, but not in *A. salisburgensis*. NIK1, a histidine kinase involved in the SLN1 branch of the HOG pathway [104], was downregulated nine-fold in *A. sclerotialis*.

3.6.2. Ion Homeostasis

The genome of *A. salisburgensis* contains most of the ion transporters reported to be involved in ion trafficking during osmotic stress in *Saccharomyces cerevisiae* and *Hortaea werneckii*, with the exception of the high affinity potassium transporter (HAK1) [104], the potassium antiporter (KHA1) [15], and the outward-rectifier potassium channel TOK1 [15]. In *A. sclerotialis*, the endosomal/prevacuolar sodium/hydrogen exchanger (NHX1) and HAK1 are missing. TOK1 and the mitochondrial exchanger system, while present in the halotolerant strain, were downregulated four-fold at 20% NaCl in this strain. PAM1, a proton ATPase involved in salt resistance in *Debaryomyces hansenii* [105], was upregulated by a factor of 13 in *A. sclerotialis*. Similarly, NHA1, the Na^+/H^+ antiporter, was upregulated 6.3-fold under high salt concentration. In the obligate halophile, ENA1, a P-type ATPase sodium pump involved in Na^+ and Li^+ efflux [106] was upregulated by a factor of 39.

3.6.3. Cellular Respiration

In *A. sclerotialis*, seven genes annotated as involved in respiration (GO:0045333) were upregulated under high salt concentration. The electron carrier protein CYC1 was upregulated by a factor of 4.62; the cytochrome BC1 complex subunit 7 (QCR7) expression increased by a factor of 3.94; and the cytochrome c oxidase subunit VIa (COX13) was induced by a factor of 3.63. Further, an ADP/ATP carrier protein (AAC) was upregulated by a factor of 6.14. Mitochondrial isocitrate dehydrogenase (IDH1), malate dehydrogenase (MDH1), and mitochondrial trans-2-enoyl-CoA reductase were upregulated 4.82-, 12.82-, and 4.62-fold, respectively. This strongly indicates that the ATP production is increased during salt stress in the halotolerant strain. In contrast, no genes related to cellular respiration were significantly regulated in *A. salisburgensis*.

3.6.4. Stress Response

A. sclerotialis had a stress response more pronounced than *A. salisburgensis*, both in terms of the number of regulated genes, as well as in the regulation intensity. Markers of oxidative stress response were strongly upregulated in *A. sclerotialis* at a high salt concentration, which is in line with the increased respiratory process seen in the halotolerant strain. The mitochondrial superoxide dismutase (SOD2), an important antioxidant defense, catalyzes the dismutation of superoxide O_2^-, a mitochondrial byproduct of respiration, into oxygen and hydrogen peroxide. This gene was upregulated by a factor of 8.3 in *A. sclerotialis* at high salinity. H_2O_2 is then reduced to H_2O by peroxiredoxin (PRX1) using electrons from the reduced form of thioredoxin (TRX) [107]. In *A. salisburgensis*, TRX was upregulated by a factor of 3.6 at a 20% salt concentration. Two mitochondrial PRX1 were upregulated by 9.5- and 6.5-fold, respectively. Further, the nuclear thioredoxin peroxidase DOT5 and the AHP1 peroxiredoxin were upregulated by a factor 4.8 and 5.9, respectively. H_2O_2 can also be scavenged through the glutathione system with the help of glutathione redoxins (GRX) [107]. A homologue to GRX1 and GRX2 and a homologue to GRX3 and GRX4 were upregulated 6.31 and 4.3 in *A. sclerotialis*, respectively. Two catalase paralogues (CTT1), an enzyme that is involved in peroxide scavenging, were upregulated by a factor of 10.1 and 6.9, respectively. The homologue to MCA1, a regulator of apoptosis upon H_2O_2 and clearance of insoluble protein aggregates, was upregulated by a factor of 3.71 [108,109]. Additional studies have shown that oxidative stress originating from intensive mitochondrial respiration [110] can pose a further threat to the survival of aerobic microorganisms living in high-salinity environments. Therefore, this stress can be one of the limiting factors for growth in such environments. In the halotolerant strain *Hortaea werneckii*, the levels of ROS degradation and resistance determine the upper limit of salt tolerance [111]. The peptidyl-prolyl cis/trans-isomerase, a gene involved in protein folding [112], underwent an upregulation under salt stress by a factor of 3.43 in the halotolerant strain.

In *A. salisburgensis*, the oxidative stress response was moderate. Similar to the halotolerant strain, CTT1 was upregulated by a factor of 8.8. Methionine sulfoxide reductase (MXR2), a gene protecting against methionine oxidation by catalyzing thio-dependent reduction of oxidized methionine residues [113], saw a 5.29-fold increase in transcription. Interestingly, the obligate halophile reduced the production of superoxide radicals by downregulating NADPH oxidase and the NADPH oxidase regulator by a factor five and 10, respectively. Beyond genes involved in oxidative stress response, *A. salisburgensis* upregulated HSP12, a chaperone regulated in osmotic stress conditions in *Saccharomyces cerevisiae*, by a factor of five [114].

3.6.5. Cell Interface

In the extracellular region, *A. salisburgensis* strongly upregulated the subtilisin peptidase (S8 x8964), two cellulose degradation enzymes (GH51, x62), a pectinolytic enzyme (EC:4.2.2.2 x23), and a cellulase (GH5 x22.62). Together with the upregulation of a chloroperoxidase, which might be involved in the degradation of lignin, and the environment where this fungus has been isolated, i.e., a wooden staircase in a salt mine, it seems reasonable to assume that salt triggers a cellulolytic response in the obligate halophile. *A. sclerotialis* strongly upregulated a hydrophobin (PF06766 x2646).

In the cell wall of the halotolerant fungus, a beta glucosidase involved in cell wall remodeling (UTH1) [115] was upregulated by a factor of 17, while a membrane-bound HSP70 involved in selective cation trafficking was upregulated by a factor of 6.33 [116]. In the halophilic fungus, the expression of a lysophospholipase increased 18-fold. This enzyme is involved in phospholipid degradation and was previously reported to be upregulated in the micro-algae *Dunaliella salina* under high salt concentration [117].

At the membrane level, besides the previously discussed transporters, *A. sclerotialis* increased the expression of a putative ammonia transporter (PKC1 x22.31), a dipeptidase (DPE1, x9.91), the SUR7 membrane protein involved in membrane organization and cell wall stress [118] (x7.78), a sulfate permease (SULP x6.77), and a cellulolytic enzyme (EggNog: 0PG84 x11.15). The halophile strain increased the transcription of two guanine nucleotide proteins (G protein) [119] (ion-translocating rhodopsin x10.63, RHEB x4.00) and one G protein-coupled receptor-like GPR1 (x12.46), a gene involved in endocytosis, the eisosome component PIL1/LSP1 (x32.11) [120], and two SUR7 membrane proteins (x4.37, x25.73).

3.6.6. Compatible Solute Management

Known genes involved in compatible solutes' management from halotolerant and obligate halophilic fungi were reviewed. Compatible solutes are known to be accumulated or synthesized during changed osmotic conditions. One of these is D-mannitol, which can be synthesized by a reduction of fructose, catalyzed by NADP-dependent mannitol dehydrogenases [13]. In *A. salisburgensis*, two homologues of NADP-dependent mannitol dehydrogenases were found in the genome, but were not regulated in high salt concentration, while one homologue was found in *A. sclerotialis* and was upregulated by a factor of 29. Still, *A. salisburgensis* upregulated GRE3, an aldole reductase involved in polyol metabolism [114], by a factor of 28.

Three homologous genes of STL1, a glycerol/proton symporter of the plasma membrane, were found in *A. salisburgensis* and one in *A. sclerotialis*. In comparison, four homologues were found in the genome of the obligate halophilic fungus *Wallemia ichthyophaga* [15]. In *A. salisburgensis*, one copy of the genes was upregulated by a factor of eight, and in *A. sclerotialis*, STL1 was upregulated by a factor of 91. In *Saccharomyces cerevisiae*, it was reported to be upregulated during hyperosmotic shock [121]. The aquaglyceroporin channel FPS1, which is opened in hypoosmotic conditions to release glycerol [122], did not receive any regulation in *A. salisburgensis* and was downregulated in *A. sclerotialis* by a factor of 30.

Glycine betaine is an osmoprotectant found in plant, animals, bacteria, and fungi and is involved in reactive oxygen scavenging [123]. In *Aspergillus fumigatus*, glycine betaine is produced by converting first choline to betaine aldehyde with a choline oxidase. Betaine aldehyde or its rapidly-forming hydrated equivalent gem-diol-choline are then converted to glycine betaine with the help of betaine aldehyde dehydrogenase or choline oxidase, respectively [123]. In *A. salisburgensis*, two choline oxidases and one betaine aldehyde dehydrogenase were found, while in *A. sclerotialis*, two copies and one copy were found, respectively. While *A. salisburgensis* did not significantly regulate these enzymes, *A. sclerotialis* increased the transcription of betaine aldehyde dehydrogenase by a factor of 4.5 in the high salinity condition.

Genes involved in polyols metabolism were also strongly upregulated. Polyols (also called sugar alcohols) compensate for differences between the extracellular and intracellular water potential without affecting the integrity and function of proteins [124]. In *A. sclerotialis*, gene jg11062 belongs to the same group as the L-arabitol 4-dehydrogenase from *Aspergillus fumigatus* and was upregulated by a factor of 1063 at high salt concentration (Supplementary Table S1). A sorbitol dehydrogenase and a ribitol dehydrogenase were upregulated by a factor of 62 and 67, respectively. The important role of polyols for the resistance of extremotolerant fungi under osmotic stress was already described by Sterflinger [125]. However, this general protection strategy is widespread also in bacteria and Archaea [126].

In the halotolerant strain, four SAM-dependent methyltransferases showed an increase in transcription by a factor of 455, 218, 205, and 52, respectively, similar to previous reports in algae [127], where sequence homologues of the SAM-dependent methyltransferase are involved in the production of the compatible solute homoserine betaine [128]. Glutamine synthetase was upregulated by a factor of 181, indicating that *A. sclerotialis* might use glutamine as an osmolyte, something previously reported for *Halobacillus halophilus* [129].

4. Conclusions

In this work, the genomic adaptation and gene regulation of the obligate halophile *A. salisburgensis*, a fungus isolated from a wooden staircase in a salt mine, were studied by employing comparative genomics and transcriptomics approaches. The halotolerant relative *A. sclerotialis*, a pathogen, was used to gain insight into the environment-specific osmoadaptation and regulation of the obligate halophile.

On the genomic level, *A. salisburgensis* exhibited a 27% decrease in genome size and gene content compared to *A. sclerotialis*. Considering the unique, extremely stable niche where the obligate halophile has been found, it can be hypothesized that the fungus optimized its genome content by dumping genes unnecessary for its survival, a known phenomenon in bacteria [130]. Niche adaptation is further seen in the enrichment of genes involved in cellulose degradation. Halophile-specific depletion of MFS transporters previously reported in *Wallemia ichthyophaga* and *Aspergillus ruber* [10,15] was also found in *A. salisburgensis* and might therefore be a strategy to survive in the high saline environment. This study further confirmed the specific amino-acid enrichment and depletion patterns found in other halophilic species compared to the control species.

The fact that the obligate halophile regulated three-times fewer genes than the halotolerant strain between both salinities further underlines the adaptation of the former to high salt concentration. Besides the regulation of a few transporters like ENA1, STL1, a hydrophobin, and an aldol-reductase, a gene reported to be involved in the production of compatible solutes, there was almost no sign revealing that the obligate halophile was under stress at a 20% salt concentration. Instead, the cellulolytic activity was an indication that the high salt concentration was beneficial to *A. salisburgensis*, as it triggered the expression of a battery of wood-degrading enzymes. Among them, the chloroperoxidase was making use of the chloride ion to chlorinate lignin and improved its degradation.

In contrast, the halotolerant strain exhibited both an osmotic- and oxidative-stress response. The link between both stresses might probably be found in the respiration processes. In fact, it is probable that under high salinity, homeostatic regulation requires an increased supply of ATP, which is produced mainly from respiration. At the transcriptome level, the increased respiration is seen in the upregulation of genes involved in the electron transport chain and the citrate cycle. ROS originating from the respiration induced an oxidative stress response, as can be seen from the increased transcription of genes involved in oxidative-stress, such as superoxide dismutase, thioredoxin peroxidases, glutathione oxidoreductases, and catalases. *A. sclerotialis* further upregulated a hydrophobin, a family of proteins that plays a role in salt resistance in *Wallemia ichthyophaga* and in pathogenicity.

Supplementary Materials: The supplementary table and the supplementary figure are available online at http://www.mdpi.com/2073-4425/10/5/381/s1.

Author Contributions: The conceptualization was done by G.P. and K.S.; methods were planned by G.P., H.T., K.L. and C.P.; Labwork was done by K.L., C.P. and G.P.; Bioinformatics and Data analysis was done by H.T.; original draft preparation and data interpretation was done by H.T; Editing, Data interpretation and reviewing was done by C.P., K.S., K.L. and G.P.; Funding aquisition was done by K.S.

Funding: This research was funded by BOKU-VIBT Equipment GmbH.

Acknowledgments: The computational results presented have been achieved using the Vienna Scientific Cluster (VSC).

Conflicts of Interest: The authors declare no conflict of interest.

References

1. Gunde-Cimerman, N.; Zalar, P. Extremely halotolerant and halophilic fungi inhabit brine in solar salterns around the globe. *Food Technol. Biotechnol.* **2014**, *52*, 170–179.

2. Gunde-Cimerman, N.; Zalar, P.; De Hoog, S.; Plemenitaš, A. Hypersaline waters in salterns—Natural ecological niches for halophilic black yeasts. *FEMS Microbiol. Ecol.* **2000**, *32*, 235–240, doi:10.1016/S0168-6496(00)00032-5. [CrossRef]

3. Northolt, M.D.; Van Egmond, H.P.; Paulsch, W.E. Patulin Production by Some Fungal Species in Relation to Water Activity and Temperature. *J. Food Prot.* **1978**, *41*, 885–890, doi:10.4315/0362-028X-41.11.885. [CrossRef] [PubMed]

4. Wheeler, K.A.; Hocking, A.D. Interactions among xerophilic fungi associated with dried salted fish. *J. Appl. Bacteriol.* **1993**, *74*, 164–169. [CrossRef]

5. Pitt, J.I.; Hocking, A.D. *Fungi and Food Spoilage*; Springer: Boston, MA, USA, 2009.

6. Gunde-Cimerman, N.; Ramos, J.; Plemenitaš, A. Halotolerant and halophilic fungi. In *Biodiversity of Fungi: Their Role in Human Life*; Deshmukh, S.K., Rai, M., Eds.; Science Publishers: Enfield, NH, USA, 2005; pp. 69–127.

7. Sklenář, F.; Jurjević, Ž.; Zalar, P.; Frisvad, J.; Visagie, C.; Kolařík, M.; Houbraken, J.; Chen, A.; Yilmaz, N.; Seifert, K.; et al. Phylogeny of xerophilic aspergilli (subgenus *Aspergillus*) and taxonomic revision of section Restricti. *Stud. Mycol.* **2017**, *88*, 161–236, doi:10.1016/j.simyco.2017.09.002. [CrossRef] [PubMed]

8. Sterflinger, K.; Voitl, C.; Lopandic, K.; Piñar, G.; Tafer, H. Big Sound and Extreme Fungi—Xerophilic, Halotolerant Aspergilli and Penicillia with Low Optimal Temperature as Invaders of Historic Pipe Organs. *Life* **2018**, *8*, 22, doi:10.3390/life8020022. [CrossRef] [PubMed]

9. Oren, A. Diversity of halophilic microorganisms: Environments, phylogeny, physiology, and applications. *J. Ind. Microbiol. Biotechnol.* **2002**, *28*, 56–63, doi:10.1038/sj/jim/7000176. [CrossRef]

10. Kis-Papo, T.; Weig, A.R.; Riley, R.; Peršoh, D.; Salamov, A.; Sun, H.; Lipzen, A.; Wasser, S.P.; Rambold, G.; Grigoriev, I.V.; et al. Genomic adaptations of the halophilic Dead Sea filamentous fungus *Eurotium rubrum*. *Nat. Commun.* **2014**, *5*, 3745. [CrossRef]

11. Martinelli, L.; Zalar, P.; Gunde-Cimerman, N.; Azua-Bustos, A.; Sterflinger, K.; Piñar, G. *Aspergillus atacamensis* and *A. salisburgensis*: Two new halophilic species from hypersaline/arid habitats with a phialosimplex-like morphology. *Extremophiles* **2017**, *21*, 755–773, doi:10.1007/s00792-017-0941-3. [CrossRef]

12. Piñar, G.; Dalnodar, D.; Voitl, C.; Reschreiter, H.; Sterflinger, K. Biodeterioration Risk Threatens the 3100 Year Old Staircase of Hallstatt (Austria): Possible Involvement of Halophilic Microorganisms. *PLoS ONE* **2016**, *11*, e0148279, doi:10.1371/journal.pone.0148279. [CrossRef]

13. Zajc, J.; Liu, Y.; Dai, W.; Yang, Z.; Hu, J.; Gostin Ar, C.; Gunde-Cimerman, N.; Gostinčar, C.; Gunde-Cimerman, N. Genome and transcriptome sequencing of the halophilic fungus Wallemia ichthyophaga: Haloadaptations present and absent. *BMC Genom.* **2013**, *14*, 617. [CrossRef]

14. Attaby, H.S.H. Influence of salinity stress on the growth, biochemical changes, and response to gamma irradiation of *Penicillium chrysogenum*. *Pak. J. Biol. Sci.* **2001**, *4*, 703–706.

15. Plemenitaš, A.; Lenassi, M.; Konte, T.; Kejžar, A.; Zajc, J.; Gostinčar, C.; Gunde-Cimerman, N. Adaptation to high salt concentrations in halotolerant/halophilic fungi: A molecular perspective. *Front. Microbiol.* **2014**, *5*, 199. [CrossRef]

16. Krauke, Y.; Sychrova, H. Four pathogenic Candida species differ in salt tolerance. *Curr. Microbiol.* **2010**, *61*, 335–339. [CrossRef]

17. Michán, C.; Martínez, J.L.; Alvarez, M.C.; Turk, M.; Sychrova, H.; Ramos, J. Salt and oxidative stress tolerance in *Debaryomyces hansenii* and *Debaryomyces fabryi*. *FEMS Yeast Res.* **2013**, *13*, 180–188. [CrossRef]

18. Salmerón-Santiago, K.G.; Pardo, J.P.; Flores-Herrera, O.; Mendoza-Hernández, G.; Miranda-Arango, M.; Guerra-Sánchez, G. Response to osmotic stress and temperature of the fungus *Ustilago maydis*. *Arch. Microbiol.* **2011**, *193*, 701–709. [CrossRef]

19. Sert, H.B.; Sterflinger, K. A new Coniosporium species from historical marble monuments. *Mycol. Prog.* **2010**, *9*, 353–359, doi:10.1007/s11557-009-0643-z. [CrossRef]

20. Poyntner, C.; Blasi, B.; Arcalis, E.; Mirastschijski, U.; Sterflinger, K.; Tafer, H. The Transcriptome of *Exophiala dermatitidis* during Ex-vivo Skin Model Infection. *Front. Cell. Infect. Microbiol.* **2016**, *6*, 136, doi:10.3389/fcimb.2016.00136. [CrossRef]

21. Simão, F.A.; Waterhouse, R.M.; Ioannidis, P.; Kriventseva, E.V.; Zdobnov, E.M. BUSCO: Assessing genome assembly and annotation completeness with single-copy orthologs. *Bioinformatics* **2015**, *31*, 3210–3212. [CrossRef] [PubMed]

22. Stanke, M.; Keller, O.; Gunduz, I.; Hayes, A.; Waack, S.; Morgenstern, B. AUGUSTUS: ab initio prediction of alternative transcripts. *Nucleic Acids Res.* **2006**, *34*, 435–439. [CrossRef]

23. Hoff, K.J.; Lange, S.; Lomsadze, A.; Borodovsky, M.; Stanke, M. BRAKER1: Unsupervised RNA-Seq-Based Genome Annotation with GeneMark-ET and AUGUSTUS. *Bioinformatics* **2016**, *32*, 767–769. [CrossRef]

24. Lomsadze, A.; Burns, P.D.; Borodovsky, M. Integration of mapped RNA-Seq reads into automatic training of eukaryotic gene finding algorithm. *Nucleic Acids Res.* **2014**, *42*, e119. [CrossRef]

25. Finn, R.D.; Attwood, T.K.; Babbitt, P.C.; Bateman, A.; Bork, P.; Bridge, A.J.; Chang, H.Y.; Dosztányi, Z.; El-Gebali, S.; Fraser, M.; et al. InterPro in 2017-beyond protein family and domain annotations. *Nucleic Acids Res.* **2017**, *45*, D190–D199. [CrossRef]

26. Huerta-Cepas, J.; Szklarczyk, D.; Forslund, K.; Cook, H.; Heller, D.; Walter, M.C.; Rattei, T.; Mende, D.R.; Sunagawa, S.; Kuhn, M.; et al. eggNOG 4.5: A hierarchical orthology framework with improved functional annotations for eukaryotic, prokaryotic and viral sequences. *Nucleic Acids Res.* **2016**, *44*, D286–D293. [CrossRef]

27. Saier, M.H.; Tran, C.V.; Barabote, R.D. TCDB: The Transporter Classification Database for membrane transport protein analyses and information. *Nucleic Acids Res.* **2006**, *34*, 181–186, doi:10.1093/nar/gkj001. [CrossRef]

28. Rawlings, N.D.; Waller, M.; Barrett, A.J.; Bateman, A. MEROPS: The database of proteolytic enzymes, their substrates and inhibitors. *Nucleic Acids Res.* **2014**, *42*, 503–509, doi:10.1093/nar/gkt953. [CrossRef]

29. Cantarel, B.L.; Coutinho, P.M.; Rancurel, C.; Bernard, T.; Lombard, V.; Henrissat, B. The Carbohydrate-Active EnZymes database (CAZy): An expert resource for Glycogenomics. *Nucleic Acids Res.* **2009**, *37*, 233–238, doi:10.1093/nar/gkn663. [CrossRef] [PubMed]

30. Altschul, S.F.; Madden, T.L.; Schäffer, A.A.; Zhang, J.; Zhang, Z.; Miller, W.; Lipman, D.J. Gapped BLAST and PSI-BLAST: A new generation of protein database search programs. *Nucleic Acids Res.* **1997**, *25*, 3389–3402. [CrossRef] [PubMed]

31. Lechner, M.; Findeiß, S.; Steiner, L.; Marz, M.; Stadler, P.F.; Prohaska, S.J. Proteinortho: Detection of (co-) orthologs in large-scale analysis. *BMC Bioinform.* **2011**, *12*, 124. [CrossRef] [PubMed]

32. Buchfink, B.; Xie, C.; Huson, D.H. Fast and sensitive protein alignment using DIAMOND. *Nat. Methods* **2015**, *12*, 59–60. [CrossRef]

33. Nguyen, L.T.; Schmidt, H.A.; von Haeseler, A.; Minh, B.Q. IQ-TREE: A fast and effective stochastic algorithm for estimating maximum-likelihood phylogenies. *Mol. Biol. Evol.* **2015**, *32*, 268–274. [CrossRef]

34. Minh, B.Q.; Nguyen, M.A.T.; von Haeseler, A. Ultrafast approximation for phylogenetic bootstrap. *Mol. Biol. Evol.* **2013**, *30*, 1188–1195. [CrossRef]

35. Kalyaanamoorthy, S.; Minh, B.Q.; Wong, T.K.F.; von Haeseler, A.; Jermiin, L.S. ModelFinder: Fast model selection for accurate phylogenetic estimates. *Nat. Methods* **2017**, *14*, 587–589. [CrossRef]

36. Tamura, K.; Battistuzzi, F.U.; Billing-Ross, P.; Murillo, O.; Filipski, A.; Kumar, S. Estimating divergence times in large molecular phylogenies. *Proc. Natl. Acad. Sci. USA* **2012**, *109*, 19333–19338. [CrossRef]

37. Tamura, K.; Tao, Q.; Kumar, S. Theoretical Foundation of the RelTime Method for Estimating Divergence Times from Variable Evolutionary Rates. *Mol. Biol. Evol.* **2018**, *35*, 1770–1782. [CrossRef]

38. Kumar, S.; Stecher, G.; Li, M.; Knyaz, C.; Tamura, K. MEGA X: Molecular Evolutionary Genetics Analysis across Computing Platforms. *Mol. Biol. Evol.* **2018**, *35*, 1547–1549. [CrossRef]

39. De Bie, T.; Cristianini, N.; Demuth, J.P.; Hahn, M.W. CAFE: A computational tool for the study of gene family evolution. *Bioinformatics* **2006**, *22*, 1269–1271. [CrossRef]

40. Almagro Armenteros, J.J.; Tsirigos, K.D.; Sønderby, C.K.; Petersen, T.N.; Winther, O.; Brunak, S.; von Heijne, G.; Nielsen, H. SignalP 5.0 improves signal peptide predictions using deep neural networks. *Nat. Biotechnol.* **2019**, *37*, 420–423. [CrossRef]

41. R Core Team. *R: A Language And Environment for Statistical Computing*; R Foundation for Statistical Computing: Vienna, Austria, 2016.

42. Bray, N.L.; Pimentel, H.; Melsted, P.; Pachter, L. Near-optimal probabilistic RNA-seq quantification. *Nat. Biotechnol.* **2016**, *34*, 888. [CrossRef]

43. Soneson, C.; Love, M.I.; Robinson, M.D. Differential analyses for RNA-seq: Transcript-level estimates improve gene-level inferences. *F1000Research* **2016**, *4*, 1521. [CrossRef]

44. Love, M.I.; Huber, W.; Anders, S. Moderated estimation of fold change and dispersion for RNA-seq data with DESeq2. *Genome Biol.* **2014**, *15*, 550. [CrossRef]

45. Wickham, H.; Chang, W.; ggplot2: An Implementation of the Grammar of Graphics. R Package Version 0.7. 2008. Available online: http://CRAN.R-project.org/package=ggplot2 (accessed on 20 November 2018).

46. Yu, G.; Smith, D.K.; Zhu, H.; Guan, Y.; Lam, T.T.Y. ggtree: An r package for visualization and annotation of phylogenetic trees with their covariates and other associated data. *Methods Ecol. Evol.* **2017**, *8*, 28–36. [CrossRef]

47. Kuncic, M.K.; Kralj Kuncic, M.; Kogej, T.; Drobne, D.; Gunde-Cimerman, N. Morphological Response of the Halophilic Fungal Genus Wallemia to High Salinity. *Appl. Environ. Microbiol.* **2010**, *76*, 329–337. [CrossRef] [PubMed]

48. Lages, F.; Silva-Graça, M.; Lucas, C. Active glycerol uptake is a mechanism underlying halotolerance in yeasts: A study of 42 species. *Microbiology* **1999**, *145*, 2577–2585. [CrossRef]

49. Klement, T.; Milker, S.; Jäger, G.; Grande, P.M.; Domínguez de María, P.; Büchs, J. Biomass pretreatment affects Ustilago maydis in producing itaconic acid. *Microb. Cell Fact.* **2012**, *11*, 43. [CrossRef]

50. Grammann, K.; Volke, A.; Kunte, H.J. New type of osmoregulated solute transporter identified in halophilic members of the bacteria domain: TRAP transporter TeaABC mediates uptake of ectoine and hydroxyectoine in Halomonas elongata DSM 2581(T). *J. Bacteriol.* **2002**, *184*, 3078–3085. [CrossRef]

51. Verghese, J.; Abrams, J.; Wang, Y.; Morano, K.A. Biology of the heat shock response and protein chaperones: Budding yeast (*Saccharomyces cerevisiae*) as a model system. *Microbiol. Mol. Biol. Rev.* **2012**, *76*, 115–158. [CrossRef]

52. Brennan, R.G.; Matthews, B.W. The helix-turn-helix DNA binding motif. *J. Biol. Chem.* **1989**, *264*, 1903–1906. [PubMed]

53. Kumar, S.; Stecher, G.; Suleski, M.; Hedges, S.B. TimeTree: A Resource for Timelines, TimeTrees, and Divergence Times. *Mol. Biol. Evol.* **2017**, *34*, 1812–1819. [CrossRef] [PubMed]

54. Jones, P.; Binns, D.; Chang, H.Y.; Fraser, M.; Li, W.; McAnulla, C.; McWilliam, H.; Maslen, J.; Mitchell, A.; Nuka, G.; et al. InterProScan 5: Genome-scale protein function classification. *Bioinformatics* **2014**, *30*, 1236–1240, doi:10.1093/bioinformatics/btu031. [CrossRef] [PubMed]

55. Ogata, H.; Goto, S.; Sato, K.; Fujibuchi, W.; Bono, H.; Kanehisa, M. KEGG: Kyoto encyclopedia of genes and genomes. *Nucleic Acids Res.* **1999**, *27*, 29–34. [CrossRef] [PubMed]

56. Finn, R.D.; Bateman, A.; Clements, J.; Coggill, P.; Eberhardt, R.Y.; Eddy, S.R.; Heger, A.; Hetherington, K.; Holm, L.; Mistry, J.; et al. Pfam: The protein families database. *Nucleic Acids Res.* **2014**, *42*, 222–230, doi:10.1093/nar/gkt1223. [CrossRef] [PubMed]

57. Yan, Q.; Cui, X.; Lin, S.; Gan, S.; Xing, H.; Dou, D. GmCYP82A3, a Soybean Cytochrome P450 Family Gene Involved in the Jasmonic Acid and Ethylene Signaling Pathway, Enhances Plant Resistance to Biotic and Abiotic Stresses. *PLoS ONE* **2016**, *11*, e0162253. [CrossRef] [PubMed]

58. Gahlot, S.; Joshi, A.; Singh, P.; Tuteja, R.; Dua, M.; Jogawat, A.; Kumar, M.; Raj, S.; Dayaman, V.; Johri, A.K.; et al. Isolation of genes conferring salt tolerance from Piriformospora indica by random overexpression in *Escherichia coli*. *World J. Microbiol. Biotechnol.* **2015**, *31*, 1195–1209, doi:10.1007/s11274-015-1867-5. [CrossRef] [PubMed]

59. Chang, P.K.; Ehrlich, K.C. Genome-wide analysis of the Zn(II)2Cys6 zinc cluster-encoding gene family in *Aspergillus flavus*. *Appl. Microbiol. Biotechnol.* **2013**, *97*, 4289–4300, doi:10.1007/s00253-013-4865-2. [CrossRef]

60. Paulsen, I.T.; Jack, D.L.; Saier, M.H. The amino acid/polyamine/organocation (APC) superfamily of transporters specific for amino acids, polyamines and organocations. *Microbiology* **2000**, *146*, 1797–1814, doi:10.1099/00221287-146-8-1797. [CrossRef]

61. Saier, M.H., Jr. Active transport in communication, protection and nutrition. *J. Mol. Microbiol. Biotechnol.* **2007**, *12*, 161–164. [CrossRef]

62. Gilkes, N.R.; Henrissat, B.; Kilburn, D.G.; Miller, R.C., Jr.; Warren, R.A. Domains in microbial beta-1, 4-glycanases: Sequence conservation, function, and enzyme families. *Microbiol. Rev.* **1991**, *55*, 303–315.

63. Sweigard, J.A.; Chumley, F.G.; Valent, B. Cloning and analysis of CUT1, a cutinase gene from *Magnaporthe grisea*. *Mol. Gen. Genet.* **1992**, *232*, 174–182.

64. Fu, H.L.; Meng, Y.; Ordóñez, E.; Villadangos, A.F.; Bhattacharjee, H.; Gil, J.A.; Mateos, L.M.; Rosen, B.P. Properties of arsenite efflux permeases (Acr3) from *Alkaliphilus metalliredigens* and *Corynebacterium glutamicum*. *J. Biol. Chem.* **2009**, *284*, 19887–19895. [CrossRef]

65. Paul, S.; Bag, S.K.; Das, S.; Harvill, E.T.; Dutta, C. Molecular signature of hypersaline adaptation: Insights from genome and proteome composition of halophilic prokaryotes. *Genome Biol.* **2008**, *9*, R70, doi:10.1186/gb-2008-9-4-r70. [CrossRef]

66. Tadeo, X.; López-Méndez, B.; Trigueros, T.; Laín, A.; Castaño, D.; Millet, O. Structural basis for the aminoacid composition of proteins from halophilic archea. *PLoS Biol.* **2009**, *7*, e1000257. [CrossRef] [PubMed]

67. De Maayer, P.; Anderson, D.; Cary, C.; Cowan, D.A. Some like it cold: Understanding the survival strategies of psychrophiles. *EMBO Rep.* **2014**, *15*, 508–517. [CrossRef] [PubMed]

68. Fukuchi, S.; Yoshimune, K.; Wakayama, M.; Moriguchi, M.; Nishikawa, K. Unique amino acid composition of proteins in halophilic bacteria. *J. Mol. Biol.* **2003**, *327*, 347–357. [CrossRef]

69. Pao, S.S.; Paulsen, I.T.; Saier, M.H., Jr. Major facilitator superfamily. *Microbiol. Mol. Biol. Rev.* **1998**, *62*, 1–34.

70. Temple, B.; Horgen, P.A.; Bernier, L.; Hintz, W.E. Cerato-ulmin, a hydrophobin secreted by the causal agents of Dutch elm disease, is a parasitic fitness factor. *Fungal Genet. Biol.* **1997**, *22*, 39–53. [CrossRef] [PubMed]

71. Sigler, L.; Sutton, D.A.; Gibas, C.F.C.; Summerbell, R.C.; Noel, R.K.; Iwen, P.C. Phialosimplex, a new anamorphic genus associated with infections in dogs and having phylogenetic affinity to the Trichocomaceae. *Med. Mycol.* **2010**, *48*, 335–345, doi:10.3109/13693780903225805. [CrossRef]

72. Liu, X.; Baird, W.M. Differential expression of genes regulated in response to drought or salinity stress in sunflower. *Crop Sci.* **2003**, *43*, 678–687.

73. Shelest, E. Transcription Factors in Fungi: TFome Dynamics, Three Major Families, and Dual-Specificity TFs. *Front. Genet.* **2017**, *8*, 53. [CrossRef]

74. Yashiroda, H.; Tanaka, K. But1 and But2 proteins bind to Uba3, a catalytic subunit of E1 for neddylation, in fission yeast. *Biochem. Biophys. Res. Commun.* **2003**, *311*, 691–695. [CrossRef]

75. Lopez, F.; Leube, M.; Gil-Mascarell, R.; Navarro-Aviñó, J.P.; Serrano, R. The yeast inositol monophosphatase is a lithium- and sodium-sensitive enzyme encoded by a non-essential gene pair. *Mol. Microbiol.* **1999**, *31*, 1255–1264. [CrossRef]

76. Hermand, D. F-box proteins: More than baits for the SCF? *Cell Div.* **2006**, *1*, 30. [CrossRef]

77. Sadeghi, M.; Dehghan, S.; Fischer, R.; Wenzel, U.; Vilcinskas, A.; Kavousi, H.R.; Rahnamaeian, M. Isolation and characterization of isochorismate synthase and cinnamate 4-hydroxylase during salinity stress, wounding, and salicylic acid treatment in Carthamus tinctorius. *Plant Signal. Behav.* **2013**, *8*, e27335. [CrossRef]

78. Bu, Y.; Kou, J.; Sun, B.; Takano, T.; Liu, S. Adverse effect of urease on salt stress during seed germination in Arabidopsis thaliana. *FEBS Lett.* **2015**, *589*, 1308–1313. [CrossRef]

79. Kovács, Z.; Simon-Sarkadi, L.; Vashegyi, I.; Kocsy, G. Different accumulation of free amino acids during short- and long-term osmotic stress in wheat. *Sci. World J.* **2012**, *2012*, 216521. [CrossRef]

80. Vale, R.D. The molecular motor toolbox for intracellular transport. *Cell* **2003**, *112*, 467–480. [CrossRef]

81. Leger, R.J.S.; Joshi, L.; Roberts, D.W. Adaptation of proteases and carbohydrases of saprophytic, phytopathogenic and entomopathogenic fungi to the requirements of their ecological niches. *Microbiology* **1997**, *143*, 1983–1992, doi:10.1099/00221287-143-6-1983. [CrossRef]

82. Muszewska, A.; Stepniewska-Dziubinska, M.M.; Steczkiewicz, K.; Pawlowska, J.; Dziedzic, A.; Ginalski, K. Fungal lifestyle reflected in serine protease repertoire. *Sci. Rep.* **2017**, *7*, 9147, doi:10.1038/s41598-017-09644-w. [CrossRef]

83. Liu, J.X.; Srivastava, R.; Che, P.; Howell, S.H. Salt stress responses in Arabidopsis utilize a signal transduction pathway related to endoplasmic reticulum stress signaling. *Plant J. Cell Mol. Biol.* **2007**, *51*, 897–909, doi:10.1111/j.1365-313X.2007.03195.x. [CrossRef]

84. Figueiredo, J.; Sousa Silva, M.; Figueiredo, A. Subtilisin-like proteases in plant defence: The past, the present and beyond. *Mol. Plant Pathol.* **2018**, *19*, 1017–1028, doi:10.1111/mpp.12567. [CrossRef]

85. Chen, J.H.; Jiang, H.W.; Hsieh, E.J.; Chen, H.Y.; Chien, C.T.; Hsieh, H.L.; Lin, T.P. Drought and salt stress tolerance of an Arabidopsis glutathione S-transferase U17 knockout mutant are attributed to the combined effect of glutathione and abscisic acid. *Plant Physiol.* **2012**, *158*, 340–351. [CrossRef]

86. Szopinska, A.; Degand, H.; Hochstenbach, J.F.; Nader, J.; Morsomme, P. Rapid response of the yeast plasma membrane proteome to salt stress. *Mol. Cell. Proteom.* **2011**, *10*. [CrossRef]

87. Ene, I.V.; Walker, L.A.; Schiavone, M.; Lee, K.K.; Martin-Yken, H.; Dague, E.; Gow, N.A.R.; Munro, C.A.; Brown, A.J.P. Cell Wall Remodeling Enzymes Modulate Fungal Cell Wall Elasticity and Osmotic Stress Resistance. *mBio* **2015**, *6*, e00986. [CrossRef]

88. Gostinčar, C.; Gunde-Cimerman, N. Overview of Oxidative Stress Response Genes in Selected Halophilic Fungi. *Genes* **2018**, *9*, 143. [CrossRef]

89. Ortiz-Bermúdez, P.; Srebotnik, E.; Hammel, K.E. Chlorination and cleavage of lignin structures by fungal chloroperoxidases. *Appl. Environ. Microbiol.* **2003**, *69*, 5015–5018. [CrossRef]

90. Dence, C.W. Halogenation and nitration. In *Lignins. Occurrence, Formation, Structure and Reactions*; Wiley-Interscience: New York, NY, USA, 1971; pp. 373–432.

91. García-Martínez, J.; Castrillo, M.; Avalos, J. The gene cutA of Fusarium fujikuroi, encoding a protein of the haloacid dehalogenase family, is involved in osmotic stress and glycerol metabolism. *Microbiology* **2014**, *160*, 26–36. [CrossRef]

92. Hara, M.; Furukawa, J.; Sato, A.; Mizoguchi, T.; Miura, K. Abiotic stress and role of salicylic acid in plants. In *Abiotic Stress Responses in Plants*; Ahmad, P., Prasad, M.N.V., Eds.; Springer: New York, NY, USA, 2012; pp. 235–251.

93. Miura, K.; Tada, Y. Regulation of water, salinity, and cold stress responses by salicylic acid. *Front. Plant Sci.* **2014**, *5*, 4, doi:10.3389/fpls.2014.00004. [CrossRef]

94. Hohmann, S. Control of high osmolarity signaling in the yeast Saccharomyces cerevisiae. *FEBS Lett.* **2009**, *583*, 4025–4029. [CrossRef]

95. Ma, D.; Li, R. Current understanding of HOG-MAPK pathway in Aspergillus fumigatus. *Mycopathologia* **2013**, *175*, 13–23. [CrossRef]

96. Lenassi, M.; Plemenitaš, A. Novel group VII histidine kinase HwHhk7B from the halophilic fungi Hortaea werneckii has a putative role in osmosensing. *Curr. Genet.* **2007**, *51*, 393–405. [CrossRef]

97. Mortimer, R.K.; Schild, D.; Contopoulou, C.R.; Kans, J.A. Genetic map of Saccharomyces cerevisiae, edition 10. *Yeast* **1989**, *5*, 321–403. [CrossRef]

98. Ono, T.; Suzuki, T.; Anraku, Y.; Iida, H. The MID2 gene encodes a putative integral membrane protein with a Ca(2+)-binding domain and shows mating pheromone-stimulated expression in Saccharomyces cerevisiae. *Gene* **1994**, *151*, 203–208. [CrossRef]

99. Hohmann, S. Osmotic stress signaling and osmoadaptation in yeasts. *Microbiol. Mol. Biol. Rev.* **2002**, *66*, 300–372. [CrossRef]

100. Stratford, M.; Steels, H.; Novodvorska, M.; Archer, D.B.; Avery, S.V. Extreme Osmotolerance and Halotolerance in Food-Relevant Yeasts and the Role of Glycerol-Dependent Cell Individuality. *Front. Microbiol.* **2019**, *9*, 3238, doi:10.3389/fmicb.2018.03238. [CrossRef] [PubMed]

101. Wurgler-Murphy, S.M.; Maeda, T.; Witten, E.A.; Saito, H. Regulation of the Saccharomyces cerevisiae HOG1 mitogen-activated protein kinase by the PTP2 and PTP3 protein tyrosine phosphatases. *Mol. Cell. Biol.* **1997**, *17*, 1289–1297. [CrossRef]

102. Smith, A.; Ward, M.P.; Garrett, S. Yeast PKA represses Msn2p/Msn4p-dependent gene expression to regulate growth, stress response and glycogen accumulation. *EMBO J.* **1998**, *17*, 3556–3564. [CrossRef] [PubMed]

103. Ferrigno, P.; Posas, F.; Koepp, D.; Saito, H.; Silver, P.A. Regulated nucleo/cytoplasmic exchange of HOG1 MAPK requires the importin beta homologs NMD5 and XPO1. *EMBO J.* **1998**, *17*, 5606–5614 [CrossRef]

104. Kruppa, M.; Calderone, R. Two-component signal transduction in human fungal pathogens. *FEMS Yeast Res.* **2006**, *6*, 149–159. [CrossRef] [PubMed]

105. Prista, C.; Loureiro-Dias, M.C.; Montiel, V.; García, R.; Ramos, J. Mechanisms underlying the halotolerant way of Debaryomyces hansenii. *FEMS Yeast Res.* **2005**, *5*, 693–701. [CrossRef]

106. Haro, R.; Garciadeblas, B.; Rodríguez-Navarro, A. A novel P-type ATPase from yeast involved in sodium transport. *FEBS Lett.* **1991**, *291*, 189–191. [CrossRef]

107. Pannala, V.R.; Dash, R.K. Mechanistic characterization of the thioredoxin system in the removal of hydrogen peroxide. *Free Radic. Biol. Med.* **2015**, *78*, 42–55. [CrossRef]

108. Madeo, F.; Herker, E.; Maldener, C.; Wissing, S.; Lächelt, S.; Herlan, M.; Fehr, M.; Lauber, K.; Sigrist, S.J.; Wesselborg, S.; et al. A caspase-related protease regulates apoptosis in yeast. *Mol. Cell* **2002**, *9*, 911–917. [CrossRef]

109. Lee, R.E.C.; Brunette, S.; Puente, L.G.; Megeney, L.A. Metacaspase Yca1 is required for clearance of insoluble protein aggregates. *Proc. Natl. Acad. Sci. USA* **2010**, *107*, 13348–13353. [CrossRef] [PubMed]

110. Gille, G.; Sigler, K. Oxidative stress and living cells. *Folia Microbiol.* **1995**, *40*, 131–152. [CrossRef]

111. Petrovic, U. Role of oxidative stress in the extremely salt-tolerant yeast Hortaea werneckii. *FEMS Yeast Res.* **2006**, *6*, 816–822, doi:10.1111/j.1567-1364.2006.00063.x. [CrossRef] [PubMed]

112. Guo, F.; Stanevich, V.; Wlodarchak, N.; Sengupta, R.; Jiang, L.; Satyshur, K.A.; Xing, Y. Structural basis of PP2A activation by PTPA, an ATP-dependent activation chaperone. *Cell Res.* **2014**, *24*, 190–203. [CrossRef] [PubMed]

113. Stadtman, E.R.; Levine, R.L. Free radical-mediated oxidation of free amino acids and amino acid residues in proteins. *Amino Acids* **2003**, *25*, 207–218. [CrossRef]

114. Saito, H.; Posas, F. Response to hyperosmotic stress. *Genetics* **2012**, *192*, 289–318. [CrossRef]

115. Kuznetsov, E.; Kučerová, H.; Váchová, L.; Palková, Z. SUN family proteins Sun4p, Uth1p and Sim1p are secreted from Saccharomyces cerevisiae and produced dependently on oxygen level. *PLoS ONE* **2013**, *8*, e73882. [CrossRef] [PubMed]

116. Arispe, N.; De Maio, A. ATP and ADP modulate a cation channel formed by Hsc70 in acidic phospholipid membranes. *J. Biol. Chem.* **2000**, *275*, 30839–30843. [CrossRef] [PubMed]

117. Katz, A.; Waridel, P.; Shevchenko, A.; Pick, U. Salt-induced changes in the plasma membrane proteome of the halotolerant alga Dunaliella salina as revealed by blue native gel electrophoresis and nano-LC-MS/MS analysis. *Mol. Cell. Proteom.* **2007**, *6*, 1459–1472. [CrossRef] [PubMed]

118. Douglas, L.M.; Wang, H.X.; Keppler-Ross, S.; Dean, N.; Konopka, J.B. Sur7 promotes plasma membrane organization and is needed for resistance to stressful conditions and to the invasive growth and virulence of Candida albicans. *mBio* **2012**, *3*. [CrossRef] [PubMed]

119. Vögler, O.; Barceló, J.M.; Ribas, C.; Escribá, P.V. Membrane interactions of G proteins and other related proteins. *Biochim. Biophys. Acta* **2008**, *1778*, 1640–1652. [CrossRef] [PubMed]

120. Walther, T.C.; Brickner, J.H.; Aguilar, P.S.; Bernales, S.; Pantoja, C.; Walter, P. Eisosomes mark static sites of endocytosis. *Nature* **2006**, *439*, 998–1003. [CrossRef] [PubMed]

121. Ferreira, C.; van Voorst, F.; Martins, A.; Neves, L.; Oliveira, R.; Kielland-Brandt, M.C.; Lucas, C.; Brandt, A. A member of the sugar transporter family, Stl1p is the glycerol/H+ symporter in Saccharomyces cerevisiae. *Mol. Biol. Cell* **2005**, *16*, 2068–2076. [CrossRef] [PubMed]

122. Luyten, K.; Albertyn, J.; Skibbe, W.F.; Prior, B.A.; Ramos, J.; Thevelein, J.M.; Hohmann, S. Fps1, a yeast member of the MIP family of channel proteins, is a facilitator for glycerol uptake and efflux and is inactive under osmotic stress. *EMBO J.* **1995**, *14*, 1360–1371. [CrossRef] [PubMed]

123. Lambou, K.; Pennati, A.; Valsecchi, I.; Tada, R.; Sherman, S.; Sato, H.; Beau, R.; Gadda, G.; Latgé, J.P. Pathway of glycine betaine biosynthesis in Aspergillus fumigatus. *Eukaryot. Cell* **2013**, *12*, 853–863. [CrossRef]

124. Brown, A.D. *Microbial Water Stress Physiology. Principles and Perspectives*; John Wiley & Sons: Chichester, UK 1990.

125. Sterflinger, K. Temperature and NaCl-tolerance of rock-inhabiting meristematic fungi. *Antonie Van Leeuwenhoek* **1998**, *74*, 271–281. [CrossRef]

126. Gunde-Cimerman, N.; Plemenitaš, A.; Oren, A. Strategies of adaptation of microorganisms of the three domains of life to high salt concentrations. *FEMS Microbiol. Rev.* **2018**, *42*, 353–375. [CrossRef]

127. Krell, A. Salt Stress Tolerance in the Psychrophilic Diatom Fragilariopsis Cylindrus. Ph.D. Thesis, Bremen University, Bremen, Germany, 2006.

128. Pade, N.; Michalik, D.; Ruth, W.; Belkin, N.; Hess, W.R.; Berman-Frank, I.; Hagemann, M. Trimethylated homoserine functions as the major compatible solute in the globally significant oceanic cyanobacterium Trichodesmium. *Proc. Natl. Acad. Sci. USA* **2016**, *113*, 13191–13196. [CrossRef] [PubMed]

129. Saum, S.H.; Sydow, J.F.; Palm, P.; Pfeiffer, F.; Oesterhelt, D.; Müller, V. Biochemical and molecular characterization of the biosynthesis of glutamine and glutamate, two major compatible solutes in the moderately halophilic bacterium *Halobacillus halophilus*. *J. Bacteriol.* **2006**, *188*, 6808–6815. [CrossRef] [PubMed]

130. Stepkowski, T.; Legocki, A.B. Reduction of bacterial genome size and expansion resulting from obligate intracellular lifestyle and adaptation to soil habitat. *Acta Biochim. Pol.* **2001**, *48*, 367–381. [PubMed]

Article

The Primary Antisense Transcriptome of *Halobacterium salinarum* NRC-1

João Paulo Pereira de Almeida [1,†], Ricardo Z. N. Vêncio [2,†], Alan P. R. Lorenzetti [1], Felipe ten-Caten [1], José Vicente Gomes-Filho [1] and Tie Koide [1,*]

[1] Department of Biochemistry and Immunology, Ribeirão Preto Medical School, University of São Paulo, São Paulo 14049-900, Brazil; jpereiradealmeida.mg32@gmail.com (J.P.P.d.A.); alan.lorenzetti@gmail.com (A.P.R.L.); ftencaten@gmail.com (F.t.-C.); vicente.gomes.filho@gmail.com (J.V.G.-F.)

[2] Department of Computation and Mathematics, Faculdade de Filosofia, Ciências e Letras de Ribeirão Preto, University of São Paulo, São Paulo 14049-900, Brazil; rvencio@usp.br

* Correspondence: tkoide@fmrp.usp.br; Tel.: +55-163-3153-107

† Authors contributed equally.

Received: 6 February 2019; Accepted: 1 April 2019; Published: 5 April 2019

Abstract: Antisense RNAs (asRNAs) are present in diverse organisms and play important roles in gene regulation. In this work, we mapped the primary antisense transcriptome in the halophilic archaeon *Halobacterium salinarum* NRC-1. By reanalyzing publicly available data, we mapped antisense transcription start sites (aTSSs) and inferred the probable 3′ ends of these transcripts. We analyzed the resulting asRNAs according to the size, location, function of genes on the opposite strand, expression levels and conservation. We show that at least 21% of the genes contain asRNAs in *H. salinarum*. Most of these asRNAs are expressed at low levels. They are located antisense to genes related to distinctive characteristics of *H. salinarum*, such as bacteriorhodopsin, gas vesicles, transposases and other important biological processes such as translation. We provide evidence to support asRNAs in type II toxin–antitoxin systems in archaea. We also analyzed public Ribosome profiling (Ribo-seq) data and found that ~10% of the asRNAs are ribosome-associated non-coding RNAs (rancRNAs), with asRNAs from transposases overrepresented. Using a comparative transcriptomics approach, we found that ~19% of the asRNAs annotated in *H. salinarum* belong to genes with an ortholog in *Haloferax volcanii*, in which an aTSS could be identified with positional equivalence. This shows that most asRNAs are not conserved between these halophilic archaea.

Keywords: antisense RNA; *Halobacterium salinarum*; transcriptome; dRNA-seq; archaea; transcription start site; post-transcriptional regulation; gene expression; type II toxin-antitoxin systems; Ribo-seq

---

1. Introduction

Antisense RNAs (asRNAs) are non-coding RNAs (ncRNAs) transcribed from the opposite strand of a given gene. Intuitively, asRNAs can be assumed to be cis-acting as they are complementary to the messenger RNA (mRNA) of the gene from which they derive. This does not restrict asRNA action to the gene on the opposite strand, they can also act in trans [1]. AsRNAs can act as regulators at different stages of gene expression [2]. They can modulate the stability and lifespan of RNAs by either occluding degradation sites or forming double-stranded complexes (dsRNAs) that are targets for RNases [3]. Furthermore, they may directly influence translation, inhibiting this process by occluding ribosome binding sites or promoting conformational alterations that might increase or decrease the frequency of translation of its target [2,4,5].

Simultaneous transcription of opposite DNA strands was first reported in λ phage [6], later it was identified in bacteria in processes related to plasmid replication, phage repression and transposases [7], and finally in eukaryotes [8]. Technical advances in transcriptome analysis allowed greater confidence in strand-specific datasets, allowing the identification of asRNAs in different organisms [9–12]. Overlapping antisense transcription is a ubiquitous phenomenon [13–15], while the functionality of these molecules is still under debate, since it can arise from spurious transcription [16]. In bacteria the percentage of protein coding genes with asRNAs varies from 2% in *Salmonella* [17] to up to 80% in *Pseudomonas aeruginosa* [18]. In archaea, asRNAs have been reported for a while [19,20]. Some studies assured that they were primary transcripts: 1244 in *Haloferax volcanii* [21], 1110 in *Methanolobus psychrophilus* [22], 1018 in *Thermococcus kodakarensis* [23], 48 in *Methanosarcina mazei* Gö1 [24], 29 in *Thermococcus onnurineus* NA1 [25], and 12 in *Methanocaldococcus jannaschii* DSM [26]. In *H. volcanii*, the only halophile represented, asRNAs are present in genes related to chemotaxis, transcription regulation, and insertion sequences [27].

Halobacterium salinarum is an extreme halophile archaeon, thriving in 4.3 M NaCl concentrations [28]. Historically known for its bacteriorhodopsin light-dependent proton pump [29], *H. salinarum* has a wide metabolic versatility for energy production such as amino acid oxidation, anaerobic respiration using different electron acceptors, and arginine fermentation [30,31]; it also shows high tolerance to diverse environmental stresses [32–34]. The plethora of transcriptional data in different environmental conditions and genetic backgrounds allowed the construction of a global gene regulatory network based primarily on mRNA levels [35,36]. Many overlapping features in *H. salinarum* genome, such as alternative promoter usage in operons [37] and internal RNAs that produce protein isoforms [38], have been reported, highlighting the complexity of transcriptional regulation. In addition, ncRNAs in intergenic regions [37], RNAs associated to transcription start sites (TSSaRNAs) [39], and associated to insertion sequences (sotRNAs) [40] have also been identified. However, there is little information on antisense RNAs. The first asRNA reported in this organism was a 151 nucleotide (nt) molecule complementary to the 5′ end of a phage related gene [41,42]; asRNAs regulating gas vesicles genes were also identified [19], but no further global analysis of asRNAs have been performed since then.

In this work, we report a genome-wide primary asRNA mapping in *H. salinarum* NRC-1. To do that, we took advantage of published differential RNA-sequencing (dRNA-seq) data for the identification of transcription start sites antisense to annotated genes (aTSS) [38]. To annotate the minimum length of asRNAs, we reanalyzed published strand-specific RNA-seq data acquired in paired-end mode. Data was manually inspected to guide asRNA annotation of probable 3′ ends. Publicly available Ribosome profiling (Ribo-seq) data (Raman et al. submitted) were reanalyzed to identify asRNAs bound to ribosomes. We investigated asRNA global properties: size distribution, location relative to the cognate genes, expression levels, and cognate gene functions. Finally, we did a comparison with dRNA-seq results of *H. volcanii* to evaluate the cross-species conservation of asRNAs.

2. Materials and Methods

2.1. Antisense Transcription Start Sites Annotation

We reanalyzed publicly available raw dRNA-seq data from *H. salinarum* NRC-1 grown in complex media (250 g/L NaCl, 20 g/L MgSO$_4$, 2 g/L KCl, 3 g/L sodium citrate, and 10 g/L bacteriological peptone (Oxoid)) over a growth curve sampled at 17 h, 37 h, and 86 h, and grown under standard conditions (37 °C, 225 rpm, constant light) sampled at mid-log phase [38]. We used our in-house workflow ("Caloi-seq", https://github.com/alanlorenzetti/frtc/), described in [38], for *H. salinarum* dRNA-seq data. Briefly, libraries were downloaded from NCBI Sequence Read Archive (SRA) [43] and trimmed using Trimmomatic [44]. Reads surviving as a pair were aligned to reference genome (NCBI Assembly ASM680v1) in paired-end mode using HISAT2 [45], suppressing alignments resulting in fragments longer than 1000 nt. Orphan R1 and R2 sequences were aligned using the single-end mode. Multimappers aligning up to 1000 times were allowed to be reported. SAM files were converted to BAM using SAMtools [46] and input in MMR to find the most likely position for each multimapper [47]. The resultant BAM files were filtered to keep only R1 reads, which serve as input for the TSS inference step. BEDTools [48] was used to compute the fragment 5′ accumulation profile, employing all the aligned R1 reads. Data visualization was performed using IGV [49] and Gaggle Genome Browser [50].

TSSs were identified from dRNA-Seq experiments using TSSAR [51] with the following parameters: p-value $p < 0.005$, a minimum of four reads, and a distance of TSS grouping of at least five nt. An antisense TSS (aTSS) was defined based on genome annotation of *H. salinarum* NRC-1 available at RefSeq updated in 2017 with additional manual curation using *H. salinarum* R1 annotation as reference [52] (Table S1). TSSs located inside genes on the opposite strand were classified as aTSS. TSSs on the opposite strand and up to 200 nt downstream of a gene 3′ end were considered downstream aTSS (daTSS).

Potentially structured regions were filtered out by calculating folding minimum free energy (MFE) along the whole genome sequence as previously described [38]. A sliding window (51 nt) with an offset of 10 nt was used to tile the genome and all subsequences were subjected to secondary structure prediction using RNAfold [53] with default parameters. The distribution of MFE obtained for the tiled genome was compared with the distribution obtained for only subsequences immediately downstream of aTSS. The 33.3% quantile in the whole-genome MFE distribution was arbitrarily chosen as cutoff for potentially forming structures and thus putative false positive aTSS.

2.2. Antisense RNA Loci Inference

To annotate asRNA loci we used aTSS positions together with all TEX- strand-specific paired-end RNA-seq libraries publicly available for *H. salinarum* NRC-1. Library accessions and number of sequenced, trimmed, and aligned reads used for asRNA loci annotation are shown in Table S2. These libraries are the control libraries for a dRNA-seq experiment, representing four replicates of *H. salinarum* NRC-1 grown under standard conditions and sampled at mid-log phase and biological duplicates of three different time points over a growth curve (17 h, 37 h, and 86 h) [38]. For these libraries, we took the MMR adjusted BAM files, cited in the last section, and computed the genome-wide coverage (transcriptional profile) using deepTools [54], considering the extension of full fragments for paired-end alignments and the proper strand orientation for alignments of R1 and R2 orphan reads. BEDTools [48] was used to compute the fragment 3′ accumulation profile, employing all the aligned R2 reads. Visual inspection was performed using IGV [49] and Gaggle Genome Browser [50] by looking for a mapped aTSS followed by a sharp decrease in read coverage to infer an asRNA locus in at least four different libraries. To aid the inference of locus ending, we used the fragment 3′ end accumulation profile, requiring at least 10 observations to demarcate the minimum end position of an asRNA. The steps for processing dRNA-seq data and computing the transcriptional profiles are depicted as a workflow in Figure S1, and asRNA loci inference is schematically shown in Figure S2.

2.3. Promoter and 3′ End Sequence Analysis

The computation of nucleotide frequency for the promoter region of aTSSs and 3′ end of asRNAs was performed using Weblogo [55]. For the promoter region we analyzed sequences 40 nt upstream and 10 nt downstream of the aTSS; for the 3′ end, we used sequences 10 nt upstream and downstream of the last nucleotide.

2.4. Gene Functions

Gene functional classification and Gene Ontology enrichment analysis were performed using the PANTHER system [56]. In the absence of an associated PANTHER annotation, MicrobesOnLine [57] and BlastCD [58] tools were used to identify conserved protein domains.

2.5. Type II Toxin-Antitoxin Systems Annotation and Antisense Transcription Start Sites Identification in Other Archaea

TA finder 2.0 [59] was used to annotate type II Toxin-Antitoxin (TA) systems in *E. coli* and archaeal organisms with available dRNA-seq data. aTSS positions were mapped as described in Section 2.1, reanalyzing dRNA-seq data from NCBI BioProject database for *Haloferax volcanii* DS2

(PRJNA324298), *Methanocaldococcus jannaschii* DSM 2661 (PRJNA342613), *Thermococcus kodakarensis* KOD1 (PRJNA242777), and *Thermococcus onnurineus* NA1 (PRJNA339284). aTSS positions for *E. coli* were obtained from Thomason et al. [60], and positions of asRNAs in type II TA systems found in an immunoprecipitation ofdsRNAs experiment were obtained from Lybecker et al. [61].

2.6. Differential Expression Analysis

A GFF file including *H. salinarum* NRC-1 annotated genes, in addition to the 846 asRNAs identified in this study, was built, generating a total of 3680 features. Matrices for read counts per feature were generated using HT-Seq [62] from BAM files. Differential expression analysis was performed using DEseq2 [63]. Genes with more than two-fold change (FC) up or down regulation ($\log_2$ FC > 1 or < −1) and adjusted *p*-values p_{adj} < 0.01 were considered differentially expressed.

2.7. Antisense Transcription Start Sites Comparison between Halobacterium salinarum and Haloferax volcanii

Ortholog genes in *H. salinarum* and *H. volcanii* were identified using OrtholugeDB 2.1 [64]. Pairs of genes with aTSSs were sub-selected and gene sizes were normalized by their length in a 0 to 100 normalized scale ($d = 0$ at start codon, $d = 100$ at stop codon). The length-normalized genes were divided in 10 partitions (tenths) depending on d: [0;10), [10;20), ... , [90;100), and aTSSs were considered conserved if present in the same partition in the orthologous genes. Differences between relative positions, $D = |d_{Hsal} - d_{Hvol}|$, were used to estimate how "equivalent" aTSS are between both organisms.

2.8. Ribo-Seq Data Analysis

H. salinarum NRC-1 Ribo-seq data is publicly available ahead of publication (Raman et al. submitted) at NCBI BioProject database under accession PRJNA413990 and was reanalyzed using the same pipeline described in Sections 2.1 and 2.2. AsRNAs presenting coverage greater than 50% and at least 20 reads were considered putative ribosome-associated ncRNAs (rancRNAs).

3. Results

3.1. Mapping Primary Antisense RNAs

We identified the primary transcription start sites (TSS) for asRNAs (aTSS) in *H. salinarum* NRC-1 by reanalyzing dRNA-seq data sampled at three time points over growth [38]. We were able to identify 2146 aTSS, located in 1231 genes. Probable false positives due to secondary structure were filtered out (see Methods 2.1), resulting in 1626 aTSSs located in 963 genes (Table S3). Figure 1a shows the position of aTSS in a length-normalized scale (0 to 100 scale) relative to the gene on the opposite strand. The overrepresentation of aTSS at the 3′ end is stronger than at the 5′ end.

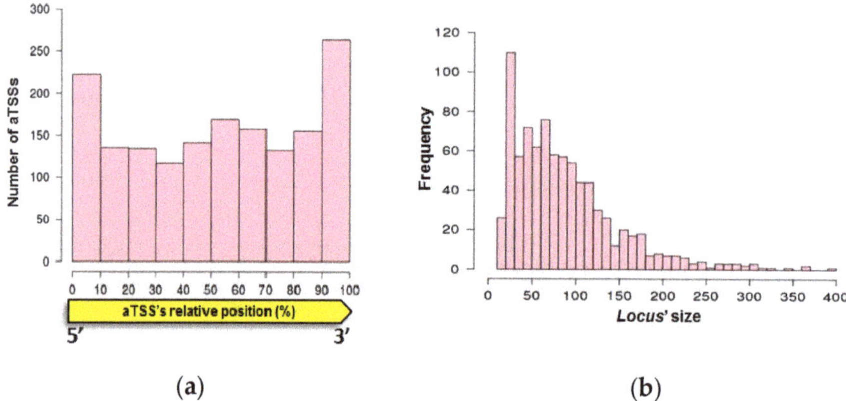

Figure 1. Antisense RNA (asRNA) properties. (**a**) Antisense transcription start site (aTSS) positions relative to cognate genes. (**b**) Size distribution of mapped asRNAs.

Given that (i) asRNAs can be located in regions downstream of the cognate gene, (ii) 3' untranslated regions (UTR) can be targeted by ncRNAs, and (iii) asRNAs from 3' UTRs are important regulatory RNAs in archaea with short 5' UTRs [65], we searched for downstream antisense TSS (daTSS). They were defined as transcript start sites located on the opposite strand of annotated genes, located up to 200 nt downstream of the 3' end of a gene. We found 80 daTSS (Table S4).

Having strand-specific paired-end data available, we could search for genome positions at which antisense transcripts preferentially terminate. Such presumed RNA 3' ends, if downstream of an aTSS (RNA 5' end), were inferred to represent the end of an asRNA locus (see Methods 2.2). We were able to define 846 asRNAs loci distributed in 613 genes, starting from aTSS or daTSS (Table S4). Most of the genes contain only one asRNA locus and 26 of the asRNAs are antisense to two genes. The size distribution of asRNAs is shown in Figure 1b, most of them being smaller than 100 nt.

We analyzed the frequency of nucleotides around aTSSs and daTSSs (Figure S3a) and observed the predominance of purines at the TSS, pyrimidines at position −1 and a BRE-TATA signature, showing a characteristic archaeal promoter region for the mapped asRNAs. The dinucleotide composition around the TSS is postulated as important for the transcription initiation among organisms from all domains [66–68], as well as the GC enrichment around the −36 position for TFIIB binding and TATA-box at −26 [69,70]. As previously shown, dRNA-seq data reliably identifies TSS genome-wide in *H. salinarum* NRC-1 [38], recapitulating known features of the promoter region. Nevertheless, this does not exclude the possibility that many of the asRNAs identified are products of spurious transcription.

For asRNAs with at least 10 fragments that corroborated the 3' end annotation, we analyzed the frequency of nucleotides 10 nt upstream and downstream of the 3' end. We observed an enrichment of pyrimidines, specially uracil at positions 0 and −1 (Figure S3b). Comparing this signature with poly-U signatures found in other archaea [71], our data shows a much shorter region, which is unlikely to form a secondary structure involved in transcription termination of asRNAs.

Most of the characterized sRNAs in prokaryotes bind to the 5' region of the mRNA, pairing with the start codon or ribosome binding site (RBS) in the 5' UTR [17], and thus inhibiting the access of the ribosomal machinery. asRNAs that overlap these regions can be candidates to act through a similar mechanism. Halophiles such as *H. salinarum* NRC-1 and *H. volcanii* have mostly leaderless transcripts [21,37,72]. However, it has been experimentally shown in *Salmonella* that ncRNAs overlapping nucleotides up to the 5th codon in the mRNA are capable of blocking the translation machinery, even without AUG or Shine–Dalgarno pairing [73]. We thus looked for asRNAs that overlap their cognate gene's start codon or at least 12 nt (four codons) downstream of it. We found 145 asRNAs in *H. salinarum* overlapping the 5' end of genes (Table S5), which could, in principle, impair mRNA translation by occluding ribosome binding. Figure 2 and Figure S4 show examples of putative RBS occlusion by an asRNA: *gcvP1*, encoding a glycine dehydrogenase subunit (VNG_RS06215); *cdcH*, encoding an AAA-type ATPase (VNG_RS06465); and *rpl1*, which encodes the 50S ribosomal protein L1 (VNG_RS04315). The gene *gcvP1* is the first from an operon and has a strong Shine–Dalgarno-like signature at −19 upstream of the start codon, colocalized with asRNA VNG_as06215_888 (Figure 2). Similarly, RBS at −11 and −13 upstream of *rpl1* and *cdcH* are colocalized with asRNA VNG_as04315_654 and VNG_as06465_925, respectively (Figure S4).

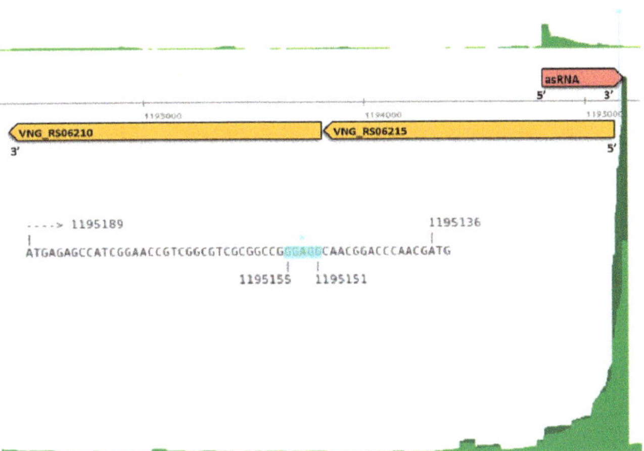

Figure 2. Antisense RNA in 5′ untranslated region (UTR). *gcvP1*, encoding a glycine dehydrogenase subunit (VNG_RS06215). VNG_RS06215 locus (orange arrow) is in reverse strand (5′→3′ right to left), neighbor gene VNG_RS06210 (orange arrow) is also in reverse strand. Differential RNA-sequencing (dRNA-seq) read coverage signal is shown in dark and light green for TEX+ and TEX− libraries, respectively. Coverage signals below and above the central axis are for reverse and forward strands, respectively. VNG_as06215_888 asRNA (pink arrow) encompasses Shine–Dalgarno-like signature (* light blue highlight in genome coordinates and zoomed in sequence).

All data are available for browsing in Gaggle Genome Browser format (interactive versions of Figure 2 and similar outputs) at http://labpib.fmrp.usp.br/~{}rvencio/asrna/.

3.2. Antisense RNAs Expression Levels

Some of the asRNAs with characterized functions are expressed at levels equivalent or higher than the mRNA on the opposite strand, which would be expected if a dsRNA is necessary for post-transcriptional regulation [74]. However, most of the asRNAs identified in eukaryotes and prokaryotes are expressed at low levels, which has challenged their identification before high-resolution sequencing methods were available [2,16,60]. Low expression levels might indicate that these asRNAs are products of spurious transcription from low complexity promoter regions [16]. However, this does not exclude the possibility of a functional RNA, given that even at low levels, asRNAs can present a buffering effect and fine-tune gene regulation [2,3].

We observe a negative correlation between asRNA and mRNA transcripts when we compare the average read counts for these transcripts (Figure S5a), as observed in *H. volcanii* [21]. Using RNA-seq data, we analyzed the relationship between the fold change of the read counts of asRNAs relative to the mRNA on the opposite strand (Figure 3). We observed that most of the asRNAs annotated in our study present low expression levels relative to the gene on the opposite strand. Only 112 asRNAs (~13%, Table S6) present expression levels equal or greater than the gene on the opposite strand in at least one of the conditions analyzed. These molecules could be candidates for *cis*-regulators of their respective cognate genes in the considered experimental conditions.

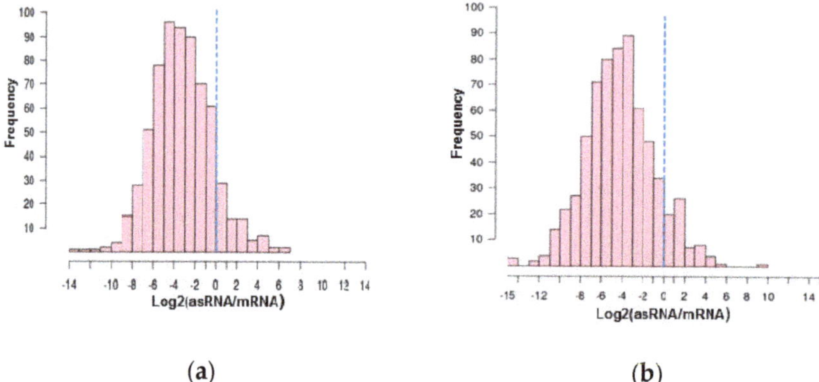

(a)   (b)

Figure 3. Relative expression levels of asRNAs and messenger RNA (mRNAs) on the opposite strand for arbitrarily selected representative libraries: (**a**) stationary phase (17 h) and (**b**) gas vesicle release phase (86 h). Vertical dotted lines mark 1:1 expression levels. Expanded version in Figure S5.

We also evaluated the expression profiles of the asRNAs and cognate genes over the growth curve of *H. salinarum* NRC-1 using RNA-seq libraries sampled at three time points [38]. We compared the expression levels from (i) the stationary phase relative to exponential phase (17 h vs. 37 h) and (ii) late exponential gas vesicle release phase relative to stationary phase (37 h vs. 86 h).

In the first transition, we observed 93 asRNA differentially expressed (Table S7). For 26 asRNAs, the gene on the opposite strand is also differentially expressed (Figure S6a), and six asRNAs overlap the 5′ UTR of the genes. In the second transition, we observed 63 asRNAs differentially expressed (Table S8). For 30 cases, the gene on the opposite strand is also differentially expressed. Twenty-seven pairs were differentially expressed in the same direction and only three in opposite directions (Figure S6b). Of them, seven asRNAs overlap the 5′ UTR region. Overall, 56 pairs of gene-asRNAs are differentially expressed in at least one of the transitions analyzed. These asRNAs can be potential *cis* regulators of the cognate genes.

3.3. Function of the Genes on the Opposite Strand of Antisense RNAs

From the total of 613 genes that present at least one asRNA locus, 198 are hypothetical proteins and 32 are nonredundant transposases (i.e., some of them are multicopy, but counted only once). Gene enrichment analysis returned no overrepresented Gene Ontology (GO) term relative to the whole genome background distribution. Genes with asRNAs are associated with different functions according to GO categorization (Figure S7). Next, we present the results for some of the gene categories.

H. salinarum NRC-1 is known to produce intracellular gas vesicles: structures composed of proteins that are filled with gas and allow floating on the media surface. Their synthesis and regulation have been studied in details and includes multiple layers of regulation [75]. We were able to identify 11 asRNAs in the *gvp* gene cluster (Table 1), including one previously identified on the opposite strand of *gvpD* [19], which encodes a repressor of gas vesicle production. Most asRNAs related to *gvp* genes showed similar expression levels relative to their cognate genes.

Table 1. asRNAs in *gvp* gene cluster, located in pNRC100. (nt: Nucleotide).

aTSS ID	asRNA ID	Strand	Start	End	Size (nt)	Locus	Annotation
aTSS_1555	VNG_as12280_1555	+	16743	16863	120	VNG_RS12280	*gvpL*
aTSS_1556	VNG_as12280_1556	+	17055	17138	83	VNG_RS12280	*gvpL*
aTSS_1557	VNG_as12290_1557	+	18092	18168	76	VNG_RS12290	*gvpJ*
aTSS_1558	VNG_as13760_1558	+	18390	18428	38	VNG_RS13760	*gvpI*
aTSS_1559	VNG_as13760_1559	+	18615	18643	28	VNG_RS13760	*gvpI*
aTSS_1560	VNG_as12295_1560	+	18698	18722	24	VNG_RS12295	*gvpH*
aTSS_1565	VNG_as12315_1565	+	21000	21165	165	VNG_RS12315	*gvpD*
aTSS_1567	VNG_as12315_1567	+	22084	22106	22	VNG_RS12315	*gvpD*
aTSS_1568	VNG_as12315_1568	+	22128	22292	164	VNG_RS12315	*gvpD*
aTSS_1569	VNG_as12325_1569	−	22865	22964	99	VNG_RS12325	*gvpC*
aTSS_1570	VNG_as12325_1570	−	23888	23940	52	VNG_RS12325	*gvpC*

AsRNAs to the *gvpD* gene have been detected by Krüger and Pfeifer [19], complementary to the 5′ and 3′ end of the gene and detected when GvpD protein was present at low levels. We were able to identify three RNAs antisense to *gvpD*, one of them (VNG_as12315_1568) overlaps the 5′ end that recapitulates previously published information. This asRNA was annotated as a 164 nt molecule in our work, presenting a similar size to the 190 nt band observed in [19]. The aTSS mapped is located 3 nt downstream to the 5′ end of the probe used in the previous work, showing the reliability of our asRNA mapping (Figure S8). We also identified primary asRNAs colocalized with genes *gvpC*, *gvpH*, *gvpI*, *gvpJ*, *gvpL*, and a strong signal antisense to the 5′ end of *gvpA*, although no aTSS was mapped (Figure S8). These data indicate asRNAs as important players in gas vesicle regulation in *H. salinarum*, as reported in cyanobacteria *Calothrix* sp. PCC 7601 [76].

Moreover, we were able to identify RNAs possibly involved in rhodopsin regulation in *H. salinarum*: asRNAs in *bop* (VNG_RS05715—bacteriorhodopsin) and its regulators *brz* (VNG_RS05710—bacteriorhodopsin regulating zinc finger protein) and *brb* (OE3105F bacteriorhodopsin regulating basic protein), in addition to an asRNA in halorhodopsin (VNG_RS00745) (Table 2). Brb fine-tunes the activation of *bop*; this activity was experimentally shown using reporter genes and mutagenesis. The Brb protein was proven to exist, but was not detected using mass spectrometry [77]. The asRNA overlapping bacteriorhodopsin regulators (VNG_da3105F_36) starts downstream of *brb*, overlaps it completely, and ends inside *brz* gene (Figure S9). In the RNA-seq libraries analyzed in this study, the number of reads for this asRNA was approximately 4-fold higher than for *brb* gene. If asRNAs should be present at high levels to post-transcriptionally regulate an mRNA, this data could account for the difficulty in Brb protein detection since asRNA–mRNA pairing could block translation, indicating a possible role for this asRNA. The presence of asRNAs in bacteriorhodopsin and halorhodopsin genes could indicate additional regulators to be studied for understanding the photobiology of this organism.

Table 2. asRNAs in rhodopsin related genes, located in the main chromosome.

aTSS ID	asRNA ID	Strand	Start	End	Size (nt)	Locus	Annotation
aTSS_175	VNG_as00745_175	+	155806	155906	100	VNG_RS00745	*halorhodopsin*
*daTSS_36	VNG_da3105F_36	−	1088797	1089100	303	VNG_RS05710 VNG_OE3105F	*brz—bacteriorhodopsin regulating zinc finger protein; brb—bacteriorhodopsin-regulating basic protein*
aTSS_824	VNG_as05715_824	−	1089545	1089615	70	VNG_RS05715	*bacteriorhodopsin*

* The asRNA VNG_da3105F_36 overlaps two genes.

In bacteria, type I TA systems are known for the presence of an asRNA acting as an antitoxin, while in type II TA systems, both toxin and antitoxin are known to be proteins [78]. In 2014, Lybecker et al. [61] found asRNAs to type II TA systems in *E. coli*, indicating a possible role for asRNAs in these systems. In *H. salinarum* NRC-1, we identified asRNAs to genes VNG_RS11240 and VNG_RS00140, which are annotated type II antitoxins.

The presence of asRNAs in type II TA systems has not been systematically explored and could indicate another layer of regulation for these systems. To verify the compatibility of TSS identification and asRNAs identified by Lybecker et al. using dsRNA immunoprecipitation [61], we reanalyzed dRNA-seq data from *E. coli* [60], and we were able to detect the aTSS corresponding to the asRNAs reported by Lybecker et al. [61] in type II TA systems. Given that aTSS identification was reliable in *E. coli*, we looked for asRNAs in type II TA systems in archaea by precisely annotating these genes using TA finder 2.0 [59] and reanalyzing available dRNA-seq data for *H. salinarum* NRC-1 (PRJNA448992) [38], *H. volcanii* DS2 (PRJNA324298) [21], *M. jannaschii* DSM 2661 (PRJNA342613) [26], *T. kodakarensis* KOD1 (PRJNA242777) [23], and *T. onnurineus* NA1 (PRJNA339284) [25]. We annotated new type II TA systems in archaea (Table S9), including nine complete pairs in *H. salinarum*. Both genes composing one of these new pairs, VNG_RS11890 (toxin) and VNG_RS11895 (antitoxin), have asRNAs displaying expression levels higher than the cognate genes (Figure S10). Most of the annotated type II TA systems have at least one aTSS in one of the genes. For *T. kodakarensis* and *M. jannaschii*, aTSSs are predominantly in the toxin genes.

We found asRNAs for 37 genes related to translation process, including ribosomal proteins, translation initiation factors, transfer RNA (tRNA) ligases and tRNAs, Asn, Lys, and Ser (Table S10). There are reports of RNAs antisense to tRNAs in *S. solfataricus* [79] and *T. kodakarensis* [23], which indicate a conserved regulatory role for these molecules.

Interestingly, 58 out of 846 asRNAs are overlapping transposase genes in *H. salinarum* NRC-1 (Table S11). Since transposases are usually encoded within repetitive elements called insertion sequences (IS), we eliminated redundancy in numbers by choosing only one representative element for multicopy entities. AsRNAs in transposases have been reported in other archaea such as *H. volcanii*, *T. kodakarensis*, *S. solfataricus*, and *M. mazei* [20]. Retrieving legacy data from tiling microarray experiments performed along *H. salinarum* growth curve [37], we verified that several transcripts antisense to IS are differentially expressed (Figure S11).

In bacteria, there are examples of asRNA inhibiting the translation of transposase mRNAs by occluding the ribosomal machinery assembly at the 5′ end of an mRNA [80]. We found 10 asRNAs that overlap the 5′ end of transposase coding gene (highlighted in Table S11), which could be potential candidates for a similar regulatory mechanism.

3.4. Ribosome-Associated Antisense RNAs

We reanalyzed Ribo-seq data, (BioProject PRJNA413990), to identify asRNAs that are putative ribosome-associated ncRNAs (rancRNAs) [81]. By looking at asRNA loci covered by at least 20 reads along at least half its extension, we identified 91 asRNAs (~11%) with relevant signal (Table S12). Recently published rancRNA data in *H. volcanii* found 68 candidates antisense to genes, ~6% of their total [82].

Interestingly, 11 asRNAs with relevant Ribo-seq signal are located in transposases (representative instances highlighted in Table S12). This might indicate that these asRNAs can be either targets for translation or regulate/interfere with the ribosomal machinery. Given that years of *H. salinarum* proteomics studies refuted a widespread colocalization of open reading frames in both strands as spurious "overprediction" [52], the regulatory ribosome binding scenario seems more plausible.

3.5. Conservation of Antisense Transcription Start Sites

Transcriptome analysis of bacteria has shown that the conservation of asRNAs even among phylogenetically close organisms is low. The comparison between *E. coli* and *Salmonella enterica* serovar Typhimurium showed that only 14% of the asRNAs are conserved [83]. The number of conserved aTSSs varies in different organisms: comparison between *Campylobacter* strains showed 45% conservation [84]; within eight different species of *Shewanella* genus, 22% [85]; and among *Synechocystis* strains, only 4% [86].

To evaluate the conservation of aTSS, we compared the identified positions in *H. salinarum* with dRNA-seq data reanalysis of *H. volcanii*. First we sub-selected pairs of orthologous genes in both halophiles (1554 pairs) [64]. Then, from these groups of orthologous genes, we sub-selected pairs with an annotated asRNA in *H. salinarum*'s genes and at least one aTSS in its correspondent ortholog in *H. volcanii*, obtaining 244 pairs. We normalized genes sizes by coding sequences (CDS) length defining $d = 0$ at start codons and $d = 100$ at stop codons. We arbitrarily partitioned CDS in 10 regions depending on d: (0;10), (10;20), and so on up to (90;100). Then, we searched for annotated asRNAs in *H. salinarum*'s genes and aTSSs in their correspondent ortholog in *H. volcanii* located in the same "equivalent" region. We were able to identify 160 asRNAs, distributed in 110 *H. salinarum*'s genes that contain at least one aTSS in the same tenth region of its ortholog gene in *H. volcanii* (Table S13), representing ~19% of the annotated *H. salinarum* asRNAs. We deemed an aTSS as a conserved feature if its relative position falls into the same tenth partition in both organisms. The distribution of differences between relative positions ($D = |d_{Hsal} - d_{Hvol}|$) shows that the majority of such 160 conserved aTSS are in fact positionally equivalent ($D < 3$) (Figure 4).

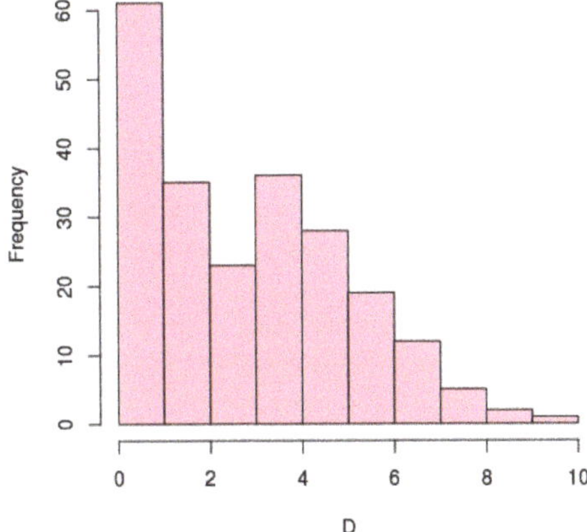

Figure 4. Distribution of differences (D) between relative positions among aTSS conserved in *H. salinarum* NRC-1 and *H. volcanii* DS2. Positions are coding sequences (CDS) length-normalized thus $D = 0$ mean same position in both organisms.

We found only seven genes in which aTSS are found exactly at the same relative position ($D = 0$): *aglM2*, *rpl10e*, *eEF1A*, *rpoB1*, *ndhG4*, *nirH*, and *atpI*. Two of them are also equal in absolute positions since they have the same length in both species (highlighted in Table S13).

Currently, *H. volcanii* is the only other halophilic archaeon for which primary transcriptome is available (dRNA-seq data). Regular strand-specific comparative transcriptome analysis (RNA-seq) is available for other halophiles but only one, in *Natrinema* sp. J7-2 (formerly *H. salinarum* J7), focused on salinity adaptation questions [87]. Since regular RNA-seq does not have the resolution power to pinpoint aTSS as dRNA-seq, we inspected differential read distributions between *Natrinema* sp. J7-2 cultures grown on relatively low vs. high salt concentrations (15% vs. 30% NaCl) guided by the *H. salinarum* and *H. volcanii* conservation results. From the seven aforementioned genes bearing conserved aTSS, only *nirH* (which encodes a sirohaem decarboxylase) is clearly upregulated in low salt relative to high salt, while its putative conserved asRNA is upregulated in high salt relative to low salt concentrations (Figure S12).

Recently published oxidative stress data in *H. volcanii* [27] was investigated to find an intersection between differentially expressed asRNAs in *H. salinarum* NRC-1. Among the 160 conserved asRNAs, four and 15 were down- and upregulated in the *H. volcanii* oxidative stress response (Table S14). From these, two were also found differentially expressed in 37 h/17 h transition: antisense to *rmeS*, encoding restriction endonuclease subunit (VNG_RS00420), and *pyrG*, encoding a CTP synthase (VNG_RS07090). Analogously, three were found in the 86 h/37 h transition: antisense to a ISH8-type transposase (VNG_RS00205); *lon*, encoding an archaeal Lon protease (VNG_RS01200); and *csg*, encoding the S-layer glycoprotein of *H. salinarum* (VNG_RS10505).

4. Discussion

In this study, we were able to define a map of the primary antisense transcriptome in *H. salinarum* NRC-1, corroborating the pervasiveness of antisense transcripts in this organism. The first step was to reanalyze dRNA-seq data, where RNAs with triphosphorylated 5′ ends were enriched by treatment with terminator exonuclease and then compared to untreated samples. It was possible to precisely identify aTSSs, which were further filtered to remove possible artifacts due to secondary structure resulting in 1626 aTSSs. We could observe a typical archaeal promoter structure upstream of the mapped aTSSs (Figure S3a), showing that we can confidently identify these positions. In *H. volcanii*, it was reported that ~30% of the genes present an aTSS during exponential growth [21], in *H. salinarum* we observed similar figure of ~34%. We also observed a similar distribution of aTSS relative to the sense gene with an accumulation at the 3′ end (Figure 1a), suggesting that in halophiles aTSS are preferentially located at this region.

Even with high resolution genomic techniques, the definition of the precise 3′ end of a transcript can be challenging since the signal decrease is usually not very sharp and termination signatures are still being discovered, especially in archaea [71]. In the studies using dRNA-seq to analyze the primary transcriptome of *E. coli* [60] and *H. volcanii* [21], the size of the asRNAs were arbitrarily defined as 50 and 100 nt, respectively. Visual inspection of the RNA-seq signal has been used as a valid approach to infer the 3′ end of diverse ncRNAs in bacteria and archaea [27,67,88,89]. In this work, we not only used the visual inspection of 10 RNA-seq libraries, but also the information of paired-end sequencing, which is routinely used to map transcriptomes and genomes in eukaryotes [90]. This approach allowed us to increase the confidence in the inference of the possible 3′ end, defining the minimum size of the asRNA. We observed a short uracil enrichment at the two last positions of the asRNAs (Figure S3b), which might indicate some sort of termination signature, although it is too short to form a termination structure as previously reported in Santangelo and Reeve [91] and Dar et al. [71]. The application of different library preparation methods such as transcription termini sequence (Term-seq) or long reads sequencing [71,92] can help infer the exact size of a transcript to study their properties in the future.

We defined 846 asRNA loci with an aTSS and a 3′ end that marks their minimum length; this is probably the minimum list of asRNA primary transcriptome of *H. salinarum*. There are several other antisense signals, such as aTSS, for which we were not able to define a 3′ end, long 3′ UTR regions of mRNAs, or RNAs that are not primary (possibly processed). An example is shown in Figure S13, where a long 3′ UTR of a coding sequence can act as an asRNA for a translation initiation factor and Figure S8 where a strong signal antisense to *gvpA* is detected, although no TSS could be mapped. Most of the characterized sRNAs in prokaryotes bind to the 5′ UTR of mRNA, inhibiting ribosome binding, and thus blocking translation [17]. However, in *H. salinarum* most of the transcripts are leaderless, which could account for the overrepresentation of asRNAs overlapping 3′ ends of mRNAs (Figure 1a). Nevertheless, we identified 145 asRNAs that could impair translation by overlapping 5′ end region of the mRNA.

Although unquestionably prevalent, asRNAs are usually expressed at low levels, which brings to discussion whether these molecules are functional or by-products of a noisy transcription process. If an asRNA has to pair with an mRNA and form a dsRNA to perform a post-transcriptional regulation, it would be expected that, given stoichiometric considerations, asRNAs would be present at similar levels of the target RNAs [74]. In our study, only 112 (~13%) of the asRNAs present expression equal to or greater than the cognate gene. Many bacterial asRNAs can be spurious transcripts that would be maintained in the genome due to the absence of a negative selection (low energy cost or no deleterious effects) [16]. However, low expression levels do not exclude the possibility of a functional molecule, given that they could interact with different proteins or present a buffering effect to fine-tune mRNA regulation [2]. Even if expressed at low levels and possibly non-functional, many of the asRNAs are detected in different conditions by transcriptomics methods, indicating that they could be a source for evolutionary processes that originate new regulatory elements, as molecular exaptations [93].

Concerning the function of genes with an asRNA, we could find many interesting examples that might present antisense regulation. We were able to recapitulate previously reported information on asRNA to *gvpD* gene, adding the information that this is a primary transcript. Another 10 asRNAs in the *gvp* gene cluster were identified, suggesting a post-transcriptional regulation of these genes yet to be described. In genes related to bacteriorhodopsin, we found a long asRNA that, if it hybridizes with *brz* and *brb* mRNAs, could block their translation, which could explain the difficulty in experimentally detecting Brb protein in physiological conditions [77]. Type II TA systems were found to present aTSS not only in *H. salinarum* but also in three other archaea with available dRNA-seq data, suggesting a conserved role of asRNAs in type II TA systems. AsRNAs were also found in 37 genes related to translation and 32 transposases. For this last group 31% has a clear overlap with the 5′ end, suggesting the mechanism of translation inhibition of these genes. We found that ~11% of the asRNAs interact with ribosomes. This could indicate that these asRNAs are either encoding unknown proteins or regulating ribosomal machinery in yet to be described ways.

Expression levels of asRNA–mRNA pairs were evaluated over a growth curve, and we found that 17% of annotated asRNAs show either positive or negative correlation with the expression of the cognate gene. Among these pairs, we found a toxin from a type II TA system (VNG_RS00140) upregulated while the asRNA (VNG_da00140_3) is downregulated (Figure S6a). This negative correlation is usually found in type I TA systems [4], where the asRNA is an antitoxin. We found 27 hypothetical proteins differentially expressed with their asRNA also differentially expressed. In *H. volcanii*, differential expression during oxidative stress identified 48% of the genes that encoded hypothetical proteins, along with their respective asRNAs [27]. This suggests a probable regulation of these genes by asRNAs in halophiles, although the impact of these genes is still unknown. Although differential expression analysis is commonly performed when studying asRNAs, it is important to note that this is a simplistic approach, since sense and antisense gene regulation might be completely independent [94] and mRNA levels are not necessarily correlated to protein levels [95]. This list of differentially expressed asRNAs–mRNAs can be a first approach to select targets for further experimental validation in *H. salinarum*.

Conservation of asRNAs is usually reported as very low, even among closely related organisms. We found that only ~19% of the aTSSs with asRNAs annotated are conserved between *H. salinarum* and *H. volcanii*, with two examples where they are found at exactly the same location. To infer the functionality of a given gene, a premise is that along the evolutionary process the sequence will be conserved between species. The low conservation of asRNAs, as well as their low expression levels, raise the question about the functionality of most asRNAs identified by high-throughput methods [85]. Higher evolution rates in ncRNAs could be related to a faster adaptation of organisms to remodel regulatory pathways to generate specific responses, which could be applied to the asRNAs [2,86].

Supplementary Materials: The following are available online at http://www.mdpi.com/2073-4425/10/4/280/s1: Figure S1: Pipeline for asRNA annotation, Figure S2: asRNA loci inference approach, Figure S3: asRNA properties, Figure S4: Examples of putative RBS occlusion by an asRNA, Figure S5: Coverage of asRNAs relative to mRNAs on the opposite strand, Figure S6: Differential expression of asRNAs and mRNAs on the opposite strand, Figure S7: Functional categorization of genes presenting asRNAs according to Gene Ontology (GO), Figure S8: asRNAs in *gvp* genes, Figure S9: asRNAs in *brb* and *brz* genes, Figure S10: asRNAs in type II TA system, Figure S11: asRNAs antisense to a transposase differentially expressed along the growth curve, Figure S12: Differential expression of putative asRNAs antisense to the *nirH* gene in *Natrinema* sp. J7-2, Figure S13: Long 3′ UTR of a coding sequence acting as an asRNA, Table S1: Genome annotation of *H. salinarum* NRC-1 (RefSeq updated in 2017) with additional manual curation using *H. salinarum* R1 annotation as reference, Table S2: RNA-seq data used to annotate asRNA loci, Table S3: aTSS mapped in *H. salinarum* NRC-1, Table S4: asRNA loci mapped in *H. salinarum* NRC-1, Table S5: asRNAs overlapping 5′ end of genes, Table S6: asRNAs with expression levels equal or greater than the gene on the opposite strand, Table S7: asRNAs differentially expressed in the stationary phase relative to exponential phase (17 h vs 37 h), Table S8: asRNAs differentially expressed in the late exponential gas vesicle release phase relative to stationary phase (37 h vs 86 h), Table S9: Annotation and aTSSs in type II TA systems in archaea, Table S10: asRNAs in genes related to translation, Table S11: asRNAs in genes encoding transposases, highlighted asRNAs overlap the 5′ end of transposases, Table S12: Ribosome associated asRNAs, Table S13: aTSSs comparison between *H. salinarum* and *H. volcanii,* Table S14: Conserved asRNAs differentially expressed in *H. volcanii* during oxidative stress

Author Contributions: Conceptualization, T.K. and R.Z.N.V.; Methodology, R.Z.N.V., J.P.P.A., A.P.R.L., F.t.-C., AND J.V.G.-F.; Software: J.P.P.A., A.P.R.L., and R.Z.N.V.; Validation, J.P.P.A. and A.P.R.L.; Formal Analysis, J.P.P.A. and R.Z.N.V.; Investigation, J.P.P.A. and R.Z.N.V.; Resources, T.K.; Data Curation, J.P.P.A., A.P.R.L., and R.Z.N.V.; Writing—Original Draft Preparation, T.K., R.Z.N.V., and J.P.P.A.; Writing—Review and Editing, T.K., R.Z.N.V., J.P.P.A and A.P.R.L; Visualization, J.P.P.A.; Supervision, T.K. and R.Z.N.V.; Project Administration, T.K.; Funding Acquisition, T.K.

Funding: This research was funded by FAPESP—Fundação de Amparo a Pesquisa do Estado de São Paulo, grant numbers [2015/21038-1], [2013/21522-5], [2017/03052-2], [2011/14455-4], and [2015/12012-9]. Fundação de Apoio ao Ensino, Pesquisa e Assistência do Hospital das Clínicas da Faculdade de Medicina de Ribeirão Preto da Universidade de São Paulo [415/2018]; Coordenação de Aperfeiçoamento de Pessoal de Nível Superior—Brasil (CAPES) [Finance Code 001].

Acknowledgments: The authors would like to thank Silvia Helena Epifânio for technical support and the Project Management of Ribeirão Preto Medical School for administrative assistance. We thank former laboratory members Livia S. Zaramela and Diego M. Salvanha for early discussions on asRNAs project. We also thank Nitin Baliga's laboratory at the Institute for Systems Biology for providing Ribo-seq data access prior to publication.

Conflicts of Interest: The authors declare no conflict of interest. The funders had no role in the design of the study; in the collection, analyses, or interpretation of data; in the writing of the manuscript, or in the decision to publish the results.

References

1. Sayed, N.; Jousselin, A.; Felden, B. A cis-antisense RNA acts in *trans* in *Staphylococcus aureus* to control translation of a human cytolytic peptide. *Nat. Struct. Mol. Biol.* **2012**, *19*, 105–113. [CrossRef]

2. Pelechano, V.; Steinmetz, L.M. Gene regulation by antisense transcription. *Nat. Rev. Genet.* **2013**, *14*, 880–893. [CrossRef]

3. Lasa, I.; Toledo-Arana, A.; Dobin, A.; Villanueva, M.; de Los Mozos, I.R.; Vergara-Irigaray, M.; Segura, V.; Fagegaltier, D.; Penadés, J.R.; Valle, J.; et al. Genome-wide antisense transcription drives mRNA processing in bacteria. *Proc. Natl. Acad. Sci. USA* **2011**, *108*, 20172–20177. [CrossRef]

4. Kawano, M.; Aravind, L.; Storz, G. An antisense RNA controls synthesis of an SOS-induced toxin evolved from an antitoxin. *Mol. Microbiol.* **2007**, *64*, 738–754. [CrossRef]

5. Carrieri, C.; Cimatti, L.; Biagioli, M.; Beugnet, A.; Zucchelli, S.; Fedele, S.; Pesce, E.; Ferrer, I.; Collavin, L.; Santoro, C.; et al. Long non-coding antisense RNA controls Uchl1 translation through an embedded SINEB2 repeat. *Nature* **2012**, *491*, 454–459. [CrossRef]

6. Bøvre, K.; Szybalski, W. Patterns of convergent and overlapping transcription within the b2 region of coliphage λ. *Virology* **1969**, *38*, 614–626. [CrossRef]

7. Inouye, M. Antisense RNA: Its functions and applications in gene regulation—A review. *Gene* **1988**, *72*, 25–34. [CrossRef]

8. Vanhée-Brossollet, C.; Vaquero, C. Do natural antisense transcripts make sense in eukaryotes? *Gene* **1998**, *211*, 1–9. [CrossRef]

9. Lasa, I.; Toledo-Arana, A.; Gingeras, T. An effort to make sense of antisense transcription in bacteria. *RNA Biol.* **2012**, *9*, 1039–1044. [CrossRef]

10. Levin, J.Z.; Yassour, M.; Adiconis, X.; Nusbaum, C.; Thompson, D.A.; Friedman, N.; Gnirke, A.; Regev, A. Comprehensive comparative analysis of strand-specific RNA sequencing methods. *Nat. Methods* **2010**, *7*, 709–715. [CrossRef]

11. Sharma, C.M.; Vogel, J. Differential RNA-seq: The approach behind and the biological insight gained. *Curr. Opin. Microbiol.* **2014**, *19*, 97–105. [CrossRef]

12. Sun, Y.; Li, D.; Zhang, R.; Peng, S.; Zhang, G.; Yang, T.; Qian, A. Strategies to identify natural antisense transcripts. *Biochimie* **2017**, *132*, 131–151. [CrossRef]

13. Beiter, T.; Reich, E.; Williams, R.W.; Simon, P. Antisense transcription: A critical look in both directions. *Cell. Mol. Life Sci.* **2009**, *66*, 94–112. [CrossRef]

14. Georg, J.; Hess, W.R. Cis-antisense RNA, another level of gene regulation in bacteria. *Microbiol. Mol. Biol. Rev.* **2011**, *75*, 286–300. [CrossRef]

15. Wade, J.T.; Grainger, D.C. Pervasive transcription: Illuminating the dark matter of bacterial transcriptomes. *Nat. Rev. Microbiol.* **2014**, *12*, 647–653. [CrossRef]

16. Lloréns-Rico, V.; Cano, J.; Kamminga, T.; Gil, R.; Latorre, A.; Chen, W.H.; Bork, P.; Glass, J.I.; Serrano, L.; Lluch-Senar, M. Bacterial antisense RNAs are mainly the product of transcriptional noise. *Sci. Adv.* **2016**, *2*, e1501363. [CrossRef]

17. Wagner, E.G.H.; Romby, P. Chapter three-small RNAs in bacteria and archaea: Who they are, what they do, and how they do it. *Adv. Genet.* **2015**, *90*, 133–208.

18. Eckweiler, D.; Häussler, S. Antisense transcription in *Pseudomonas aeruginosa*. *Microbiology* **2018**, *164*, 889–895. [CrossRef]

19. Krüger, K.; Pfeifer, F. Transcript analysis of the *c-vac* region and differential synthesis of the two regulatory gas vesicle proteins GvpD and GvpE in *Halobacterium salinarium* PHH4. *J. Bacteriol.* **1996**, *178*, 4012–4019. [CrossRef]

20. Gelsinger, D.R.; DiRuggiero, J. The non-coding regulatory RNA revolution in archaea. *Genes* **2018**, *9*, 141. [CrossRef]

21. Babski, J.; Haas, K.A.; Näther-Schindler, D.; Pfeiffer, F.; Förstner, K.U.; Hammelmann, M.; Hilker, R.; Becker, A.; Sharma, C.M.; Marchfelder, A.; et al. Genome-wide identification of transcriptional start sites in the haloarchaeon *Haloferax volcanii* based on differential RNA-Seq (dRNA-Seq). *BMC Genom.* **2016**, *17*, 629. [CrossRef]

22. Li, J.; Qi, L.; Guo, Y.; Yue, L.; Li, Y.; Ge, W.; Wu, J.; Shi, W.; Dong, X. Global mapping transcriptional start sites revealed both transcriptional and post-transcriptional regulation of cold adaptation in the methanogenic archaeon *Methanolobus psychrophilus*. *Sci. Rep.* **2015**, *5*, 9209. [CrossRef] [PubMed]

23. Jäger, D.; Förstner, K.U.; Sharma, C.M.; Santangelo, T.J.; Reeve, J.N. Primary transcriptome map of the hyperthermophilic archaeon *Thermococcus kodakarensis*. *BMC Genom.* **2014**, *15*, 684. [CrossRef] [PubMed]

24. Jäger, D.; Sharma, C.M.; Thomsen, J.; Ehlers, C.; Vogel, J.; Schmitz, R.A. Deep sequencing analysis of the *Methanosarcina mazei* Go1 transcriptome in response to nitrogen availability. *Proc. Natl. Acad. Sci. USA* **2009**, *106*, 21878–21882. [CrossRef]

25. Cho, S.; Kim, M.-S.; Jeong, Y.; Lee, B.-R.; Lee, J.-H.; Kang, S.G.; Cho, B.-K. Genome-wide primary transcriptome analysis of H2-producing archaeon *Thermococcus onnurineus* NA1. *Sci. Rep.* **2017**, *7*, 43044. [CrossRef]

26. Smollett, K.; Blombach, F.; Reichelt, R.; Thomm, M.; Werner, F. A global analysis of transcription reveals two modes of Spt4/5 recruitment to archaeal RNA polymerase. *Nat. Microbiol.* **2017**, *2*, 17021. [CrossRef] [PubMed]

27. Gelsinger, D.R.; DiRuggiero, J. Transcriptional landscape and regulatory roles of small noncoding RNAs in the oxidative stress response of the Haloarchaeon *Haloferax volcanii*. *J. Bacteriol.* **2018**, *200*, e00779-17. [CrossRef] [PubMed]

28. Gunde-Cimerman, N.; Plemenitaš, A.; Oren, A. Strategies of adaptation of microorganisms of the three domains of life to high salt concentrations. *FEMS Microbiol. Rev.* **2018**, *42*, 353–375. [CrossRef]

29. Oesterhelt, D.; Stoeckenius, W. Rhodopsin-like protein from the purple membrane of *Halobacterium halobium*. *Nat. New Biol.* **1971**, *233*, 149–152. [CrossRef] [PubMed]

30. Oren, A.; Trüper, H. Anaerobic growth of halophilic archaeobacteria by reduction of dimethysulfoxide and trimethylamine N-oxide. *FEMS Microbiol. Lett.* **1990**, *70*, 33–36. [CrossRef]

31. Ruepp, A.; Soppa, J. Fermentative arginine degradation in *Halobacterium salinarium* (Formerly *Halobacterium halobium*): Genes, gene products, and transcripts of the arcRACB gene cluster. *J. Bacteriol.* **1996**, *178*, 4942–4947. [CrossRef] [PubMed]

32. Kaur, A.; Pan, M.; Meislin, M.; Facciotti, M.T.; El-Gewely, R.; Baliga, N.S. A systems view of haloarchaeal strategies to withstand stress from transition metals. *Genome Res.* **2006**, *16*, 841–854. [CrossRef]

33. Coker, J.A.; DasSarma, P.; Kumar, J.; Müller, J.A.; DasSarma, S. Transcriptional profiling of the model Archaeon *Halobacterium sp.* NRC-1: Responses to changes in salinity and temperature. *Saline Syst.* **2007**, *3*, 6. [CrossRef]

34. Baliga, N.S.; Bjork, S.J.; Bonneau, R.; Pan, M.; Iloanusi, C.; Kottemann, M.C.H.; Hood, L.; DiRuggiero, J. Systems level insights into the stress response to UV radiation in the halophilic archaeon *Halobacterium* NRC-1. *Genome Res.* **2004**, *14*, 1025–1035. [CrossRef] [PubMed]

35. Bonneau, R.; Facciotti, M.T.; Reiss, D.J.; Schmid, A.K.; Pan, M.; Kaur, A.; Thorsson, V.; Shannon, P.; Johnson, M.H.; Bare, J.C.; et al. A predictive model for transcriptional control of physiology in a free living cell. *Cell* **2007**, *131*, 1354–1365. [CrossRef] [PubMed]

36. Brooks, A.N.; Reiss, D.J.; Allard, A.; Wu, W.-J.; Salvanha, D.M.; Plaisier, C.L.; Chandrasekaran, S.; Pan, M.; Kaur, A.; Baliga, N.S. A system-level model for the microbial regulatory genome. *Mol. Syst. Biol.* **2014**, *10*, 740. [CrossRef]

37. Koide, T.; Reiss, D.J.; Bare, J.C.; Pang, W.L.; Facciotti, M.T.; Schmid, A.K.; Pan, M.; Marzolf, B.; Van, P.T.; Lo, F.Y.; et al. Prevalence of transcription promoters within archaeal operons and coding sequences. *Mol. Syst. Biol.* **2009**, *5*, 285. [CrossRef]

38. Ten-Caten, F.; Vêncio, R.Z.N.; Lorenzetti, A.P.R.; Zaramela, L.S.; Santana, A.C.; Koide, T. Internal RNAs overlapping coding sequences can drive the production of alternative proteins in archaea. *RNA Biol.* **2018**, *15*, 1119–1132. [CrossRef]

39. Zaramela, L.S.; Vêncio, R.Z.N.; Ten-Caten, F.; Baliga, N.S.; Koide, T. Transcription start site associated RNAs (TSSaRNAs) are ubiquitous in all domains of life. *PLoS ONE* **2014**, *9*, e107680. [CrossRef]

40. Gomes-Filho, J.V.; Zaramela, L.S.; da Silva Italiani, V.C.; Baliga, N.S.; Vêncio, R.Z.N.; Koide, T. Sense overlapping transcripts in IS1341-type transposase genes are functional non-coding RNAs in archaea. *RNA Biol.* **2015**, *12*, 490–500. [CrossRef]

41. Stolt, P.; Zillig, W. Structure specific ds/ss-RNase activity in the extreme halophile *Halobacterium salinarium*. *Nucleic Acids Res.* **1993**, *21*, 5595–5599. [CrossRef]

42. Wagner, E.G.H.; Simons, R.W. Antisense RNA control in bacteria, phages, and plasmids. *Annu. Rev. Microbiol.* **1994**, *48*, 713–742. [CrossRef]

43. Leinonen, R.; Sugawara, H.; Shumway, M. The sequence read archive. *Nucleic Acids Res.* **2010**, *39*, D19–D21. [CrossRef]

44. Bolger, A.M.; Lohse, M.; Usadel, B. Trimmomatic: A flexible trimmer for Illumina sequence data. *Bioinformatics* **2014**, *30*, 2114–2120. [CrossRef]

45. Kim, D.; Langmead, B.; Salzberg, S.L. HISAT: A fast spliced aligner with low memory requirements. *Nat. Methods* **2015**, *12*, 357–360. [CrossRef]

46. Li, H.; Handsaker, B.; Wysoker, A.; Fennell, T.; Ruan, J.; Homer, N.; Marth, G.; Abecasis, G.; Durbin, R. The sequence alignment/map format and SAMtools. *Bioinformatics* **2009**, *25*, 2078–2079. [CrossRef]

47. Kahles, A.; Behr, J.; Rätsch, G. MMR: A tool for read multi-mapper resolution. *Bioinformatics* **2015**, *32*, 770–772. [CrossRef]

48. Quinlan, A.R. BEDTools: The Swiss-army tool for genome feature analysis. *Curr. Protoc. Bioinform.* **2014**, *47*, 11–12. [CrossRef]

49. Thorvaldsdóttir, H.; Robinson, J.T.; Mesirov, J.P. Integrative genomics viewer (IGV): High-performance genomics data visualization and exploration. *Brief. Bioinform.* **2013**, *14*, 178–192. [CrossRef]

50. Bare, J.C.; Koide, T.; Reiss, D.J.; Tenenbaum, D.; Baliga, N.S. Integration and visualization of systems biology data in context of the genome. *BMC Bioinform.* **2010**, *11*, 382. [CrossRef]

51. Amman, F.; Wolfinger, M.T.; Lorenz, R.; Hofacker, I.L.; Stadler, P.F.; Findeiß, S. TSSAR: TSS annotation regime for dRNA-seq data. *BMC Bioinform.* **2014**, *15*, 89. [CrossRef]

52. Pfeiffer, F.; Oesterhelt, D. A manual curation strategy to improve genome annotation: Application to a set of Haloarchael genomes. *Life* **2015**, *5*, 1427–1444. [CrossRef]

53. Lorenz, R.; Bernhart, S.H.; Zu Siederdissen, C.H.; Tafer, H.; Flamm, C.; Stadler, P.F.; Hofacker, I.L. ViennaRNA Package 2.0. *Algorithms Mol. Biol.* **2011**, *6*, 26. [CrossRef]

54. Ramírez, F.; Ryan, D.P.; Grüning, B.; Bhardwaj, V.; Kilpert, F.; Richter, A.S.; Heyne, S.; Dündar, F.; Manke, T. DeepTools2: A next generation web server for deep-sequencing data analysis. *Nucleic Acids Res.* **2016**, *44*, W160–W165. [CrossRef]

55. Crooks, G.E.; Hon, G.; Chandonia, J.M.; Brenner, S.E. WebLogo: A sequence logo generator. *Genome Res.* **2004**, *14*, 1188–1190. [CrossRef]

56. Mi, H.; Muruganujan, A.; Casagrande, J.T.; Thomas, P.D. Large-scale gene function analysis with the PANTHER classification system. *Nat. Protoc.* **2013**, *8*, 1551–1566. [CrossRef]

57. Dehal, P.S.; Joachimiak, M.P.; Price, M.N.; Bates, J.T.; Baumohl, J.K.; Chivian, D.; Friedland, G.D.; Huang, K.H.; Keller, K.; Novichkov, P.S.; et al. MicrobesOnline: An integrated portal for comparative and functional genomics. *Nucleic Acids Res.* **2009**, *38*, D396–D400. [CrossRef]

58. Marchler-Bauer, A.; Bo, Y.; Han, L.; He, J.; Lanczycki, C.J.; Lu, S.; Chitsaz, F.; Derbyshire, M.K.; Geer, R.C.; Gonzales, N.R.; et al. CDD/SPARCLE: Functional classification of proteins via subfamily domain architectures. *Nucleic Acids Res.* **2016**, *45*, D200–D203. [CrossRef]

59. Xie, Y.; Wei, Y.; Shen, Y.; Li, X.; Zhou, H.; Tai, C.; Deng, Z.; Ou, H.-Y. TADB 2.0: An updated database of bacterial type II toxin-antitoxin loci. *Nucleic Acids Res.* **2018**, *46*, D749–D753. [CrossRef]

60. Thomason, M.K.; Bischler, T.; Eisenbart, S.K.; Förstner, K.U.; Zhang, A.; Herbig, A.; Nieselt, K.; Sharma, C.M.; Storz, G. Global transcriptional start site mapping using differential RNA sequencing reveals novel antisense RNAs in *Escherichia coli*. *J. Bacteriol.* **2015**, *197*, 18–28. [CrossRef]

61. Lybecker, M.; Zimmermann, B.; Bilusic, I.; Tukhtubaeva, N.; Schroeder, R. The double-stranded transcriptome of *Escherichia coli*. *Proc. Natl. Acad. Sci. USA* **2014**, *111*, 3134–3139. [CrossRef] [PubMed]

62. Anders, S.; Pyl, P.T.; Huber, W. HTSeq-A Python framework to work with high-throughput sequencing data. *Bioinformatics* **2015**, *31*, 166–169. [CrossRef]

63. Love, M.I.; Huber, W.; Anders, S. Moderated estimation of fold change and dispersion for RNA-seq data with DESeq2. *Genome Biol.* **2014**, *15*, 550. [CrossRef] [PubMed]

64. Whiteside, M.D.; Winsor, G.L.; Laird, M.R.; Brinkman, F.S.L. OrtholugeDB: A bacterial and archaeal orthology resource for improved comparative genomic analysis. *Nucleic Acids Res.* **2012**, *41*, D366–D376. [CrossRef] [PubMed]

65. Babski, J.; Maier, L.-K.; Heyer, R.; Jaschinski, K.; Prasse, D.; Jäger, D.; Randau, L.; Schmitz, R.A.; Marchfelder, A.; Soppa, J. Small regulatory RNAs in Archaea. *RNA Biol.* **2014**, *11*, 484–493. [CrossRef]

66. Kim, D.; Hong, J.S.-J.; Qiu, Y.; Nagarajan, H.; Seo, J.H.; Cho, B.K.; Tsai, S.F.; Palsson, B.Ø. Comparative analysis of regulatory elements between *Escherichia coli* and *Klebsiella pneumoniae* by genome-wide transcription start site profiling. *PLoS Genet.* **2012**, *8*, e1002867. [CrossRef]

67. Wurtzel, O.; Sapra, R.; Chen, F.; Zhu, Y.; Simmons, B.A.; Sorek, R. A single-base resolution map of an archaeal transcriptome. *Genome Res.* **2010**, *20*, 133–141. [CrossRef]

68. Carninci, P.; Sandelin, A.; Lenhard, B.; Katayama, S.; Shimokawa, K.; Ponjavic, J.; Semple, C.A.M.; Taylor, M.S.; Engström, P.G.; Frith, M.C.; et al. Genome-wide analysis of mammalian promoter architecture and evolution. *Nat. Genet.* **2006**, *38*, 626–635. [CrossRef]

69. Bell, S.D.; Jackson, S.P. Mechanism and regulation of transcription in archaea. *Curr. Opin. Microbiol.* **2001**, *4*, 208–213. [CrossRef]

70. Seitzer, P.; Wilbanks, E.G.; Larsen, D.J.; Facciotti, M.T. A Monte Carlo-based framework enhances the discovery and interpretation of regulatory sequence motifs. *BMC Bioinform.* **2012**, *13*, 317. [CrossRef]

71. Dar, D.; Prasse, D.; Schmitz, R.A.; Sorek, R. Widespread formation of alternative 3′ UTR isoforms via transcription termination in archaea. *Nat. Microbiol.* **2016**, *1*, 16143. [CrossRef]

72. Brenneis, M.; Hering, O.; Lange, C.; Soppa, J. Experimental characterization of cis-acting elements important for translation and transcription in halophilic archaea. *PLoS Genet.* **2007**, *3*, e229. [CrossRef]

73. Bouvier, M.; Sharma, C.M.; Mika, F.; Nierhaus, K.H.; Vogel, J. Small RNA binding to 5′ mRNA coding region inhibits translational initiation. *Mol. Cell* **2008**, *32*, 827–837. [CrossRef]

74. Fozo, E.M.; Hemm, M.R.; Storz, G. Small toxic proteins and the antisense RNAs that repress them. *Microbiol. Mol. Biol. Rev.* **2008**, *72*, 579–589. [CrossRef]

75. Pfeifer, F.; Krüger, K.; Röder, R.; Mayr, A.; Ziesche, S.; Offner, S. Gas vesicle formation in halophilic Archaea. *Arch. Microbiol.* **1997**, *167*, 259–268. [CrossRef]

76. Csiszàr, K.; Houmard, J.; Damerval, T.; de Marsac, N.T. Transcriptional analysis of the cyanobacterial *gvpABC* operon in differentiated cells: Occurrence of an antisense RNA complementary to three overlapping transcripts. *Gene* **1987**, *60*, 29–37. [CrossRef]

77. Tarasov, V.; Schwaiger, R.; Furtwängler, K.; Dyall-Smith, M.; Oesterhelt, D. A small basic protein from the *brz-brb* operon is involved in regulation of bop transcription in *Halobacterium salinarum*. *BMC Mol. Biol.* **2011**, *12*, 42. [CrossRef]

78. Harms, A.; Brodersen, D.E.; Mitarai, N.; Gerdes, K. Toxins, targets, and triggers: An overview of toxin-antitoxin biology. *Mol. Cell* **2018**, *70*, 768–784. [CrossRef]

79. Tang, T.H.; Polacek, N.; Zywicki, M.; Huber, H.; Brugger, K.; Garrett, R.; Bachellerie, J.P.; Hüttenhofer, A. Identification of novel non-coding RNAs as potential antisense regulators in the archaeon *Sulfolobus solfataricus*. *Mol. Microbiol.* **2005**, *55*, 469–481. [CrossRef]

80. Ellis, M.J.; Trussler, R.S.; Haniford, D.B. A *cis*-encoded sRNA, Hfq and mRNA secondary structure act independently to suppress IS*200* transposition. *Nucleic Acids Res.* **2015**, *43*, 6511–6527. [CrossRef]

81. Pircher, A.; Gebetsberger, J.; Polacek, N. Ribosome-associated ncRNAs: An emerging class of translation regulators. *RNA Biol.* **2014**, *11*, 1335–1339. [CrossRef]

82. Wyss, L.; Waser, M.; Gebetsberger, J.; Zywicki, M.; Polacek, N. mRNA-specific translation regulation by a ribosome-associated ncRNA in *Haloferax volcanii*. *Sci. Rep.* **2018**, *8*, 12502. [CrossRef]

83. Raghavan, R.; Sloan, D.B.; Ochman, H.C. Antisense transcription is pervasive but rarely conserved in enteric bacteria. *MBio* **2012**, *3*, e00156-12. [CrossRef]

84. Dugar, G.; Herbig, A.; Förstner, K.U.; Heidrich, N.; Reinhardt, R.; Nieselt, K.; Sharma, C.M. High-resolution transcriptome maps reveal strain-specific regulatory features of multiple *Campylobacter jejuni* isolates. *PLoS Genet.* **2013**, *9*, e1003495. [CrossRef]

85. Shao, W.; Price, M.N.; Deutschbauer, A.M.; Romine, M.F.; Arkin, A.P. Conservation of transcription start sites within genes across a bacterial genus. *MBio* **2014**, *5*, e01398-14. [CrossRef]

86. Kopf, M.; Klähn, S.; Scholz, I.; Hess, W.R.; Voß, B. Variations in the non-coding transcriptome as a driver of inter-strain divergence and physiological adaptation in bacteria. *Sci. Rep.* **2015**, *5*, 9560. [CrossRef]

87. Mei, Y.; Liu, H.; Zhang, S.; Yang, M.; Hu, C.; Zhang, J.; Shen, P.; Chen, X. Effects of salinity on the cellular physiological responses of *Natrinema sp.* J7-2. *PLoS ONE* **2017**, *12*, e0184974. [CrossRef]

88. Irnov, I.; Sharma, C.M.; Vogel, J.; Winkler, W.C. Identification of regulatory RNAs in *Bacillus subtilis*. *Nucleic Acids Res.* **2010**, *38*, 6637–6651. [CrossRef]

89. Van der Meulen, S.B.; de Jong, A.; Kok, J. Transcriptome landscape of *Lactococcus lactis* reveals many novel RNAs including a small regulatory RNA involved in carbon uptake and metabolism. *RNA Biol.* **2016**, *13*, 353–366. [CrossRef]

90. Garber, M.; Grabherr, M.G.; Guttman, M.; Trapnell, C. Computational methods for transcriptome annotation and quantification using RNA-seq. *Nat. Methods* **2011**, *8*, 469. [CrossRef]

91. Santangelo, T.J.; Reeve, J.N. Archaeal RNA polymerase is sensitive to intrinsic termination directed by transcribed and remote sequences. *J. Mol. Biol.* **2006**, *355*, 196–210. [CrossRef] [PubMed]

92. Garalde, D.R.; Snell, E.A.; Jachimowicz, D.; Sipos, B.; Lloyd, J.H.; Bruce, M.; Pantic, N.; Admassu, T.; James, P.; Warland, A.; et al. Highly parallel direct RNA sequencing on an array of nanopores. *Nat. Methods* **2018**, *15*, 201–206. [CrossRef]

93. Brosius, J.; Gould, S.J. On "genomenclature": A comprehensive (and respectful) taxonomy for pseudogenes and other "junk DNA". *Proc. Natl. Acad. Sci. USA* **1992**, *89*, 10706–10710. [CrossRef]

94. Goyal, A.; Fiškin, E.; Gutschner, T.; Polycarpou-Schwarz, M.; Groß, M.; Neugebauer, J.; Gandhi, M.; Caudron-Herger, M.; Benes, V.; Diederichs, S. A cautionary tale of sense-antisense gene pairs: Independent regulation despite inverse correlation of expression. *Nucleic Acids Res.* **2017**, *45*, 12496–12508. [CrossRef]

95. Vogel, C.; Marcotte, E.M. Insights into regulation of protein abundance from proteomic and transcriptomic analyses. *Nat. Rev. Genet.* **2012**, *13*, 227–232. [CrossRef] [PubMed]

Article

Genomic Evidence of Recombination in the Basidiomycete *Wallemia mellicola*

Xiaohuan Sun [1,2,†], Cene Gostinčar [3,4,*,†], Chao Fang [1,2], Janja Zajc [3,5], Yong Hou [1,2], Zewei Song [1,2,‡] and Nina Gunde-Cimerman [3,‡]

1   China National GeneBank, BGI-Shenzhen, Jinsha Road, Shenzhen 518120, China; sunxiaohuan@genomics.cn (X.S.); fangchao@genomics.cn (C.F.); houyong@genomics.cn (Y.H.); songzewei@genomics.cn (Z.S.)
2   BGI-Shenzhen, Beishan Industrial Zone, Shenzhen 518083, China
3   Department of Biology, Biotechnical Faculty, University of Ljubljana, 1000 Ljubljana, Slovenia; janja.zajc@bf.uni-lj.si (J.Z.); nina.Gunde-Cimerman@bf.uni-lj.si (N.G.-C.)
4   Lars Bolund Institute of Regenerative Medicine, BGI-Qingdao, Qingdao 266555, China
5   Department of Biotechnology and Systems biology, National Institute of Biology, 1000 Ljubljana, Slovenia
*   Correspondence: cene.gostincar@bf.uni-lj.si or cgostincar@gmail.com
†   These authors contributed equally to this work as first authors.
‡   These authors contributed equally to this work.

Received: 25 April 2019; Accepted: 30 May 2019; Published: 4 June 2019

Abstract: One of the most commonly encountered species in the small basidiomycetous sub-phylum Wallemiomycotina is *Wallemia mellicola*, a xerotolerant fungus with a widespread distribution. To investigate the population characteristics of the species, whole genomes of twenty-five strains were sequenced. Apart from identification of four strains of clonal origin, the distances between the genomes failed to reflect either the isolation habitat of the strains or their geographical origin. Strains from different parts of the world appeared to represent a relatively homogenous and widespread population. The lack of concordance between individual gene phylogenies and the decay of linkage disequilibrium indicated that *W. mellicola* is at least occasionally recombining. Two versions of a putative mating-type locus have been found in all sequenced genomes, each present in approximately half of the strains. *W. mellicola* thus appears to be capable of (sexual) recombination and shows no signs of allopatric speciation or specialization to specific habitats.

Keywords: population genomics; halotolerance; xerotolerance; basidiomycete; allergenic fungus; recombination

---

1. Introduction

Towards the end of the 19th century, fish inspector Wallem was trying to tackle the problem of salted drying fish being spoiled by microbial growth [1]. From his samples in 1887, mycologist Johan Olav Olsen isolated and described the fungus *Wallemia ichthyophaga* [2]. More than a century later, and after several nomenclature changes, the only recognized species of *Wallemia* was *Wallemia sebi*. In 2005, the name *W. ichthyophaga* was resurrected for a group of *Wallemia* spp. strains able to grow only in media with substantially lowered water activity and an additional species—*W. muriae*—was described [3]. In 2015, a multi-locus phylogenetic analysis led to the description of additional species, *W. mellicola*, *W. canadensis*, *W. tropicalis* [4], followed by a description of *W. hederae* the following year [5] and finally *W. peruviensis* a year later [6]. In the resulting taxonomy *W. sebi* s. str. and *W. mellicola* were the most commonly isolated and most ubiquitous species of the genus. In addition to differences in molecular taxonomic markers, *W. mellicola* can be recognized by the larger size of conidia compared to *W. sebi*, while it is also less salt-tolerant and chaotolerant [4].

Due to their unusual morphology, *Wallemia* spp. long evaded reliable positioning into the fungal tree of life. The use of molecular phylogenetics showed that the genus is distant from all other known fungi, but its exact phylogenetic position remained uncertain. The first comprehensive molecular study by Zalar et al. [3] placed the *Wallemia* spp. into a new order (Wallemiales) and class (Wallemiomycetes) at the base of the Basidiomycota phylogenetic tree. Additional molecular analyses based on six genes confirmed a basal position of Wallemiomycetes to all of Pucciniomycotina, Ustilaginomycotina and Agaricomycotina [7]. Following the genome sequencing of *W. mellicola* and *W. ichthyophaga*, the analyses based on larger datasets positioned Wallemiomycetes as a sister group of Agaricomycotina [8,9]. Finally, the class Wallemiomycetes was accommodated in a new sub-phylum Wallemiomycotina, which was estimated to have emerged almost half a billion years ago, while its position in this study (as a sister group of just Agaricomycotina or basal to all three major subphyla of Basidiomycota) was again unclear and depended on the dataset used for inferring the phylogenetic relationships [10].

Wallemia spp. used to be known mainly as contaminants of food preserved with low-water-activity [3,11,12]. Later it became clear that they are frequent in both indoor and outdoor environments. They have been found in indoor air and house dust [13,14] and were reported to represent a large share of the microbiome of some species of house dust mites [15]. In natural environments *Wallemia* spp. are isolated particularly often from habitats characterized by low water activity [5]. While only a few isolates are known for some of the species of the genus, *W. mellicola* is encountered much more frequently. It can be found in different habitats around the world, among them air and house dust, hypersaline water of solar salterns, soil, salted, food preserved with low water activity, plant surface and pollen, straw and seeds [1]. These habitats reflect the extremotolerant character of *Wallemia* spp. Although tolerance of low water activity, especially if induced by high concentrations of salt, is rare among basidiomycetes, *Wallemia* spp. are among the most xerotolerant fungal taxa described to date, and some of them are even xerophilic—requiring low water activity to grow—an exceedingly rare trait in the fungal kingdom [3,5,16]. While *W. mellicola* is not the most extreme of *Wallemia* spp. in terms of halotolerance, the upper salinity levels supporting its growth are still high: 4.1 M NaCl and 1.4 M $MgCl_2$ [1]. However, even though its growth optimum is at water activity of 0.97 to 0.92, *W. mellicola* also grows well in regular mycological media without additional osmolytes and is therefore considered to be xerotolerant/halotolerant rather than xerophilic/halophilic [4].

Strains of *W. mellicola* are known to produce secondary metabolites, namely tricyclic dihydroxysesquiterpenes wallimidione, walleminone, walleminol, and two azasteroids with antitumor activity, UCA 1064-A and UCA 1064-B [17]. Unusually, the production of wallimidione increases with increasing concentration of salt up to 2.6 M NaCl. This trait raises questions about the safety of salt-preserved food contaminated with mycotoxigenic *Wallemia mellicola* and other *Wallemia* spp. [17]. Walleminol (known also as walleminol A) was detected in food [18]. There are also sporadic reports of human infections by *Wallemia* spp. [19], although these may be underreported due to slow growth of the species [1].

Despite the above, the major threat posed by *Wallemia* spp. appears to be their allergenic potential, either through exposure by inhalation or, as shown by recent research, by the overgrowth of *W. mellicola* in the gastrointestinal tract. *Wallemia* spp. have long been associated with the development of farmer's lung disease, a type of bronchial asthma or hypersensitivity pneumonitis (reviewed in [1]). A survey of air in animal and hay barns detected propagules of *Wallemia* spp. reaching up to 500×10^6 colony forming units (CFU)/m^3, while only 20 to 500 CFU/m^3 were found in residential buildings [5]. Immune sensitization to *Wallemia* spp. is frequently observed in asthmatic patients. Species of *Wallemia* were among the few fungi that increased the risk of asthma for inhabitants of homes damaged by water [20,21].

Wallemia spp. are often found in the human (and mice) gastrointestinal mycobiota. In mice the eradication of *Candida* spp. with antifungals leads to gastrointestinal overgrowth of *W. mellicola*, *Aspergillus amsteoldami*, and *Epicoccum nigrum*. While feeding healthy mice with these fungi did not lead to changes in their gut mycobiota, oral administration of *W. mellicola* after transient antibiotic therapy led to expansion of *W. mellicola* in the gut (a phenomenon not observed for either *A. amstelodami* or *E. nigrum*). This expansion in turn led to altered pulmonary immune responses to inhaled aeroallergens–without *Wallemia* present in the lungs [22,23].

The genome of *W. mellicola* (strain CBS 633.66, isolated from date honey and at the time classified as *W. sebi*) was published in 2012 [8]. The genome turned out to be unusually compact for a basidiomycete (9.8 Mbp) and contained a putative mating-type locus, even though sexual reproduction in *W. mellicola* has not been described to date.

To investigate the intraspecific relationships between strains of *W. mellicola* isolated from various indoor and outdoor environments in different parts of the world, we sequenced the whole genomes of 25 strains and analysed them using population and comparative genomic tools.

2. Materials and Methods

2.1. Culture, Medium, Growth Conditions and DNA Isolation

Twenty-five strains of *W. mellicola* (Table 1) were obtained from the Ex Culture Collection of the Department of Biology, Biotechnical Faculty, University of Ljubljana (Slovenia). They were cultivated and their DNA was isolated as described previously [24]. All strains used in this study are publicly available in the Ex Culture Collection under their EXF numbers (Table 1).

Table 1. Strains sequenced in this study.

Culture Collection Strain Number *	Number in This Study	Isolation Habitat	Sampling Site Location
EXF-277	1	hypersaline saltern water	Spain
EXF-757	2	hypersaline saltern water	Dominican Republic
EXF-1274 (CBS 110588)	3	peanuts	Indonesia
EXF-1277 (CBS 110589)	4	*Channa striata* dried salted fish	Indonesia
EXF-1279 (CBS 110593)	5	straw hat	Philippines
EXF-5677	6	air	Slovenia
EXF-5829	7	chocolate	Slovenia
EXF-6156 (UAMH 2651)	8	moldy white bread	United Kingdom
EXF-6157 (UAMH 2757)	9	soil	Canada
EXF-6158 (UAMH 6689)	10	maple syrup	Canada
EXF-8738	11	house dust	Uruguay
EXF-8740	12	house dust	Micronesia
EXF-8747	13	house dust	Indonesia
EXF-8749	14	house dust	Thailand
EXF-8757	15	house dust	Mexico
EXF-10633	16	dry common fig	Slovenia
EXF-483	17	hypersaline saltern water	Spain
EXF-1262 (CBS 213.34)	18	Unknown	Italy
EXF-1443 (IBT 19078)	19	Unknown	Denmark
EXF-5828	20	chocolate	Slovenia
EXF-5830	21	chocolate	Slovenia
EXF-5922	22	chocolate	Slovenia
EXF-6152 (MUCL 45613)	23	forest plant (*Clusia rosea*)	Cuba
EXF-6151 (MUCL 45615)	24	forest plant (*Verbena officinalis*)	Cuba
EXF-8741	25	house dust	Micronesia

* EXF strain numbers (Ex Culture Collection of the Department of Biology, Biotechnical Faculty, University of Ljubljana, Slovenia); other culture collection numbers are in parentheses (CBS—CBS-KNAW culture collection, Netherlands; UAMH—UAMH Centre For Global Microfungal Biodiversity, Canada; IBT—IBT culture collection, DTU, Denmark; MUCL—BCCM/MUCL Agro-food & Environmental Fungal Collection, Belgium).

2.2. Genome Sequencing

The genome sequencing was performed using the platform BGISEQ-500, with 2 × 150 bp libraries, prepared as described previously [25]. Multiplexing of the samples was used, and after demultiplexing, the quality of the reads was investigated using FastQC. Trimming the reads for adaptors and quality (removal of bases with Q < 20) was performed with the 'bbduk' script (https://jgi.doe.gov/data-and-tools/bbtools/).

The sequencing reads, assembly and annotation data have been deposited in Genbank under BioProject PRJNA527769 and in CNGB Nucleotide Sequence Archive (CNSA) (https://db.cngb.org/cnsa/) of China National GeneBank DataBase (CNGBdb) with accession code CNP0000446.

2.3. Variant Calling

Sequencing reads were mapped to the reference *W. mellicola* genome of strain CBS 633.66 (GenBank AFQX00000000.1) [8] with 'bwa mem', using the default parameters. This was followed by sorting with Samtools 1.6 [26], and identification of duplicates with Picard 2.10.2. Variant calling was performed with Genome Analysis Toolkit 4.1 [27]. 'Genome Analysis Toolkit (GATK) Best Practices' were modified by using the 'hard filtering' and haploid ploidy.

2.4. Assembly and Annotation

IDBA-Hybrid 1.1.3 [28] was used to assemble the genomes. The process was guided by the *W. mellicola* CBS 633.66 reference genome [8]. The other parameters were: maximum k-value 120, seed kmer 20, minimum support 2, similarity for alignment 0.95, maximum allowed gap in the reference 100, minimum size of contigs 500.

Protein-coding genes were annotated with MAKER 2.31.8 [29]. The published predicted proteome of *W. mellicola* CBS 633.66 [8] and the fungal proteins of the Swissprot database (downloaded on 12.06.2018) were used as evidence. Semi-HMM-based Nucleic Acid Parser (SNAP) [30] was trained in three bootstrap iterations (*W. mellicola* CBS 633.66 proteins were used as evidence in the first iteration, *W. mellicola* and Swissprot database in the second and third), using protein-alignment-derived gene models following the workflow of Campbell et al. (2014). Augustus predictions with the *Laccaria bicolor* training parameters was also used [31].

BUSCO 3 software [32] in proteomic mode and with the Basidiomycota protein dataset [33] was used to investigate the genome assembly and gene prediction completeness. All of the parameters were used as the default values.

Genome Annotation Generator (GAG) 2.0.1 software [34] was used to prepare the files for submission to GenBank. All of the gene models with a coding region <150 bp or with introns <10 bp were removed.

2.5. Variant-Based Analysis

Principal component analysis of the Single Nucleotide Polymorphism (SNP) data was performed with the 'glPca' function from the 'adgenet' package [35]. Linkage Disequilibrium (LD) was estimated on a dataset of biallelic SNP loci. For each pair of loci, the normalized coefficient of LD (D') and the squared correlation coefficient (r^2) were calculated using 'vcftools' [36]. To investigate LD decay, D' and r^2 of loci within 10,000 nucleotides from each other were plotted as a function of distance and a generalized additive model fitted curve was added using 'ggplot2' in R [37,38]. The LD decay range was determined as the interval outside which all of the arithmetic means of D' or r^2 were either higher (left interval border) or lower (right interval border) than half of the maximum observed D' or r^2 means.

2.6. Phylogenetic Analysis

Gene phylogenetic trees were constructed from the predicted coding sequences of complete and single-copy BUSCOs. Alignment was calculated with MAFFT 7.407 in '–auto' mode [39]. Gblocks

0.91 was employed to optimize the alignment, with the options '-b3 = 10 -b4 = 3 -b5 = n' [40]; if the resulting alignment length was > 200 nt and the mean number of different nucleotides between the sequence pairs was larger than 15 (as counted by the 'infoalign' tool included in EMBOSS 6.6.0.0 [41]), phylogeny was inferred from the alignment with PhyML 3.1 [42]. The nucleotide substitution model was Hasegawa-Kishino-Yano 85 [43], the proportion of invariable sites and the alpha parameter of the gamma distribution of substitution rate categories were estimated by PhyML. Trees were visualized in DensiTree 2.2.5 [44]. A majority rule consensus tree was constructed in R with the function 'consensus.edges' (package 'phytools'), using the default parameters [38,45].

The phylogenetic network was reconstructed from the SNP data as described previously [24].

2.7. Core Genome, GO Enrichment

The core genome *W. mellicola* was estimated with the pipeline GET_HOMOLOGUES 3.0.8 [46] from the predicted proteomes of all here sequenced strains and the reference strain *W. mellicola* CBS 633.66 [8] as a consensus of COGtriangle and OrthoMCL algorithms using default parameters. Representative sequences of each protein cluster were annotated using the PANTHER HMM scoring tools 2.1 and the HMM library version 13.1 [47]. Statistically significant enrichment of GO-Slim Biological Process terms was investigated at www.pantherdb.org for the lists of core gene clusters (present in all 26 genomes) and soft core gene clusters (in at least 24 genomes) with a list of all gene clusters used as a reference. Fisher's Exact test and the False Discovery Rate correction were used.

2.8. Mating-Type Loci

BLAST was used to search for mating genes in the assembled and annotated *W. mellicola* genomes and predicted proteomes, using homologues of putative mating genes identified in the reference genome [8] as queries. The functional annotations of the genes were assigned according to Padamsee et al. [8]. Putative mating loci and their flanking regions were visualized in R with 'ggplot2' [37,38]. The corresponding regions of the genomes were aligned and the alignments visualized using Mummer 3.23 [48].

3. Results

Wallemia mellicola has a worldwide distribution and, unlike most other species in the genus, it is frequently isolated. To investigate its population structure, 25 genomes of *W. mellicola* were sequenced and compared. Strains were selected to cover a variety of habitats (from hypersaline water and various low water activity food to house dust, air, soil, plants and soil) and isolation locations (14 countries), as listed in Table 1.

Genomes were sequenced at 318× average coverage and the minimum coverage was 194× in case of genome 5. Using the reference *W. mellicola* to guide the assembly process, the genomes were assembled into 202 to 422 contigs (average 239 ± 43 SD). The size of the genomes was small and similar between the strains (average 9.75 Mbp ± 0.05 Mbp SD). Nevertheless, the completeness of the genome was high, with 88.18% (±0.50% SD) complete basidiomycetous Benchmarking Universal Single-Copy Orthologues (BUSCOs) present in the predicted proteomes, most of them in a single copy, and only 5.89% (±0.38% SD) of BUSCOs missing entirely. Between 4327 and 4509 (average 4475 ± 37 SD) predicted genes covered 66.64% (±0.32% SD) of the assemblies. The average intron length of 64 and the average GC content of 39.95% were very similar between the individual genomes (Table 2, Supplementary Table S1).

Table 2. Statistics for the sequenced *Wallemia mellicola* genomes.

Statistic *	Minimum **	Mean **	Maximum **	Standard Deviation**
Coverage	194	318	558	92
Genome assembly size (Mbp)	9.68	9.75	9.95	0.05
Number of contigs	202	239	422	43
Contig N50	115375	144560	170540	13888
GC content (%)	39.91%	39.95%	39.97%	0.01%
CDS total length (Mbp)	6.44	6.50	6.53	0.02
CDS total length (% of genome)	66.45%	66.64%	67.08%	0.32%
Gene models (n)	4317	4475	4509	37
CDS average length (bp)	1438	1453	1512	13
Exons per gene (average)	3.98	4.02	4.17	0.04
Intron average length (bp)	63	64	66	0.57
Complete BUSCOs	87.40%	88.18%	89.80%	0.50%
Complete and single-copy BUSCOs	85.90%	87.91%	88.60%	0.56%
Complete and duplicated BUSCOs	0.10%	0.27%	3.90%	0.76%
Fragmented BUSCOs	5.50%	5.94%	6.50%	0.31%
Missing BUSCOs	4.60%	5.89%	6.50%	0.38%
SNP density	0.41%	0.52%	0.60%	0.04%

* Complete data for each genome is available in Supplementary Table S1; ** Calculated from 25 here sequenced *W. mellicola* genomes; CDS, Coding Sequence; BUSCOs, Benchmarking Universal Single-Copy Orthologues; SNP, Single Nucleotide Polymorphism.

All 25 genomes and the reference *W. mellicola* genome shared 2845 genes (the core genome, identified by the GET_HOMOLOGUES pipeline). The softcore genome (genes present in at least 24 of 26 genomes) contained additional 611 genes. When the genomes were classified with the PANTHER classification system, the following categories were identified as significantly overrepresented in the core genome: Molecular Function: nucleoside-triphosphatase activity, oxidoreductase activity, protein binding, transporter activity, transferase activity, nucleic acid binding; Biological Process: nucleic acid metabolic process, cellular localization, transport, cellular component organization, gene expression, regulation of biological process; Cellular Component: chromosomal part, endomembrane system, cytosol, plasma membrane, vacuole, protein-containing complex, nuclear lumen.

The density of single nucleotide polymorphisms (SNPs) when compared to the reference *W. mellicola* genome (Table 2, Supplementary Table S1) was very similar for all strains (0.41–0.60%). No strains with very high similarity to the reference genome were found. There was also little clustering of the strains in the principal component analysis (PCA) of SNP data (Figure 1). The first two axes explained 23.7% and 8.80% of the variation. A cluster of four strains isolated from food (three from Slovenia, one from United Kingdom) were observed and were most likely of clonal origin (Supplementary Table S2). Two pairs of strains from the same country clustered closely together (4 and 13, 6 and 16) but apart from that the strains from the same habitats or from the same geographic locations were spread relatively far from each other on the PCA plot.

Phylogenetic analysis of core BUSCOs (40 single-copy BUSCOs present in all sequenced genomes and with a minimum average of 15 different nucleotides between gene pairs) returned very different phylogenetic trees (Figure 2). There was little concordance between the tree topologies, resulting in a majority rule consensus tree with an extreme, star-like topology. The phylogenetic network analysis of the SNP data detected a fair amount of reticulation in the network.

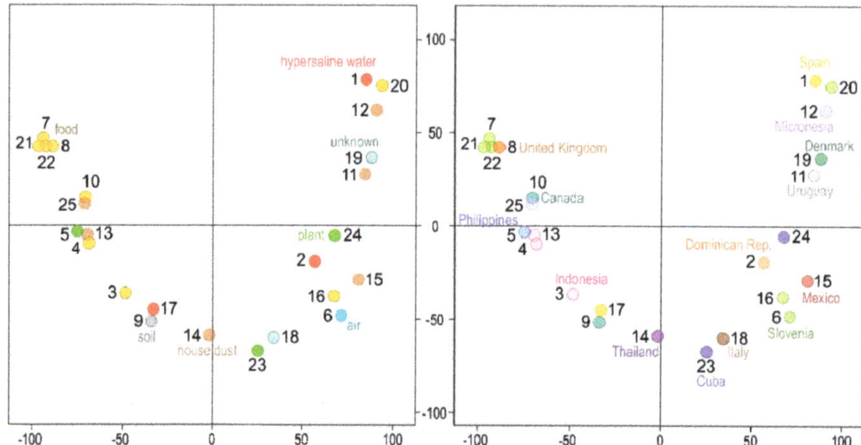

Figure 1. Clustering of the *Wallemia mellicola* genomes. Principal component analysis of single nucleotide polymorphisms (SNP) data estimated by comparing 25 sequenced genomes to the reference genome. The genomes are represented by circles, the color of which corresponds to the habitat (left) or sampling location (right) of the sequenced strains. The first two axes explain 23.7% (*x* axis) and 8.80% (*y* axis) of variation.

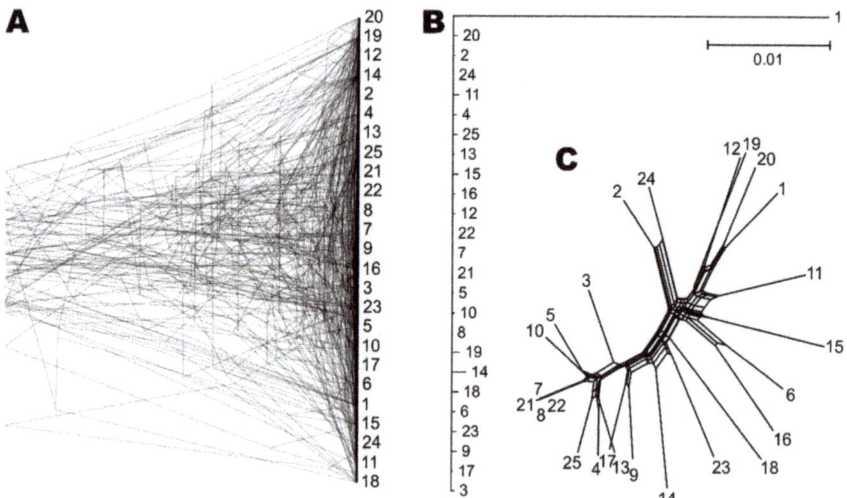

Figure 2. Phylogeny of *W. mellicola* strains. (**A**) Overlay of 40 core Benchmarking Universal Single-Copy Orthologue (BUSCO) gene trees estimated by PhyML 3.1 using the Hasegawa-Kishino-Yano 85 nucleotide substitution model and estimating the alpha parameter of the gamma distribution of the substitution rate categories and the proportion of invariable sites. (**B**) Majority rule consensus tree of 40 core gene trees described above. (**C**) Phylogenetic network reconstructed with the Neighbor-Net algorithm based on the dissimilarity distance matrix calculated from the SNP data.

A lack of concordance between phylogenetic histories of individual parts of the genome can be best explained by recombination shuffling these parts of the genome between individual organisms. A well-recognized estimator of the amount of recombination is the influence of the physical distance between the loci on the linkage disequilibrium (LD) between these loci. In non-recombining organisms the linkage between the loci should be absolute, no matter the distance between them, while in recombining populations the linkage between two loci is expected to decrease as a function of the distance between the two loci, approaching (but not necessarily reaching) equilibrium. In *W. mellicola* the LD decay, the average distance over which the LD falls to half of its maximum value, was 1291–3064 bp (intersect of the fitted curve was at 1990) when estimated with the squared correlation coefficient (r^2; Figure 3) and 224–860 when estimated with the normalized coefficient of LD (D'; data not shown).

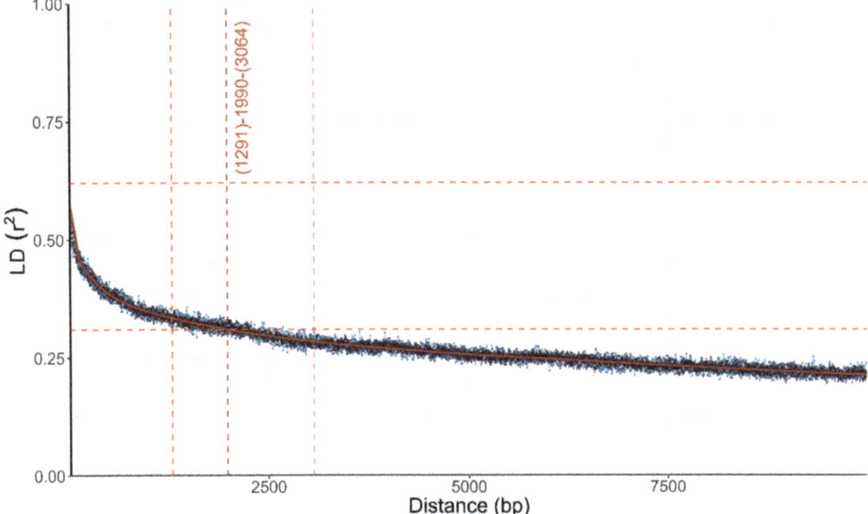

Figure 3. Linkage disequilibrium (LD) decay in *Wallemia mellicola* estimated by calculation of the squared correlation coefficient (r^2) between pairs of biallelic loci. r^2 is plotted against the physical distance of the loci in the genome. Horizontal lines mark the maximum observed value and half of the maximum observed value. Vertical lines mark the interval of the physical distance delimited by the first point of the curve under half of the maximum r^2 value (left vertical line), the last point above half of the maximum r^2 value (right vertical line) and the point where the fitted curve intersects with half of the maximum r^2 value (middle vertical line).

A putative mating-type locus of *W. mellicola* that was identified by Padamsee et al. [8] was also found in all 25 here sequenced genomes (Figure 4, Supplementary File S1). Two groups of strains (containing 12 and 13 strains each) were recognized by comparing the gene annotations in the locus—the two groups differed in some of the genes and in the orientation of the locus. No other chromosomal inversions were identified in the corresponding contig (Figure 4).

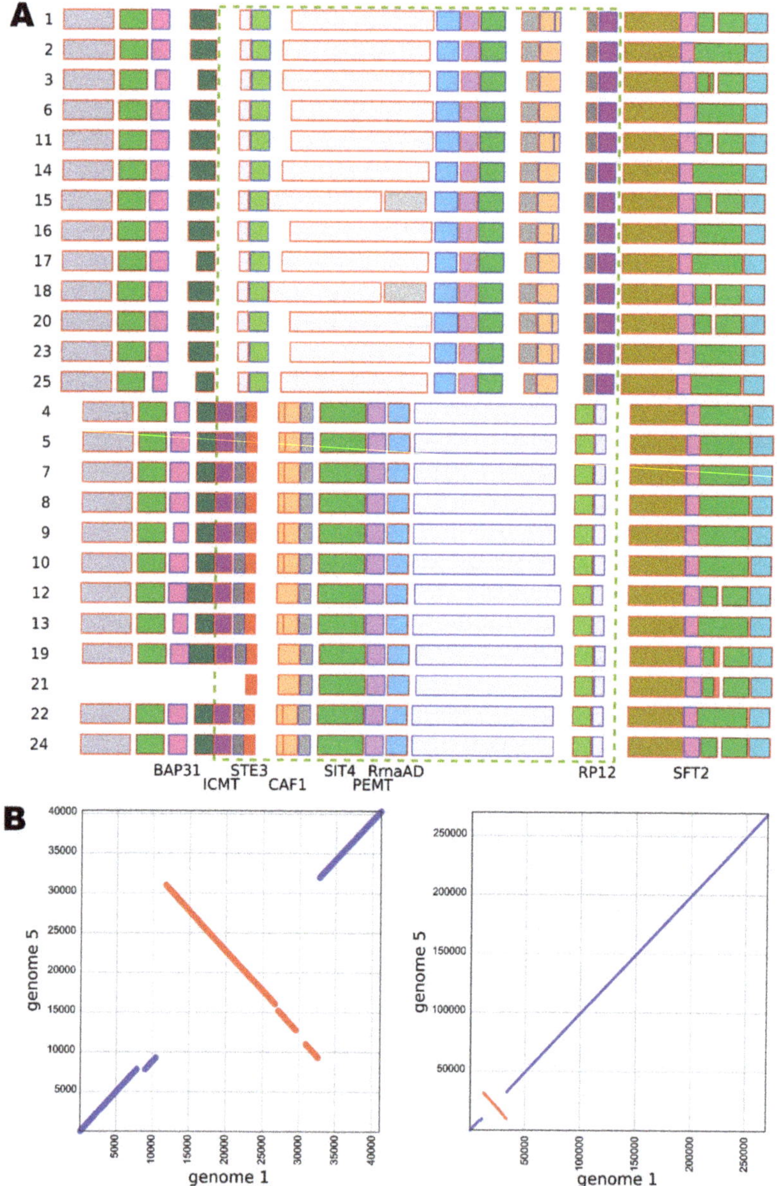

Figure 4. Putative mating-type loci in different strains of *Wallemia mellicola*. (**A**) Annotated mating-type loci and their flanking regions with putative gene functions assigned according to Padamsee et al. [8]. The chromosomal inversion is marked with a dashed green rectangle. Genome numbers are marked on the left. The blue or red outline of the rectangles representing the genes shows the gene orientation: left to right (blue) or right to left (red). (**B**) Alignment of the contigs containing the putative mating-type loci from the genomes 1 (*x* axis) and 5 (*y* axis) at different magnifications.

4. Discussion

Some extremotolerant fungal species are found in a large number of diverse environments, presumably using their stress tolerance to endure the often challenging conditions in their chosen habitats [49–51]. Although an infrequently mentioned example of such species, *Wallemia mellicola*, a species described upon a taxonomic revision of *Wallemia sebi*, can be isolated from habitats as different as air and house dust, hypersaline water of solar salterns, soil, salted, food preserved with low water activity, plant surface and pollen, straw and seeds [1] and is even a common part of the human gastrointestinal mycobiota [23]. To compare *W. mellicola* isolates from a variety of habitats and locations, twenty-five strains (Table 1) were genome sequenced and compared to investigate the possible existence of local subpopulations or cryptic specialization towards specific habitats and to check for evidence of recombination within the species. Such result would not be unusual—in recent years population genomics enabled the discovery of several cases of cryptic diversification or specialization in fungal species, which had previously appeared homogenous due to the lower phylogenetic resolution of standard taxonomic markers [52–55].

The sequenced genomes showed high similarity to the reference *W. mellicola* genome, which was reported to be very compact and with a large gene density [8]. The genomes of *Wallemia* spp. thus remain among the smallest in Basidiomycota [56,57]. The twenty-five genomes sequenced here were on average 9.75 Mbp large (compared to 9.8 Mbp of the reference genome) and 4475 protein coding genes were annotated per genome—with very little variation between the genomes (Table 2, Supplementary Table S1). This was less than 5284 genes annotated in the reference genome, which might be a consequence of different approaches to the genome assembly (de novo for the reference genome and reference-guided from short reads here) and/or to the gene annotation. At the same time 94% of the Benchmarking Universal Single-Copy Orthologues (BUSCOs) were found in the genome on average, indicating that the genomes were assembled to a high degree of completeness.

The variation between genomes (0.52% average SNP density compared to the reference genome (Table 2, Supplementary Table S1)) was comparable to variation observed in other fungal species, for example 0.55% in wild strains of *Saccharomyces cerevisiae* [58], 0.41% in *Neurospora crassa* [59] or up to 0.66% in *Candida glabrata* [60]. Only one clonal cluster of nearly identical genomes was observed on the PCA plot of SNP data, representing strains 7, 8, 21 and 22, which differed by less than 500 SNPs per genome pair (Supplementary Table S2). All four strains were isolated from food, three from chocolate in Slovenia, one from bread in UK. The next most similar pair of genomes (4 and 13, both isolated in Indonesia) differed by 17223 SNPs. However, despite this there was no general trend of higher similarity based on geographical proximity—although only a rough estimate, the average number of SNPs between the isolates isolated in different countries and the average difference between the genomes from the same country differed by only 2.4%. Similarly, the PCA did not uncover any clustering of strains based on either their habitat or geographical origin (Figure 1). This indicates that adaptation of localized populations or cryptic specialization for specific (types of) habitats is not widespread in *W. mellicola*. Considering the many habitats *W. mellicola* is able to inhabit, such generalization is not self-evident. To name just a few examples, using genomic data, it was discovered that the mushroom *Suillus brevipes* shows strong population differentiation and environmental adaptation [61], the pathogen *Cryptococcus neoformans* diversified into lineages with different pathogenic potential [62], the arbuscular mycorrhizal fungus *Rhizophagus irregularis* diverged into four main genetic groups, which were not related to the geographical origin of the strains [63], and the yeast *Metschnikowia reukaufii* diverged into lineages, which were again not related to geographical origin, but were shown to be metabolically distinct [53].

The general absence of clear clonal clusters (apart from the above-mentioned exception) raised a question about signs of recombination within *W. mellicola* genomes. This would be in line with the proposal made by Padamsee et al. [8] that *W. mellicola* is capable of sexual reproduction and their description of a putative mating locus in its genome. The availability of the genome sequences of *W. mellicola* for the first time enabled us to check for evidence of recombination in the species. If an organism is recombining (either by meiotic recombination or through other mechanisms), different parts of its genome are expected to have different phylogenetic histories. Two main approaches are used to check this in practice. The first is to reconstruct the phylogenies of individual genes and compare them. If, on the one hand, the gene trees share the same topology, they fulfil the "strong phylogenetic signal" criterion for clonality [64]. If, on the other hand, there is a lack of concordance between the phylogenies, this can be interpreted as a good indication of recombination. In the case of *W. mellicola* the latter scenario was observed and the differences between the gene trees were so numerous that the majority rule tree had an extreme star-shaped topology (Figure 2). The second approach for investigating recombination on the genome level is to look at the inheritance of loci on the same DNA molecule. In clonal organisms two loci will always be inherited together—they will be linked and in maximum linkage disequilibrium (LD). In sexual organisms two loci can segregate randomly and even if they are on the same DNA molecule, the linkage between them can be broken by chromosomal crossover—which is more likely to happen the further apart the loci are located. Such decrease of LD with distance, and particularly the distance over which half of the maximum observed LD is reached (a value known as LD decay distance), is a good measure of the amount of recombination in the population [65]. In the sequenced strains of *W. mellicola* the LD decay distance calculated from biallelic SNPs was around 1990 bp (Figure 3). This value is higher than observed in many other fungi [55,58,66], but much lower than in highly clonal species, where LD decay distances can be larger by two orders of magnitude [66]. This indicates that *W. mellicola* is at least occasionally recombining, but also extensively combining this with clonal reproduction—an observation in line with the identification of four clones in the sequenced dataset.

The indications for recombination in *W. mellicola* appear to support its proposed capability for sexual reproduction [8]. Indeed, the putative mating-type locus has been found in all 25 here sequenced genomes. Furthermore, the locus was found to exist in two variants, which share several regions of high similarity, but differ in their orientations relative to the rest of the contigs on which they are located, and which contain no other chromosomal inversions (Figure 4). These two variants possibly represent two different mating types of *W. mellicola*—a hypothesis to be tested in the future.

5. Conclusions

Wallemia mellicola is an extremotolerant basidiomycetous fungus from the sub-phylum Wallemiomycotina, with a distinct phylogenetic position and capable of inhabiting a wide variety of habitats. Sequencing and analysis of twenty-five *W. mellicola* genomes showed that the strains form no clusters based on the habitat or geographical location from which they were isolated. The sequenced strains appear to represent a relatively homogenous and widespread population with only one clonal lineage detected in the dataset. The lack of concordance between individual gene phylogenies and the decay of linkage disequilibrium indicated that *W. mellicola* is at least occasionally recombining. The mechanism of recombination could be sexual reproduction—two versions of a putative mating-type locus have been found in all sequenced genomes, each present in approximately half of the strains.

Supplementary Materials: The following are available online at http://www.mdpi.com/2073-4425/10/6/427/s1, Table S1: Statistics of the sequenced W. mellicola genomes, Table S2: Number of different SNP loci between pairs of W. mellicola genomes, File S1: Sequence of the putative mating-type locus and flanking genomic regions.

Author Contributions: Conceptualization, N.G.-C., Z.S. and C.G.; methodology, C.G., X.S. and J.Z.; investigation, X.S., J.Z. and C.G.; resources, Z.S. and N.G.-C.; data curation, C.G.; writing—original draft preparation, C.G.; writing—review and editing, C.G., N.G.-C., X.S. and Z.S.; visualization, C.G.; supervision, Z.S. and N.G.-C.; software, F.C.; funding acquisition, Y.H., N.G.-C. and Z.S.; Z.S. and N.G.C. contributed equally to this work as leading authors.

Funding: The authors acknowledge the financial support from the Slovenian Research Agency to the Infrastructural Centre Mycosmo (MRIC UL), to the programs P1-0170 and P1-0198, and to the Postdoctoral Project Z7-7436 J. Zajc. The authors would like to thank Chris Berrie for language editing assistance.

Acknowledgments: The authors would like to thank Yuchong Tang for his invaluable help in project management, organization of scientific visits and facilitation of collaboration between the project partners. We thank Anja

Černoša and Teja Lavrin for their valuable assistance in preparation of DNA samples. The authors would also like to thank Keith A. Seifert for providing the strains EXF-6156 (UAMH 2651), EXF-6157 (UAMH 2757) and EXF-6158 (UAMH 6689); and Jens Frisvad for the strain EXF-1443 (IBT 19078).

Conflicts of Interest: The authors declare no conflict of interest.

References

1. Zajc, J.; Gunde-Cimerman, N. The Genus *Wallemia*—From Contamination of Food to Health Threat. *Microorganisms* **2018**, *6*, 46. [CrossRef] [PubMed]

2. Johan-Olsen, O. *Om Sop på Klipfisk den Såkaldte Mid*; Dybwad: Christiania, Norway, 1887.

3. Zalar, P.; Sybren de Hoog, G.; Schroers, H.-J.; Frank, J.M.; Gunde-Cimerman, N. Taxonomy and phylogeny of the xerophilic genus *Wallemia* (Wallemiomycetes and Wallemiales, cl. et ord. nov.). *Antonie Van Leeuwenhoek* **2005**, *87*, 311–328. [CrossRef] [PubMed]

4. Jančič, S.; Nguyen, H.D.T.; Frisvad, J.C.; Zalar, P.; Schroers, H.-J.; Seifert, K.A.; Gunde-Cimerman, N. A Taxonomic Revision of the *Wallemia sebi* Species Complex. *PLoS ONE* **2015**, *10*, e0125933. [CrossRef] [PubMed]

5. Jančič, S.; Zalar, P.; Kocev, D.; Schroers, H.J.; Džeroski, S.; Gunde-Cimerman, N. Halophily reloaded: New insights into the extremophilic life-style of *Wallemia* with the description of *Wallemia hederae* sp. nov. *Fungal Divers.* **2016**, *76*, 97–118. [CrossRef]

6. Díaz-Valderrama, J.R.; Nguyen, H.D.T.; Aime, M.C. *Wallemia peruviensis* sp. nov., a new xerophilic fungus from an agricultural setting in South America. *Extremophiles* **2017**, *21*, 1017–1025. [CrossRef]

7. Matheny, P.B.; Gossmann, J.A.; Zalar, P.; Kumar, T.K.A.; Hibbett, D.S. Resolving the phylogenetic position of the Wallemiomycetes: An enigmatic major lineage of Basidiomycota. *Can. J. Bot.* **2006**, *84*, 1794–1805. [CrossRef]

8. Padamsee, M.; Kumar, T.K.A.; Riley, R.; Binder, M.; Boyd, A.; Calvo, A.M.; Furukawa, K.; Hesse, C.; Hohmann, S.; James, T.Y.; et al. The genome of the xerotolerant mold *Wallemia sebi* reveals adaptations to osmotic stress and suggests cryptic sexual reproduction. *Fungal Genet. Biol.* **2012**, *49*, 217–226. [CrossRef] [PubMed]

9. Zajc, J.; Liu, Y.; Dai, W.; Yang, Z.; Hu, J.; Gostinčar, C.; Gunde-Cimerman, N. Genome and transcriptome sequencing of the halophilic fungus *Wallemia ichthyophaga*: Haloadaptations present and absent. *BMC Genom.* **2013**, *14*, 617. [CrossRef]

10. Zhao, R.L.; Li, G.J.; Sánchez-Ramírez, S.; Stata, M.; Yang, Z.L.; Wu, G.; Dai, Y.C.; He, S.H.; Cui, B.K.; Zhou, J.L.; et al. A six-gene phylogenetic overview of Basidiomycota and allied phyla with estimated divergence times of higher taxa and a phyloproteomics perspective. *Fungal Divers.* **2017**, *84*, 43–74. [CrossRef]

11. Pitt, J.I.; Hocking, A.D. *Fungi and Food Spoilage*, 2nd ed.; Aspen Publishers, Inc.: Gaithersburg, MD, USA, 1999; ISBN 978-0-387-92206-5.

12. Samson, R.A.; Hoekstra, E.S.; Frisvad, J.C.; Filtenborg, O. *Introduction to Food- and Airborne Fungi*, 6th ed.; Centraalbureau voor Schimmelcultures: Baarn, The Netherlands, 2002.

13. Takahashi, T. Airborne fungal colony-forming units in outdoor and indoor environments in Yokohama, Japan. *Mycopathologia* **1997**, *139*, 23–33. [CrossRef]

14. Fröhlich-Nowoisky, J.; Pickersgill, D.A.; Despres, V.R.; Poschl, U. High diversity of fungi in air particulate matter. *Proc. Natl. Acad. Sci. USA* **2009**, *106*, 12814–12819. [CrossRef] [PubMed]

15. Hubert, J.; Nesvorna, M.; Kopecky, J.; Erban, T.; Klimov, P. Population and Culture Age Influence the Microbiome Profiles of House Dust Mites. *Microb. Ecol.* **2019**, *77*, 1048–1066. [CrossRef] [PubMed]

16. Zajc, J.; Kogej, T.; Ramos, J.; Galinski, E.A.; Gunde-Cimerman, N. The osmoadaptation strategy of the most halophilic fungus *Wallemia ichthyophaga*, growing optimally at salinities above 15% NaCl. *Appl. Environ. Microbiol.* **2014**, *80*, 247–256. [CrossRef] [PubMed]

17. Jančič, S.; Frisvad, J.C.; Kocev, D.; Gostinčar, C.; Džeroski, S.; Gunde-Cimerman, N. Production of secondary metabolites in extreme environments: Food- and airborne *Wallemia* spp. produce toxic metabolites at hypersaline conditions. *PLoS ONE* **2016**, *11*, e0169116. [CrossRef] [PubMed]

18. Moss, M.O. Recent studies of mycotoxins. *J. Appl. Microbiol.* **1998**, *84*, 62–76. [CrossRef]

19. Guarro, J.; Gugnani, H.C.; Sood, N.; Batra, R.; Mayayo, E.; Gene, J.; Kakkar, S. Subcutaneous phaeohyphomycosis caused by *Wallemia sebi* in an immunocompetent host. *J. Clin. Microbiol.* **2008**, *46*, 1129–1131. [CrossRef]

20. Sakamoto, T.; Urisu, A.; Yamada, M.; Matsuda, Y.; Tanaka, K.; Torii, S. Studies on the Osmophilic Fungus *Wallemia sebi* as an Allergen Evaluated by Skin Prick Test and Radioallergosorbent Test. *Int. Arch. Allergy Immunol.* **1989**, *90*, 368–372. [CrossRef]

21. Vesper, S.J.; McKinstry, C.; Yang, C.; Haugland, R.A.; Kercsmar, C.M.; Yike, I.; Schluchter, M.D.; Kirchner, H.L.; Sobolewski, J.; Allan, T.M.; et al. Specific molds associated with asthma in water-damaged homes. *J. Occup. Environ. Med.* **2006**, *48*, 852–858. [CrossRef]

22. Wheeler, M.L.; Limon, J.J.; Bar, A.S.; Leal, C.A.; Gargus, M.; Tang, J.; Brown, J.; Funari, V.A.; Wang, H.L.; Crother, T.R.; et al. Immunological Consequences of Intestinal Fungal Dysbiosis. *Cell Host Microbe* **2016**, *19*, 865–873. [CrossRef]

23. Skalski, J.H.; Limon, J.J.; Sharma, P.; Gargus, M.D.; Nguyen, C.; Tang, J.; Coelho, A.L.; Hogaboam, C.M.; Crother, T.R.; Underhill, D.M. Expansion of commensal fungus *Wallemia mellicola* in the gastrointestinal mycobiota enhances the severity of allergic airway disease in mice. *PLoS Pathog.* **2018**, *14*, e1007260. [CrossRef]

24. Gostinčar, C.; Stajich, J.E.; Zupančič, J.; Zalar, P.; Gunde-Cimerman, N. Genomic evidence for intraspecific hybridization in a clonal and extremely halotolerant yeast. *BMC Genom.* **2018**, *19*, 364. [CrossRef] [PubMed]

25. Fang, C.; Zhong, H.; Lin, Y.; Chen, B.; Han, M.; Ren, H.; Lu, H.; Luber, J.M.; Xia, M.; Li, W.; et al. Assessment of the cPAS-based BGISEQ-500 platform for metagenomic sequencing. *Gigascience* **2018**, *7*, 1–8. [CrossRef] [PubMed]

26. Li, H.; Handsaker, B.; Wysoker, A.; Fennell, T.; Ruan, J.; Homer, N.; Marth, G.; Abecasis, G.; Durbin, R. The Sequence Alignment/Map format and SAMtools. *Bioinformatics* **2009**, *25*, 2078–2079. [CrossRef] [PubMed]

27. Alkan, C.; Coe, B.P.; Eichler, E.E. GATK toolkit. *Nat. Rev. Genet.* **2011**, *12*, 363–376. [CrossRef]

28. Peng, Y.; Leung, H.C.M.; Yiu, S.M.; Chin, F.Y.L. IDBA-UD: A *de novo* assembler for single-cell and metagenomic sequencing data with highly uneven depth. *Bioinformatics* **2012**, *28*, 1420–1428. [CrossRef] [PubMed]

29. Campbell, M.S.; Holt, C.; Moore, B.; Yandell, M. Genome Annotation and Curation Using MAKER and MAKER-P. *Curr. Protoc. Bioinform.* **2014**, *48*, 1–39. [CrossRef]

30. Korf, I. Gene finding in novel genomes. *BMC Bioinform.* **2004**, *5*, 59. [CrossRef] [PubMed]

31. Stanke, M.; Morgenstern, B. AUGUSTUS: A web server for gene prediction in eukaryotes that allows user-defined constraints. *Nucleic Acids Res.* **2005**, *33*, 465–467. [CrossRef] [PubMed]

32. Simão, F.A.; Waterhouse, R.M.; Ioannidis, P.; Kriventseva, E.V.; Zdobnov, E.M. BUSCO: Assessing genome assembly and annotation completeness with single-copy orthologs. *Bioinformatics* **2015**, *31*, 3210–3212. [CrossRef]

33. Kriventseva, E.V.; Tegenfeldt, F.; Petty, T.J.; Waterhouse, R.M.; Simão, F.A.; Pozdnyakov, I.A.; Ioannidis, P.; Zdobnov, E.M. OrthoDB v8: Update of the hierarchical catalog of orthologs and the underlying free software. *Nucleic Acids Res.* **2015**, *43*, 250–256. [CrossRef]

34. Geib, S.M.; Hall, B.; Derego, T.; Bremer, F.T.; Cannoles, K.; Sim, S.B. Genome Annotation Generator: A simple tool for generating and correcting WGS annotation tables for NCBI submission. *Gigascience* **2018**, *7*, 4. [CrossRef] [PubMed]

35. Jombart, T.; Ahmed, I. adegenet 1.3-1: New tools for the analysis of genome-wide SNP data. *Bioinformatics* **2011**, *27*, 3070–3071. [CrossRef] [PubMed]

36. Danecek, P.; Auton, A.; Abecasis, G.; Albers, C.A.; Banks, E.; DePristo, M.A.; Handsaker, R.E.; Lunter, G.; Marth, G.T.; Sherry, S.T.; et al. The variant call format and VCFtools. *Bioinformatics* **2011**, *27*, 2156–2158. [CrossRef] [PubMed]

37. Wickham, H. *ggplot2*; Springer New York: New York, NY, USA, 2009; ISBN 978-0-387-98140-6.

38. R Development Core Team R: A Language and Environment for Statistical Computing. 2019. Available online: ftp://ftp.uvigo.es/CRAN/web/packages/dplR/vignettes/intro-dplR.pdf (accessed on 23 March 2019).

39. Katoh, K.; Toh, H. Recent developments in the MAFFT multiple sequence alignment program. *Brief. Bioinform.* **2008**, *9*, 286–298. [CrossRef] [PubMed]

40. Talavera, G.; Castresana, J. Improvement of phylogenies after removing divergent and ambiguously aligned blocks from protein sequence alignments. *Syst. Biol.* **2007**, *56*, 564–577. [CrossRef]

41. Rice, P.; Longden, I.; Bleasby, A. EMBOSS: The European Molecular Biology Open Software Suite. *Trends Genet.* **2000**, *16*, 276–277. [CrossRef]

42. Guindon, S.; Dufayard, J.F.; Lefort, V.; Anisimova, M.; Hordijk, W.; Gascuel, O. New algorithms and methods to estimate maximum-likelihood phylogenies: Assessing the performance of PhyML 3.0. *Syst. Biol.* **2010**, *59*, 307–321. [CrossRef]

43. Hasegawa, M.; Kishino, H.; Yano, T. aki Dating of the human-ape splitting by a molecular clock of mitochondrial DNA. *J. Mol. Evol.* **1985**, *22*, 160–174. [CrossRef]

44. Bouckaert, R.R. DensiTree: Making sense of sets of phylogenetic trees. *Bioinformatics* **2010**, *26*, 1372–1373. [CrossRef]

45. Revell, L.J. Phytools: An R package for phylogenetic comparative biology (and other things). *Methods Ecol. Evol.* **2012**, *3*, 217–223. [CrossRef]

46. Vinuesa, P.; Contreras-Moreira, B. Robust identification of orthologues and paralogues for microbial pan-genomics using GET_HOMOLOGUES: A case study of pIncA/C plasmids. *Methods Mol. Biol.* **2015**, *1231*, 203–232. [CrossRef] [PubMed]

47. Thomas, P.D. PANTHER: A Library of Protein Families and Subfamilies Indexed by Function. *Genome Res.* **2003**, *13*, 2129–2141. [CrossRef] [PubMed]

48. Kurtz, S.; Phillippy, A.; Delcher, A.L.; Smoot, M.; Shumway, M.; Antonescu, C.; Salzberg, S.L. Versatile and open software for comparing large genomes. *Genome Biol.* **2004**, *5*, 2. [CrossRef] [PubMed]

49. Gostinčar, C.; Grube, M.; De Hoog, S.; Zalar, P.; Gunde-Cimerman, N. Extremotolerance in fungi: Evolution on the edge. *FEMS Microbiol. Ecol.* **2010**, *71*, 2–11. [CrossRef]

50. Gostinčar, C.; Grube, M.; Gunde-Cimerman, N. Evolution of fungal pathogens in domestic environments? *Fungal Biol.* **2011**, *115*, 1008–1018. [CrossRef] [PubMed]

51. Gostinčar, C.; Gunde-Cimerman, N.; Grube, M. 10 Polyextremotolerance as the fungal answer to changing environments. In *Microbial Evolution under Extreme Conditions*; Bakermans, C., Ed.; DE GRUYTER: Berlin/München,Germany; Boston, MA, USA, 2015; pp. 185–208. ISBN 9783110335064.

52. Silva, D.N.; Várzea, V.; Paulo, O.S.; Batista, D. Population genomic footprints of host adaptation, introgression and recombination in coffee leaf rust. *Mol. Plant Pathol.* **2018**, *19*, 1742–1753. [CrossRef] [PubMed]

53. Dhami, M.K.; Hartwig, T.; Letten, A.D.; Banf, M.; Fukami, T. Genomic diversity of a nectar yeast clusters into metabolically, but not geographically, distinct lineages. *Mol. Ecol.* **2018**, *27*, 2067–2076. [CrossRef] [PubMed]

54. Branco, S.; Gladieux, P.; Ellison, C.E.; Kuo, A.; LaButti, K.; Lipzen, A.; Grigoriev, I.V.; Liao, H.L.; Vilgalys, R.; Peay, K.G.; et al. Genetic isolation between two recently diverged populations of a symbiotic fungus. *Mol. Ecol.* **2015**, *24*, 2747–2758. [CrossRef]

55. Ellison, C.E.; Hall, C.; Kowbel, D.; Welch, J.; Brem, R.B.; Glass, N.L.; Taylor, J.W. Population genomics and local adaptation in wild isolates of a model microbial eukaryote. *Proc. Natl. Acad. Sci. USA* **2011**, *108*, 2831–2836. [CrossRef]

56. Gregory, T.R.; Nicol, J.A.; Tamm, H.; Kullman, B.; Kullman, K.; Leitch, I.J.; Murray, B.G.; Kapraun, D.F.; Greilhuber, J.; Bennett, M.D. Eukaryotic genome size databases. *Nucleic Acids Res.* **2007**, *35*, D332–D338. [CrossRef]

57. Mohanta, T.K.; Bae, H. The diversity of fungal genome. *Biol. Proced. Online* **2015**, *17*, 8. [CrossRef] [PubMed]

58. Peter, J.; De Chiara, M.; Friedrich, A.; Yue, J.-X.; Pflieger, D.; Bergström, A.; Sigwalt, A.; Barre, B.; Freel, K.; Llored, A.; et al. Genome evolution across 1,011 *Saccharomyces cerevisiae* isolates. *Nature* **2018**, *556*, 339–344. [CrossRef] [PubMed]

59. Pomraning, K.R.; Smith, K.M.; Freitag, M. Bulk Segregant Analysis Followed by High-Throughput Sequencing Reveals the *Neurospora* Cell Cycle Gene, ndc-1, To Be Allelic with the Gene for Ornithine Decarboxylase, spe-1. *Eukaryot. Cell* **2011**, *10*, 724–733. [CrossRef] [PubMed]

60. Carreté, L.; Ksiezopolska, E.; Pegueroles, C.; Gómez-Molero, E.; Saus, E.; Iraola-Guzmán, S.; Loska, D.; Bader, O.; Fairhead, C.; Gabaldón, T. Patterns of Genomic Variation in the Opportunistic Pathogen Candida glabrata Suggest the Existence of Mating and a Secondary Association with Humans. *Curr. Biol.* **2018**, *28*, 15–27. [CrossRef] [PubMed]

61. Branco, S.; Bi, K.; Liao, H.-L.; Gladieux, P.; Badouin, H.; Ellison, C.E.; Nguyen, N.H.; Vilgalys, R.; Peay, K.G.; Taylor, J.W.; et al. Continental-level population differentiation and environmental adaptation in the mushroom *Suillus brevipes*. *Mol. Ecol.* **2017**, *26*, 2063–2076. [CrossRef] [PubMed]

62. Desjardins, C.A.; Giamberardino, C.; Sykes, S.M.; Yu, C.-H.; Tenor, J.L.; Chen, Y.; Yang, T.; Jones, A.M.; Sun, S.; Haverkamp, M.R.; et al. Population genomics and the evolution of virulence in the fungal pathogen *Cryptococcus neoformans*. *Genome Res.* **2017**, *27*, 1207–1219. [CrossRef] [PubMed]

63. Savary, R.; Masclaux, F.G.; Wyss, T.; Droh, G.; Cruz Corella, J.; Machado, A.P.; Morton, J.B.; Sanders, I.R. A population genomics approach shows widespread geographical distribution of cryptic genomic forms of the symbiotic fungus *Rhizophagus irregularis*. *ISME J.* **2018**, *12*, 17–30. [CrossRef] [PubMed]

64. Tibayrenc, M.; Ayala, F.J. Reproductive clonality of pathogens: A perspective on pathogenic viruses, bacteria, fungi, and parasitic protozoa. *Proc. Natl. Acad. Sci. USA* **2012**, *109*, 3305–3313. [CrossRef]

65. Taylor, J.W.; Hann-Soden, C.; Branco, S.; Sylvain, I.; Ellison, C.E. Clonal reproduction in fungi. *Proc. Natl. Acad. Sci. USA* **2015**, *112*, 8901–8908. [CrossRef]

66. Nieuwenhuis, B.P.S.; James, T.Y. The frequency of sex in fungi. *Philos. Trans. R. Soc. B Biol. Sci.* **2016**, *371*, 20150540. [CrossRef]

Article

The Patchy Distribution of Restriction–Modification System Genes and the Conservation of Orphan Methyltransferases in Halobacteria

Matthew S. Fullmer [1,2,†], Matthew Ouellette [1,†], Artemis S. Louyakis [1,†], R. Thane Papke [1,*] and Johann Peter Gogarten [1,*]

[1] Department of Molecular and Cell Biology, University of Connecticut, Storrs, CT 06269-3125, USA; Matthew.Fullmer@uconn.edu (M.S.F.); matthew.ouellette@uconn.edu (M.O.); artemis.louyakis@uconn.edu (A.S.L.)
[2] Bioinformatics Institute, School of Biological Sciences, The University of Auckland, Auckland 1010, New Zealand
[*] Correspondence: thane@uconn.edu (R.T.P.); gogarten@uconn.edu (J.P.G.)
[†] These authors contributed equally.

Received: 15 February 2019; Accepted: 14 March 2019; Published: 19 March 2019

Abstract: Restriction–modification (RM) systems in bacteria are implicated in multiple biological roles ranging from defense against parasitic genetic elements, to selfish addiction cassettes, and barriers to gene transfer and lineage homogenization. In bacteria, DNA-methylation without cognate restriction also plays important roles in DNA replication, mismatch repair, protein expression, and in biasing DNA uptake. Little is known about archaeal RM systems and DNA methylation. To elucidate further understanding for the role of RM systems and DNA methylation in Archaea, we undertook a survey of the presence of RM system genes and related genes, including orphan DNA methylases, in the halophilic archaeal class Halobacteria. Our results reveal that some orphan DNA methyltransferase genes were highly conserved among lineages indicating an important functional constraint, whereas RM systems demonstrated patchy patterns of presence and absence. This irregular distribution is due to frequent horizontal gene transfer and gene loss, a finding suggesting that the evolution and life cycle of RM systems may be best described as that of a selfish genetic element. A putative target motif (CTAG) of one of the orphan methylases was underrepresented in all of the analyzed genomes, whereas another motif (GATC) was overrepresented in most of the haloarchaeal genomes, particularly in those that encoded the cognate orphan methylase.

Keywords: HGT; restriction; methylation; gene transfer; selfish genes; archaea; haloarchaea; DNA methylase; epigenetics

---

1. Introduction

DNA methyltransferases (MTases) are enzymes which catalyze the addition of a methyl group to a nucleotide base in a DNA molecule. These enzymes will methylate either adenine, producing $N6$-methyladenine (6mA), or cytosine, producing either $N4$-methylcytosine (4mC) or $C5$-methylcytosine (5mC), depending on the type of MTase enzyme [1]. DNA methyltransferases typically consist of three types of protein domains: an S-adenosyl-L-methionine (AdoMet) binding domain which obtains the methyl group from the cofactor AdoMet, a target recognition domain (TRD) that binds the enzyme to the DNA strand at a short nucleotide sequence known as the recognition sequence, and a catalytic domain that transfers the methyl group from AdoMet to a nucleotide at the recognition sequence [2]. The order in which these domains occur in a MTase varies and can be used to classify the enzymes into the subtypes of α, β, γ, δ, ε, and ζ MTases [3–5].

In bacteria and archaea, MTases are often components of restriction–modification (RM) systems, in which an MTase works alongside a cognate restriction endonuclease (REase) that targets the same recognition site. The REase will cleave the recognition site when it is unmethylated, but the DNA will escape cutting when the site has been methylated by the MTase; this provides a self-recognition system to the host where it differentiates between its own methylated DNA and that of unmethylated, potentially harmful foreign DNA that is then digested by the host's REase [6–8]. RM systems have also been described as addiction cassettes akin to toxin-antitoxin systems, in which postsegregational killing occurs when the RM system is lost since the MTase activity degrades more quickly than REase activity, resulting in digestion of the host genome at unmodified recognition sites [9,10]. RM systems have been hypothesized to act as barriers to genetic exchange and drive population diversification [11,12]. In *Escherichia coli* for example, conjugational uptake of plasmids is reduced by the RM system EcoKI when the plasmids contain EcoKI recognition sequences [13]. However, transferred DNA that is digested by a cell's restriction endonuclease can still effectively recombine with the recipient's chromosomal DNA [7,14,15]; the effect of DNA digestion serves to limit homologous recombinant DNA fragment size [16]. Restriction thus advantages its host by decreasing transfer of large mobile genetic elements and infection with phage originating in organisms without the cognate MTase [8], while also reducing linkage between beneficial and slightly deleterious mutations [17].

There are four major types of RM systems which have been classified in bacteria and archaea [18,19]: Type I RM systems consist of three types of subunits: REase (R) subunits, MTase (M) subunits, and site specificity (S) subunits which contain two tandem TRDs. These subunits form pentamer complexes of two R subunits, two M subunits, and one S subunit, and these complexes will either fully methylate recognition sites which are modified on only one DNA strand (hemimethylated) or cleave the DNA several bases upstream or downstream of recognition sites which are unmethylated on both strands [20,21]. The MTases and REases of Type II RM systems have their own TRDs and operate independently of each other, but each one targets the same recognition site [22]. There are many different subclasses of Type II RM system enzymes, such as Type IIG enzymes which contain both REase and MTase domains and are therefore capable of both methylation and endonuclease activity [23]. Type III RM systems consist of REase (Res) and MTase (Mod) subunits which work together as complexes, with the Mod subunit containing the TRD which recognizes asymmetric target sequences [24]. Type IV RM systems are made up of only REases, but unlike in other RM systems, these REases will target and cleave methylated recognition sites [20,25].

MTases can also exist in bacterial and archaeal hosts as orphan MTases, which occur independently of cognate restriction enzymes and typically have important physiological functions [26]. In *E. coli*, the orphan MTase, Dam, an adenine MTase which targets the recognition sequence GATC, is involved in regulating the timing of DNA replication by methylating the GATC sites present at the origin of replication (*oriC*) [27]. The protein SeqA binds to hemimethylated GATC sites at *oriC*, which prevents reinitiation of DNA replication at *oriC* after a new strand has been synthesized [28,29]. Dam methylation is also important in DNA repair in *E. coli*, where the methylation state of GATC sites is used by the methyl-directed mismatch repair (MMR) system to identify the original DNA strand in order to make repairs to the newly-synthesized strand [30–32]. In *Cauldobacter crescentus*, the methylation of target sites in genes such as *ctrA* by orphan adenine MTase CcrM helps regulate the cell cycle of the organism [33–35]. The importance of orphan MTases in cellular processes is likely the reason why they are more widespread and conserved in bacteria compared to MTases associated with RM systems [36,37].

MTases and RM systems have been well-studied in Bacteria, but less research has been performed in Archaea, with most studies focused on characterizing RM systems of thermophilic species [38–42]. Recent research into the halophilic archaeal species, *Haloferax volcanii*, has demonstrated a role for DNA methylation in DNA metabolism and probably uptake: cells could not grow on wild type *E. coli* DNA as a phosphorous source, whereas unmethylated *E. coli* was metabolized completely [43,44]. In an effort to better understand this phenomenon, we characterized the genomic methylation patterns (methylome) and MTases in the halophilic archaeal species *Haloferax volcanii* [45,46]. However, the distribution of RM systems and MTases among the Archaea has not been extensively studied, and thus their life histories and impact on host evolution are unclear.

To that end we surveyed the breadth of available genomes from public databases representing the class, Halobacteria, also known as the Haloarchaea, for RM system and MTase candidate genes. We further sequenced additional genomes from the genus *Halorubrum*, which provided an opportunity to examine patterns among very closely related strains. Upon examining their patterns of occurrence, we discovered orphan methyltransferases widely distributed throughout the Haloarchaea. In contrast, RM system candidate genes had a sparse and spotty distribution indicating frequent gene transfer and loss. Even individuals from the same species isolated from the same environment and at the same time differed in the RM system complement.

2. Materials and Methods

2.1. Search Approach

The starting data consists of 217 Halobacteria genomes from NCBI and 14 in-house sequenced genomes (Table S1). We note that some of these genomes were assembled from shotgun metagenome sequences and not from individual cultured strains. Genome completion was determined through identification of 371 Halobacteriaceae marker genes using CheckM v1.0.7 [47]. Queries for all available restriction-methylation-specificity genes were obtained from the Restriction Enzyme dataBASE (REBASE) website [48,49]. As methylation genes are classified by function rather than by homology [48], the protein sequences of each category were clustered into homologous groups (HGs) via the *uclust* function of the USEARCH v9.0.2132 package [50] at a 40 percent identity. The resulting ~36,000 HGs were aligned with MUSCLE v3.8.31 [51]. HMMs were then generated from the alignments using the *hmmbuild* function of HMMER3 v3.1b2 [52]. The ORFs of the 217 genomes were searched against the profiles via the *hmmsearch* function of HMMER3. Top hits were extracted and cross hits filtered with in-house Perl scripts available at the Gogarten-lab's GitHub repository rms_analysis [53]. Steps were taken to collapse and filter HGs. First, the hits were searched against the arCOG database [54] using BLAST [55] to assign arCOG identifiers to the members of each group. Second, the R package *igraph* v1.2.2 [56] was used to create a list of connected components from the arCOG identifications. All members of a connected component were collapsed into a single collapsed HG (cHG).

Because REBASE is a database of all methylation-restriction-related activities there are many members of the database outside our interest. At this point, we made a manual curation of our cHGs attempting to identify known functions that did not apply to our area of interest. Examples include protein methylation enzymes, exonucleases, cell division proteins, etc. The final tally of this clustering and filtering yielded 1696 hits across 48 total candidate cHGs. arCOG annotations indicate DNA methylase activity, restriction enzyme activity, or specificity module activity as part of an RM system for 26 cHGs. The remaining 22 cHGs had predominant arCOG annotations matching other functions that may reasonably be excluded from conservative RM system-specific analyses. For a graphical representation of the search strategy (Figure S1). The putative Type IV methyl-directed restriction enzyme gene *mrr*, which is known to be present in *Haloferax volcanii*, had not been assembled into a cHG. We assembled a cluster of *mrr* homologs and determined their presence and absence using Mrr from *Haloferax volcanii* DS2 (accession: ADE02322.1) as query in BLASTP searches against each genome (E-value cut-off 10^{-10}).

2.2. Reference Phylogeny

A reference tree was created using the full complement of ribosomal proteins. The ribosomal protein set for *Halorubrum lacusprofundi* ATCC 49239 was obtained from the BioCyc website [57]. Each protein open reading frame (ORF) was used as the query in a BLAST [55] search against each genome. Hits for each gene were aligned with MUSCLE v3.8.31 [51] and then concatenated with in-house scripting. The concatenated alignment was subjected to maximum likelihood phylogenetic inference in the IQ-TREE v1.6.1 suite with ultrafast bootstrapping and automated model selection [58,59]. The final model selection was LG + F + R9.

2.3. Presence–absence Phylogeny

It is desirable to use maximum-likelihood methodology rather than simple distance measures. To realize this, the matrix was converted to an A/T alignment by replacing each present with an "A" and absent with a "T." This allowed the use of an F81 model with empirical base frequencies. This confines the base parameters to only A and T while allowing all of the other advantages of an ML approach. IQ-TREE was employed to infer the tree with 100 bootstraps [59].

2.4. Horizontal Gene Transfer Detection

Gene trees for each of the cHGs were inferred using RAxML v8.2.11 [60] under PROTCATLG models with 100 bootstraps. The gene trees were then improved by resolving their poorly supported in nodes to match the species tree using TreeFix-DTL [61]. Optimized gene tree rootings were inferred with the *OptRoot* function of Ranger-DTL. Reconciliation costs for each gene tree were computed against the reference tree using Ranger-DTL 2.0 [62] with default DTL costs. One-hundred reconciliations, each using a different random seed, were calculated for each cHG. After aggregating these with the *AggregateRanger* function of Ranger-DTL, the results were summarized and each prediction and any transfer inferred in 51% or greater of cases was counted as a transfer event.

2.5. Data Analysis and Presentation

The presence–absence matrix of cHGs was plotted as a heatmap onto the reference phylogeny using the *gheatmap* function of the R Bioconductor package *ggtree* v1.14.4 [63,64]. The rarefaction curve was generated with the *specaccum* function of the *vegan* v2.5-3 package in R [65], and the number of genomes per homologous group was plotted with *ggplot2* v3.1.0 [66]. Spearman correlations and significances between the presence–absence of cHGs was calculated with the *rcorr* function of the *hmisc* v4.1-1 package in R [67]. A significance cutoff of $p < 0.05$ was used with a Bonferroni correction. All comparisons failing this criterion were set to correlation = 0. These data were plotted into a correlogram via the *corrplot* function of the R package *corrplot* v0.84. To compare the Phylogeny calculated from presence–absence data to the ribosomal protein reference, the bootstrap support set of the presence–absence phylogeny was mapped onto the ribosomal protein reference tree using the *plotBS* function in *phangorn* v2.4.0 [68]. Support values equal to or greater than 10% are displayed. To compare phylogenies using Internode Certainty, scores were calculated using the IC/TC score calculation algorithm implemented in RAxML v8.2.11 [60,69].

Genomes were searched for location of cHGs. Proximity was used to determine synteny of groups of cHGs frequently identified on the same genomes.

Jaccard distances between presence–absence of taxa were calculated using the *distance* function of the R package *philentropy* v0.2.0 [70]. The PCoA was generated using the *wcmdscale* function in *vegan* v2.5-3 [65]. The two best sampled genera—*Halorubrum* (orange) and *Haloferax* (red)—are colored distinctively.

To determine the most likely recognition sites, each member of each cHG was searched against the REBASE Gold Standard set using BLASTp. The REBASE gold standard set was chosen over the individual gene sets on account of it having a much higher density of recognition site annotation. This simplifies the need to search for secondary hits to find predicted target sites. After applying an e-value cut-off of 10^{-20}, the top hit was assigned to each ORF.

CTAG and GATC motifs were counted with an inhouse perl script available at the Gogarten-lab's GitHub [71].

Sets of Gene Ontology (GO) terms were identified for each cHG using Blast2GO [72]. Annotations were checked against the UniProt database [73] using arCOG identifiers.

3. Results

3.1. RM-System Gene Distribution

Analysis of 217 haloarchaeal genomes and metagenome-assembled genomes yielded 48 total candidate collapsed homologous groups (cHGs) of RM-system components. Out of these 48 cHGs, 26 had arCOG annotation suggesting DNA methylase activity, restriction enzyme activity, or specificity module activity as part of an RM system. We detected 22 weaker candidates with predominant arCOG annotations matching other functions (Table 1). Our analysis shows that nearly all of the cHGs are found more than once. (Figure 1A). Indeed, 16 families are found in 20 or more genomes each (>9%), and this frequency steadily increases culminating in five families being conserved in greater than 80 genomes each (>37%) with one cHG being in ~80% of all Haloarchaea surveyed. Though these genes appear frequently in taxa across the haloarchaeal class, the majority of each candidate RM system cHG is present in fewer than half the genomes—the second most abundantly recovered cHG is found in only ~47% of all taxa surveyed. We note that the cHGs with wide distribution are annotated as MTases without an identifiable coevolving restriction endonuclease: Groups U DNA_methylase-022; W dam_methylase-031; Y dcm_methylase-044; and AT Uncharacterized-032 (members of this cHG are also annotated as methylation subunit and N6-Adenine MTase). Rarefaction analysis indicates that ~50% of the genomes assayed contain seven dominant cHGs, and that all taxa on average are represented by half of the cHGs (Figure 1B). Together, the separate analyses indicate extensive gene gain and loss of RM-system genes. In contrast, orphan MTases in cHG U and W, and to a lesser extent Y (Figure 2) have a wider distribution in some genera.

Table 1. Collapsed homologous group descriptions [$].

Alpha Code	Numerical Code	Annotated arCOG Function [$$]	arCOG Number
A	cHG_021	T_I_M	arCOG02632
B	cHG_024	T_I_M	arCOG05282
C	cHG_018	T_I_R	arCOG00880
D	cHG_034	T_I_R	arCOG00879
E	cHG_045	T_I_R	arCOG00878
F	cHG_006	T_I_S	arCOG02626
G	cHG_025	T_I_S	arCOG02628
H	cHG_036	probable_T_II_M	arCOG00890
I	cHG_001	T_II_M	arCOG02635
J	cHG_003	T_II_M	arCOG02634
K	cHG_011	T_II_M	arCOG04814
L	cHG_033	T_II_M	arCOG03521
M	cHG_007	T_II_R	arCOG11279
N	cHG_013	T_II_R	arCOG11717
O	cHG_023	T_II_R	arCOG03779
P	cHG_029	T_II_R	arCOG08993
Q	cHG_042	Adenine_DNA_methylase_probable_T_III_M	arCOG00108
R	cHG_008	T_III_R	arCOG06887
S	cHG_009	T_III_R_probable	arCOG07494
T	cHG_014	Adenine_DNA_methylase	arCOG00889
U	cHG_022	DNA_methylase	arCOG00115
V	cHG_027	DNA_methylase	arCOG00129
W	cHG_031	dam_methylase	arCOG03416
X	cHG_035	probable_RMS_M	arCOG08990
Y	cHG_044	dcm_methylase	arCOG04157
Z	cHG_048	Adenine_DNA_methylase	arCOG02636
AA	cHG_010	RNA_methylase	arCOG00910
AB	cHG_040	SAM-methylase	arCOG01792

Table 1. Cont.

Alpha Code	Numerical Code	Annotated arCOG Function [$$]	arCOG Number
AC	cHG_012	RestrictionEndonuclease	arCOG05724
AD	cHG_038	PredictedRestrictionEndonuclease	arCOG06431
AE	cHG_015	HNH_endonuclease	arCOG07787
AF	cHG_019	Endonuclease	arCOG02782
AG	cHG_020	Endonuclease	arCOG02781
AH	cHG_004	HNH_endonuclease	arCOG09398
AI	cHG_037	HNH_nuclease	arCOG05223
AJ	cHG_039	HNH_nuclease	arCOG03898
AK	cHG_041	HNH_nuclease	arCOG08099
AL	cHG_046	MBF1	arCOG01863
AM	cHG_028	CBS_domain	arCOG00608
AN	cHG_005	MarR	arCOG03182
AO	cHG_030	ParB-like nuclease	arCOG01875
AP	cHG_016	GVPC	arCOG06392
AQ	cHG_002	ASCH domain RNA binding	arCOG01734
AR	cHG_017	Uncharacterized	arCOG10082
AS	cHG_026	Uncharacterized	arCOG13171
AT	cHG_032	Uncharacterized	arCOG08946
AU	cHG_043	Uncharacterized	arCOG08856
AV	cHG_047	Uncharacterized	arCOG04588

[$]: A listing of associated Gene Ontology terms and gene family descriptions is available in Table S2. [$$]: T_I and T_II denote type I and type II restriction enzymes, respectively. M, R, and S denote the methylase, restriction endonuclease, and specificity subunits, respectively.

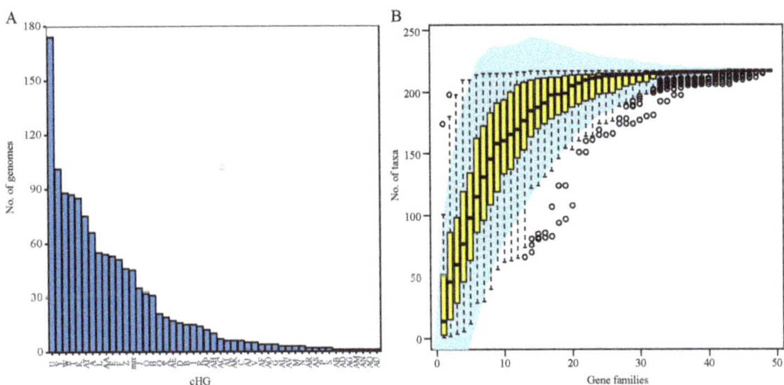

Figure 1. Distribution of collapsed Homologous Group (cHG) among haloarchaeal genomes. (**A**) The number of genomes present in each collapsed Homologous Group (cHG). No cHG contains a representative from every genome used in this study. With the exception of one cHG, all contain members from fewer than half of the genomes. The cHGs are ordered by number of genomes they contain. (**B**) Rarefaction plot of the number of genomes represented as cHGs accumulate. A 95% confidence interval is shown in shaded blue area and the yellow box whisker plots give the number of taxa from random subsamples (permutations = 100) over 48 gene families.

The phylogeny of the class Halobacteria inferred from concatenated ribosomal proteins (Figure 2) was largely comparable to prior work [74], and with a taxonomy based on concatenations of conserved proteins [75,76]. For instance, in our phylogeny, the *Halorubracaea* group with the *Haloferacaceae* recapitulating the order Haloferacales, and the families, *Halobacteriaceae, Haloarculaceae,* and *Halococcaceae*, group within the order Halobacteriales. Our genome survey in search of RM-system genes encompassed a broad taxonomic sampling, and it explores in depth the genus *Halorubrum* because it is a highly speciated genus, and because the existence of many genomes from the same species allows within species distribution assessment.

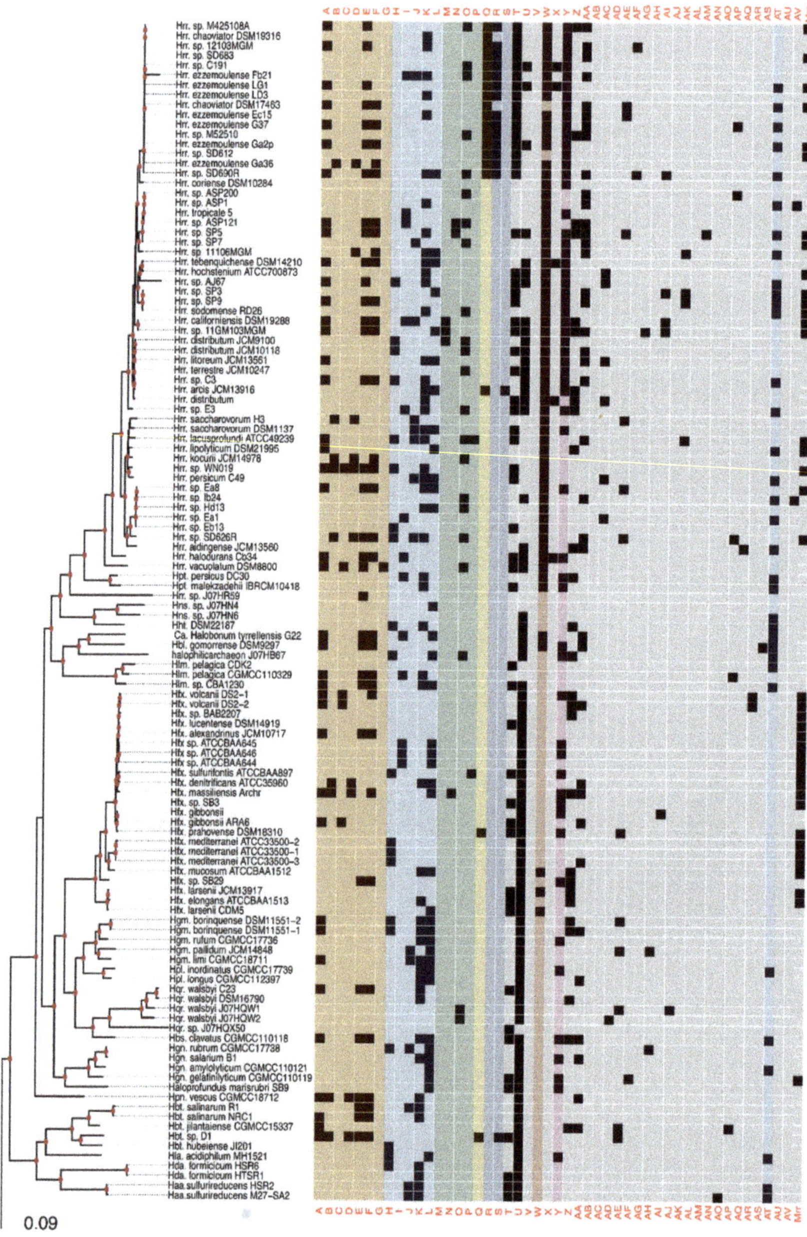

Figure 2. *Cont.*

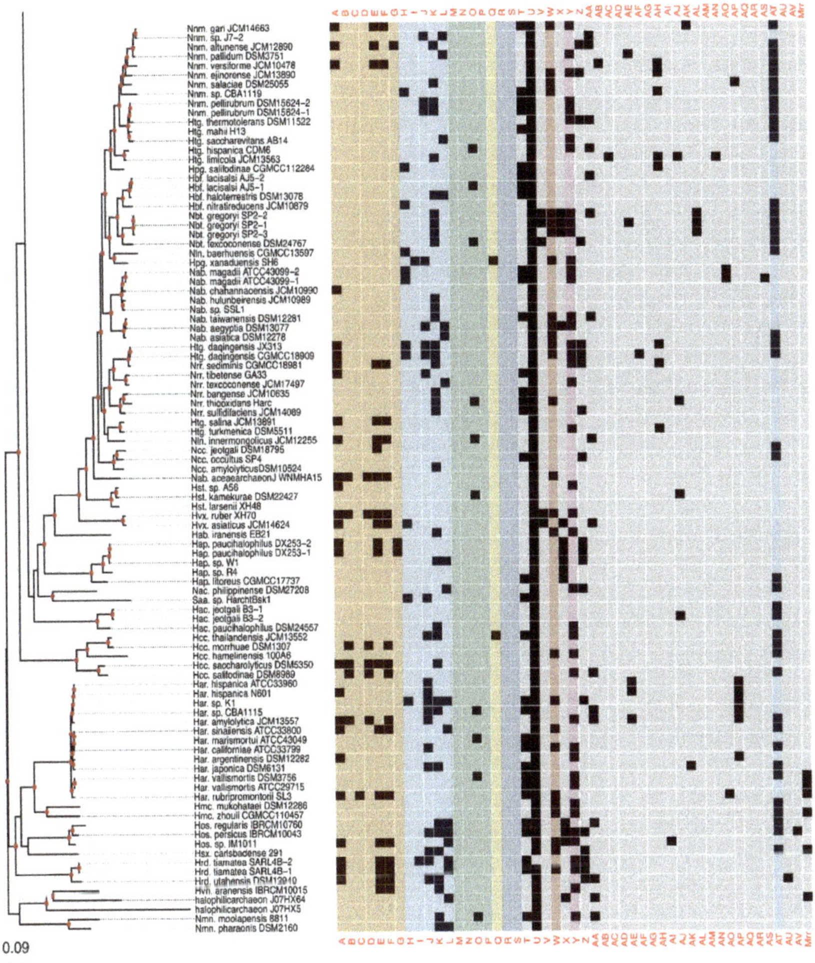

Figure 2. Presence–absence matrix of the 48 candidate RMS cHGs plotted against the reference phylogeny. For most cHGs the pattern of presence–absence does not match the reference phylogeny (compare Figures S2–S5). RMS-candidate cHGs are loosely ordered by system type and with the ambiguously assigned RM candidates at the end. Table 1 gives a key relating the column names to the majority functional annotation.

Comparison of the phylogeny in Figure 2 to the heatmap giving the presence/absence of RM system cHG candidates demonstrates that the cHG distribution is highly variable (Figure 2). The one glaring exception is cHG U, a DNA methylase found in 174 of the 217 genomes analyzed. Since it is not coupled with a restriction enzyme of equal abundance, it is assumed to be an orphan MTase. The MTase from *Hfx. volcanii* (gene HVO_0794), which recognizes the CTAG motif [45], is a member of this cHG. Though U is widely distributed, within the genus *Halorubrum* it is only found in ~37.5% (21/56) of the genomes. While U's phylogenetic profile is compatible with vertical inheritance over much of the phylogeny, the presence absence data also indicate a few gene transfer and loss events within *Halorubrum*. cHG U is present in *Hrr. tebenquichense* DSM14210, *Hrr. hochstenium* ATCC700873, *Hrr.* sp. AJ767, and in strains from related species *Hrr. distributum*, *Hrr. arcis*, *Hrr. litoreum*, and *Hrr. terrestre*, suggesting an acquisition in the ancestor of this group.

Instead of U, another orphan MTase is abundantly present in *Halorubrum* spp., cHG W. It was found in ~95% of all *Halorubrum* strains, with three exceptions—an assembled genome from the metagenome sequence data and two from incomplete draft genomes of the species *Halorubrum ezzemoulense*. Interestingly, when U is present in a *Halorubrum* sp. genome, so too is W (Figure 2). In a complementary fashion, analysis of W outside of the genus *Halorubrum* shows that it is found patchily distributed throughout the rest of the class Halobacteria (~20% −32/158), and always as a second orphan MTase with cHG U. When the members of cHG W were used to search the uniprot database, the significant matches included the *E. coli* Dam MTase, a very well-characterized GATC MTase, which provides strong evidence that this cHG is a GATC orphan MTase family. The presence and absence of cHG U and W in completely sequenced genomes is given in Table S3, together with the frequency of the CTAG and GATC motifs in the main chromosome.

The rest of the RM cHGs are much more patchily distributed (Figure 2). For instance, the cHGs that make up columns A–G represent different gene families within the Type I RM system classification: two MTases (A,B), three REases (C,D,E), and two site specificity units (SSUs) (F,G). Throughout the Haloarchaea, cHGs from columns A, E, and F, representing an MTase, an REase, and an SSU, respectively, are found co-occurring 35 times. In a subset of genomes studied for synteny, A, E, and F are encoded next to one another in *Natrinema gari*, *Halorhabdus utahensis*, *Halorubrum* SD690R, *Halorubrum ezzemoulense* G37, and *Haloorientalis* IM1011 (Figure 3). These genes probably represent a single transcriptional unit of genes working together for restriction and modification purposes. Since the Type I RM system is a five-component system, the likely stoichiometry is 2:2:1. These three cHGs co-occur four times within the species *Halorubrum ezzemoulense*, and two of these cHGs (A and E) co-occur an additional three more times, suggesting either a loss of the SSU, or an incomplete genome sequence for those strains. If it is due to incomplete sequencing, then 7/16 (43%) of the *Hrr. ezzemoulense* genomes have this set of co-occurring genes, while half do not have an identified Type I system. This is particularly stunning since strains FB21, Ec15, G37, and Ga2p were all cultivated at the same time from the same sample, a hypersaline lake in Iran. Furthermore, one strain—Ga36—has a different identified Type I RM system composed of substituted cHGs A and E with B and D, respectively, while maintaining the same SSU. This suggests the same DNA motif may be recognized by the different cHGs and that these cHGs are therefore functionally interchangeable. Members of cHGs B, F, and D were found as likely cotranscribed units in *Halococcus salifodinae*, *Natronolimnobius aegyptiacus*, *Halorubrum kocurii*, and *Haloarcula amylolytica* (Figure 3). In *Halorubrum* DL and *Halovivax ruber* XH70 genomes that contained members from cHGs A, B, D, E, and F, these genes were not found in a single unit, suggesting that they do not form a single RM system. Together, these analyses suggest this Type I RM system has a wide but sporadic distribution, that this RM system is not required for individual survival, and that functional substitutions occur for cHGs.

Type II RM systems contain an MTase and an REase that target the same motif but do not require an associated SSU because each enzyme has its own TRD. The Type II RM system cHGs are in columns H-L for the MTases, and M-P for the REases. Memberships to the Type II MTase cHGs are far more numerous in the Haloarchaea than their REase counterpart, as might be expected when witnessing decaying RM systems through the loss of the REase. The opposite result—more REases—is a more difficult scenario because an unmethylated host genome would be subject to restriction by the remaining cognate REase (e.g., addiction cassettes). There are 14 "orphan" Type II REases in Figure 2, but their cognate MTase's absence could be explained by incomplete genome sequence data.

Type III RM systems have been identified in cHGs Q (MTase) and R and S (REases). Type III MTases and REases (cHGs Q and R) co-occur almost exclusively in the species *Halorubrum ezzemoulense*, our most highly represented taxon. Furthermore, these Type III RM systems are highly restricted in their distribution to that species, with cHGs co-occurring only twice more throughout the Haloarchaea, and with a different REase cHG (S); once in *Halorubrum arcis* and another in *Halobacterium* D1. Orphan MTases occurred twice in cHG Q. Of particular interest is that closely related strains also cultivated from Lake Bidgol in Iran but which are in a different but closely related *Halorubrum* species (e.g., Ea8,

IB24, Hd13, Ea1, and Eb13) do not have a Type III RM system, implying though exposed to the same halophilic viruses, they do not rely on this system for avoiding virus infection.

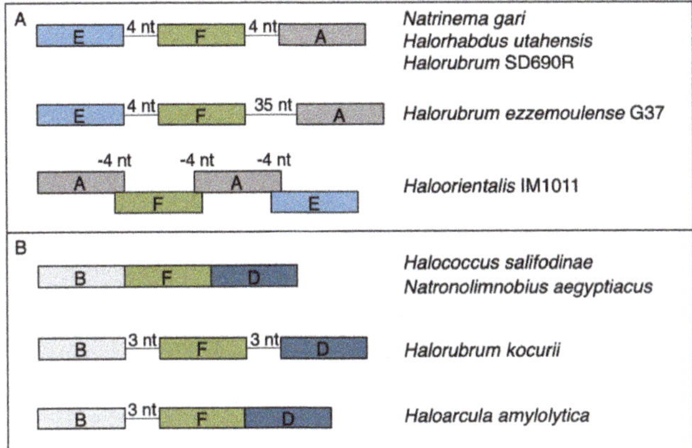

Figure 3. Gene maps for syntenic clusters of gene families (**A**) EFA and (**B**) BFD found in a subset of organisms identified to the right of each map. Genes are colored by gene families with Type I methylases (families A and B) in grays, Type I restriction endonucleases (DE) in blues, and Type I site specificity unit (F) in green.

Mrr is a Type IV REase that was suggested to cleave methylated GATC sites [77,78]. Mrr homologs are identified in most *Haloferax* sp., they have a sporadic distribution among other *Haloferacaceae* and in the *Halobacteriaceae* and are absent in the *Natrialbaceae* (Figure 2). cHGs Z-AV are not sufficiently characterized to pinpoint their role in DNA RM systems or as MTase. These cHGs likely include homing endonucleases or enzymes modifying nucleotides in RNA molecules; however, their function as orphan MTases or restriction endonucleases can, at present, not be excluded.

3.2. Horizontal Gene Transfer Explains Patchy Distribution

The patchy appearance of RM system candidates was further investigated by plotting the Jaccard distance of the presence–absence data against the alignment distance of the reference tree (Figure S2). If the presence–absence data followed vertical descent one would expect the best-fit line to move from the origin with a strong positive slope. Instead, the best fit line is close to horizontal with an r-squared value of 0.0047, indicating negligible relationship between the overall genome phylogeny and RM system complement per genome. The presence–absence clustering patterns were visualized by plotting a principle coordinate analysis (Figure S3). The high degree of overlap between the ranges of the three groups illustrates that there are few RM system genes unique to a given group and a large amount of overlap in repertoires.

To further evaluate the lack of long-term vertical descent for RM system genes, a phylogeny was inferred from the presence–absence pattern of cHGs. The resultant tree (Figure S4) is largely in disagreement with the reference phylogeny. The bootstrap support set from the presence–absence phylogeny was mapped onto the ribosomal topology (Figure S5). The resulting support values demonstrate an extremely small degree of agreement between the two methods. The few areas where there is even 10% support are near the tips of the ribosomal phylogeny and correspond to parts of established groups, such as *Haloferax*, *Natronobacterium*, and *Halorubrum*. Internode Certainty (IC) scores are another way to compare phylogenies. An average IC score of 1 represents complete agreement between the two phylogenies, and score of −1 complete disagreement. The average IC scores for the reference tree using the support set from the F81 tree was −0.509, illustrating that the presence absence data do not support the topology of the reference phylogeny.

The patchy distribution of the RM system candidate genes and their lack of conformity to the reference phylogeny suggests frequent horizontal gene transfer combined with gene loss events as the most probable explanation for the observed data. To quantify the amount of transfer, the TreeFix-Ranger pipeline was employed. TreeFix-DTL resolves poorly supported areas of gene trees to better match the concatenated ribosomal protein gene tree used as reference. Ranger-DTL resolves optimal gene tree rooting against the species tree and then computes a reconciliation estimating the number of duplications, transfers, and losses that best explains the data (Table 2). For almost every cHG with four or more taxa, our analysis infers several HGT events. Only cHG R, a putative Type III restriction enzyme found only in a group of closely related *Halorubrum ezzemoulense* strains, has not been inferred to undergo at least one transfer event.

Table 2. Important traits of cHGs with four or more open reading frames (ORFs).

Alpha (Numeric) cHG	No. of Taxa	No. of Transfers [a]	Function [b]	Predicted Recognition Sites [c]	Frequency [e]
I (001)	16	9	T_II_M	GAAGGC	31%
				GGRCA	31%
J (003)	38	21	T_II_M	CANCATC	53%
				TAGGAG	21%
AH (004)	12	4	HNH_endonuclease	GGCGCC	89%
				GATC	11%
F (006)	61	44	T_I_S	GGAYNNNNNNNTGG	24%
				CAGNNNNNNNTGCT	16%
R (008)	14	0	T_III_R	NA [d]	100%
AA (010)	55	15	RNA_methylase	ATTAAT	33%
K (011)	137	97	T_II_M	GCAAGG	49%
				GKAAYG	28%
				GCGAA	29%
AC (012)	8	5	Restriction Endonuclease	CAACNNNNNTC	29%
				CTGGAG	29%
T (014)	130	93	Adenine_DNA_methylase	GCAGG	45%
				AAGCTT	32%
AE (015)	21	13	HNH_endonuclease	GGCGCC	70%
				YSCNS	15%
AP (016)	12	6	GVPC	CANCATC	83%
C (018)	7	4	T_I_R	AACNNNNNNGTGC	73%
				CTANNNNNNRTTC	27%
AF (019)	4	3	Endonuclease	NA [d]	100%
A (021)	88	58	T_I_M	GGAYNNNNNNNTGG	37%
				GTCANNNNNNRTCA	12%
				CTCGAG	9%
U (022)	290	120	DNA_methylase	CTAG	59%
				CATTC	14%
				CCCGGG	7%
O (023)	37	28	T_II_R	NA [d]	100%
B (024)	16	8	T_I_M	GAGNNNNNNNVTGAC	75%
				GACNNNNNNNRTAC	19%
G (025)	4	2	T_I_S	GAGNNNNRTAA	75%
				GAGNNNNNTAC	25%
V (027)	5	1	DNA_methylase	CATTC	100%
AO (030)	4	2	ParB-like_nuclease	GATC	75%
				CTAG	25%
W (031)	153	70	dam_methylase	GATC	70%
				AB/SAAM	22%
AT (032)	116	60	Uncharacterized	GCAAGG	43%
				GKAAYG	26%
				GGTTAG	14%
L (033)	66	38	T_II_M-033	CAARCA	40%
				CTGAAG	36%
D (034)	16	11	T_I_R-034	GCANNNNNRTTA	69%
				GGCANNNNNNNTTC	19%

Table 2. *Cont.*

Alpha (Numeric) cHG	No. of Taxa	No. of Transfers [a]	Function [b]	Predicted Recognition Sites [c]	Frequency [e]
X (035)	19	9	probable_RMS_M	GGGAC	83%
				CCWGG	42%
H (036)	38	24	probable_T_II_M	CCSGG	18%
				GTAC	16%
AI (037)	6	4	HNH_nuclease	NA [d]	100%
AJ (039)	5	4	HNH_nuclease	GGCGCC	100%
AK (041)	6	4	HNH_nuclease	NA [d]	100%
Q (042)	21	8	Adenine_DNA_methylase probable_T_III_M	RGTAAT NA [d]	71% 19%
Y (044)	179	110	dcm_methylase	CGGCCG GTCGAC ACGT	24% 13% 11%
E (045)	58	42	T_I_R	CCCNNNNNRTTGY GCANNNNNRTTA	63% 28%
Z (048)	54	35	Adenine_DNA_methylase	CCRGAG GTMKAC	36% 30%

[a] Number of estimated horizontal gene transfer events, [b] T_I and T_II denote type I and type II restriction enzyme, respectively. M, R, and S denote the methylase, restriction endonuclease, and specificity subunits, respectively. [c] Top predicted recognition sites [d] No predicted recognition site [e] Frequency of predictions within the cHG.

RM systems usually function as cooperative units [9,19,48]. It stands to reason that some of the RM system candidates may be transferred as units, maintaining their cognate functionality. This possibility was examined by a correlation analysis. A spearman correlation was made between all pairs of cHGs. Those with a significant result at a Bonferroni-corrected $p < 0.05$ were plotted in a correlogram (Figure 4). As illustrated in Figure 3, cHGs with significant similar phylogenetic profiles often are near to one another in the genomes.

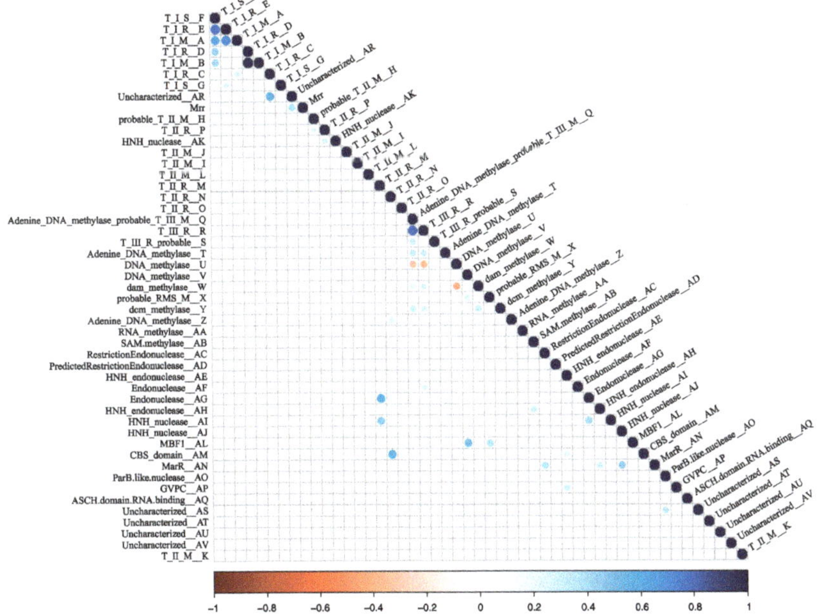

Figure 4. Heatmap of co-occurrence between the 48 RMS-candidate cHGs. Positive correlation indicates the cHGs co-occur while negative indicates that the presence of one means the other will not be present. Significance level is $p < 0.05$ with a Bonferroni correction applied for multiple tests. Blue indicates significant positive correlation; red indicates a significant negative correlation.

4. Discussion

A striking result of our study is the irregular distribution of the RM system gene candidates throughout not just the haloarchaeal class, but also within its orders, genera, species, and even communities and populations. The patchy distribution is almost certainly the result of frequent HGT and gene loss. RM system genes are well known for their susceptibility to HGT and loss, and their presence almost never define a clade or an environmental source [36,79]. Frequent acquisition of RM system genes through HGT is illustrated by their sporadic distribution. For example, *Halorubrum* genomes encode many candidate RM system cHGs that are absent from the remainder of the Halobacteria (e.g., cHG M, R, S, AC, AG, and AM). Only one of these (cHG R) is found in more than three genomes, a Type III restriction protein found in 14 of 57 *Halorubrum* genomes. Mrr homologs have a sporadic distribution among *Haloferacaceae* and *Halobacteriaceae* and are absent in *Natrialbaceae* (Figure 2). Gene loss undoubtedly contributed to the sparse cHGs distribution; however, without invoking frequent gene transfer, many independent and parallel gene losses need to be postulated. We also observed that a number haloarchaeal species possess multiple Type I subunit genes, allowing for functional substitution of the different subunits in the RM system. The existence of multiple Type I subunits has also been observed in *Helicobacter pylori*, in which 4 different SSU loci are used by the organism's Type I system to target different recognition sequences; these SSUs can even exchange TRDs, resulting in variation in the methylome of *H. pylori* [80–82]. In our results, however, we observed multiple MTase and REase subunits alongside a single SSU, suggesting the functional substitution of the subunits in these haloarchaeal organisms does not result in variation in detected recognition sequences.

Mrr is a Type IV REase that cleaves methylated target sites. Studies have demonstrated that this gene reduces transformation efficiency of GATC-methylated plasmids in *H. volcanii*, and that deletion of the *mrr* gene increases transformation efficiency on GATC-methylated plasmids, suggesting that this Type IV REase can target GATC-methylated sites for cleavage [77,78]. However, we find no anticorrelation between the presence of Mrr homologs and members of cHG W, which is homologous to the *E. coli* Dam MTase, a very well-characterized GATC MTase (Figure 2; Figure 4). This suggests that some members of cHG W or the Mrr homologs either are dysfunctional of have a site specificity different from the GATC motif.

It seems counterintuitive that RM systems are not more conserved as cellular countermeasures against commonly occurring viruses. It may be that cells do not require extensive protection via RM systems, because they use multiple defensive systems some of which might be more effective. For example, another well-known defense against viruses is the CRISPR-Cas system [83]. CRISPR recognizes short (~40 bp) regions of invading DNA that the host has been exposed to previously and degrades it. While it can be very useful against virus infection, our prior work indicated that CRISPR-Cas was also sporadically distributed within communities of closely related haloarchaeal species [84], indicating they are not required for surviving virus infection.

Both the RM and CRISPR-Cas systems are only important countermeasures after external fortifications have failed to prevent a virus from infiltrating and, therefore, their limited distributions also indicate that the cell's primary defense would be in preventing virus infection altogether, which is accomplished by different mechanisms. By altering surfaces via glycosylation, cells can avoid virus predation prior to infection. In *Haloferax* species, there are two pathways which control glycosylation of external features. One is relatively conserved and could have functions other than virus avoidance, while the other is highly variable and shows hallmarks of having genes mobilized by horizontal transfer [85]. At least one halovirus has been found to require glycosylation by its host in order to infect properly [86]. Comparison of genomes and metagenomes from hypersaline environments showed widespread evidence for distinct "genomic" islands in closely related halophiles [87] that contain a unique mixture of genes that contribute to altering the cell's surface structure and virus docking opportunities. Thus, selective pressure on postinfection, cytosolic, and nucleic acid-based virus defenses is eased, allowing them to be lost randomly in populations.

A major consideration in understanding RM system diversity is that viruses, or other infiltrating selfish genetic elements, might gain access to the host's methylation after a successful infection that was not stopped by the restriction system. Indeed, haloviruses are known to encode DNA methyltransferases in their genomes [88]. In this case, RM systems having a limited within population distribution would then be an effective defense for that part of the population possessing a different RM system. Under this scenario, a large and diverse pool of mobilized RM system genes could offer a stronger defense for the population as a whole. A single successful infection would no longer endanger the entire group of potential hosts.

Group selection may be invoked to explain the within population diversity of RM systems; a sparse distribution of RM systems may provide a potential benefit to the population as a whole, because a virus cannot easily infect all members of the population. However, often gene-level selection is a more appropriate alternative to group selection [89,90]. Under a gene centered explanation, RM systems are considered as selfish addiction cassettes that may be of little benefit to its carrier. While RM systems may be difficult to delete as a whole, stepwise deletion that begins with inactivation of the REase activity can lead to their loss from a lineage. Their long-term survival thus may be a balance of gain through gene transfer, persistence through addiction, and gene loss. This gene centered explanation is supported by a study from Seshasayee et al. [36], which examined the distribution of MTase genes in ~1000 bacterial genomes. They observed, similar to our results in the Halobacteria, that MTases associated with RM systems are poorly conserved, whereas orphan MTases share conservation patterns similar to average genes. They also demonstrated that many RM-associated and orphan MTases are horizontally acquired, and that a number of orphan MTases in bacterial genomes neighbor degraded REase genes, suggesting that they are the product of degraded RM systems that have lost functional REases [36]. Similarly, Kong et al. [79] studying genome content variation in *Neisseria meningitidis*, found an irregular distribution of RM systems, suggesting that these systems do not form an effective barrier to homologous recombination within the species. They also observed that the RM systems themselves had been frequently transferred within the species [79]. We conclude that RM genes in bacteria as well as archaea appear to undergo significant horizontal transfer and are not well-conserved. Only when these genes pick up additional functions do parts of these systems persist for longer periods of time, as exemplified in the distribution of orphan MTases. However, the transition from RM system MTase to orphan MTase is an infrequent event. A study of 43 pangenomes by Oliveira et al. [91] suggests that orphan MTases occur more frequently from transfer via large mobile genetic elements (MGEs) such as plasmids and phages rather than arise de novo from RM degradation.

The distribution of orphan methylase cHG U and W, and their likely target motifs, CTAG and GATC, respectively, suggests different biological functions for these two methylases. The widespread conservation of the CTAG MTase family cHG U supports the findings of Blow et al. [37] who identified a well-conserved CTAG orphan MTase family in the Halobacteria. Similar to other bacterial and archaeal genomes [92], the CTAG motif—the likely target for methylases in cHG U—is underrepresented in all haloarchaeal genomes (Table S3). The low frequency of occurrence, only about once per 4000 nucleotides, suggests that this motif and the cognate orphan methylase are not significantly involved in facilitating mismatch repair. The underrepresented CTAG motif was found to be less underrepresented near rRNA genes [92] and on plasmids; the CTAG motif is also a known target sequence for some Insertion Sequence (IS) elements [93] and it may be involved in repressor binding, where the CTAG motif was found to be associated with kinks in the DNA when bound to the repressor [94,95]. Interestingly, CTAG and GATC motifs are either absent or underrepresented in several haloarchaeal viruses [88,96,97]. Both the presence of the cHG U methylase and the underrepresentation of the CTAG motif appear to be maintained by selection; however, at present, the reasons for the underrepresentation of the motif in chromosomal DNA, and the role that the methylation of this motif may play remain open questions.

5. Conclusions

RM systems have a sporadic distribution in Halobacteria, even within species and populations. In contrast, orphan methylases are more persistent in lineages, and the targeted motifs are under selection for lower (in case of CTAG) or higher (in case of GATC) than expected frequency. In the case of the GATC motif, the cognate orphan MTase was found only in genomes where this motif occurs with high frequency.

Supplementary Materials: The following are available online at http://www.mdpi.com/2073-4425/10/3/233/s1, Figure S1: Workflow of RMS-candidate gene search strategy, Figure S2: Plot of alignment distance as a function of presence–absence distance, Figure S3: PCoA plot of the distances between the RMS presence–absence profiles of the 217 analyzed Halobacterial genomes, Figure S4: Maximum-likelihood phylogeny of cHG presence–absence matrix, Figure S5: Bootstrap support values of the presence–absence phylogeny mapped onto the ribosomal protein reference tree, Table S1: Basic statistics for *Halobacteriaceae* complete and draft genomes, Table S2: Gene Ontology (GO) terms for each collapsed homologous group, Table S3: Distribution of orphan methylases cHGs U and W and frequency of their putative recognition motifs in completely sequenced halobacterial chromosomes.

Author Contributions: Conceptualization, T.R.P. and J.P.G.; Data Curation, M.S.F., M.O., and A.S.L.; Formal Analysis, M.S.F., M.O., and A.S.L.; Funding Acquisition, T.R.P. and J.P.G.; Investigation, M.S.F. and M.O.; Methodology, M.S.F., A.S.L., and J.P.G.; Project Administration, T.R.P. and J.P.G.; Software, M.S.F. and A.S.L.; Supervision, T.R.P. and J.P.G.; Validation, M.S.F.; Visualization, M.S.F. and A.S.L.; Writing—Original Draft, M.S.F. and Johann Peter Gogarten; Writing—Review & Editing, M.O., AS.L., T.R.P., and J.P.

Funding: This work was supported through grants from the Binational Science Foundation (BSF 2013061), the National Science Foundation (NSF/MCB 1716046) within the BSF-NSF joint research program, and NASA exobiology (NNX15AM09G, and 80NSSC18K1533).

Acknowledgments: The Computational Biology Core, the Institute for Systems Genomics, and the University of Connecticut provided computational resources. The authors thank the anonymous reviewer for pointing out the presence of Mrr in *Haloferax volcanii*.

Conflicts of Interest: The authors declare no conflicts of interest.

References

1. Korlach, J.; Turner, S.W. Going beyond five bases in DNA sequencing. *Curr. Opin. Struct. Biol.* **2012**, *22*, 251–261. [CrossRef]

2. Bheemanaik, S.; Reddy, Y.V.R.; Rao, D.N. Structure, function and mechanism of exocyclic DNA methyltransferases. *Biochem. J.* **2006**, *399*, 177–190. [CrossRef]

3. Malone, T.; Blumenthal, R.M.; Cheng, X. Structure-guided analysis reveals nine sequence motifs conserved among DNA mmino-methyl-transferases, and suggests a catalytic mechanism for these enzymes. *J. Mol. Biol.* **1995**, *253*, 618–632. [CrossRef]

4. Bujnicki, J.M.; Radlinska, M. Molecular evolution of DNA-(cytosine-N4) methyltransferases: evidence for their polyphyletic origin. *Nucleic Acids Res.* **1999**, *27*, 4501–4509. [CrossRef]

5. Bujnicki, J.M. Sequence permutations in the molecular evolution of DNA methyltransferases. *BMC Evol. Biol.* **2002**, *2*, 3. [CrossRef]

6. Tock, M.R.; Dryden, D.T. The biology of restriction and anti-restriction. *Curr. Opin. Microbiol.* **2005**, *8*, 466–472. [CrossRef]

7. Vasu, K.; Nagaraja, V. Diverse functions of restriction-modification systems in addition to cellular defense. *Microbiol. Mol. Biol. Rev.* **2013**, *77*, 53–72. [CrossRef] [PubMed]

8. Pleška, M.; Qian, L.; Okura, R.; Bergmiller, T.; Wakamoto, Y.; Kussell, E.; Guet, C.C. Bacterial autoimmunity due to a restriction-modification system. *Curr. Biol. CB* **2016**, *26*, 404–409. [CrossRef]

9. Ohno, S.; Handa, N.; Watanabe-Matsui, M.; Takahashi, N.; Kobayashi, I. Maintenance forced by a restriction-modification system can be modulated by a region in its modification enzyme not essential for methyltransferase activity. *J. Bacteriol.* **2008**, *190*, 2039–2049. [CrossRef] [PubMed]

10. Kobayashi, I. Behavior of restriction-modification systems as selfish mobile elements and their impact on genome evolution. *Nucleic Acids Res.* **2001**, *29*, 3742–3756. [CrossRef]

11. Budroni, S.; Siena, E.; Hotopp, J.C.D.; Seib, K.L.; Serruto, D.; Nofroni, C.; Comanducci, M.; Riley, D.R.; Daugherty, S.C.; Angiuoli, S.V.; Covacci, A.; Pizza, M.; Rappuoli, R.; Moxon, E.R.; Tettelin, H.; Medini, D. Neisseria meningitidis is structured in clades associated with restriction modification systems that modulate homologous recombination. *Proc. Natl. Acad. Sci. USA* **2011**, *108*, 4494–4499. [CrossRef]

12. Erwin, A.L.; Sandstedt, S.A.; Bonthuis, P.J.; Geelhood, J.L.; Nelson, K.L.; Unrath, W.C.T.; Diggle, M.A.; Theodore, M.J.; Pleatman, C.R.; Mothershed, E.A.; Sacchi, C.T.; Mayer, L.W.; Gilsdorf, J.R.; Smith, A.L. Analysis of genetic relatedness of *Haemophilus influenzae* isolates by multilocus sequence typing. *J. Bacteriol.* **2008**, *190*, 1473–1483. [CrossRef]

13. Roer, L.; Aarestrup, F.M.; Hasman, H. The EcoKI type I restriction-modification system in *Escherichia coli* affects but Is not an absolute barrier for conjugation. *J. Bacteriol.* **2015**, *197*, 337–342. [CrossRef]

14. McKane, M.; Milkman, R. Transduction, restriction and recombination patterns in *Escherichia coli*. *Genetics* **1995**, *139*, 35–43. [PubMed]

15. Chang, S.; Cohen, S.N. In vivo site-specific genetic recombination promoted by the EcoRI restriction endonuclease. *Proc. Natl. Acad. Sci. USA* **1977**, *74*, 4811–4815. [CrossRef] [PubMed]

16. Lin, E.A.; Zhang, X.-S.; Levine, S.M.; Gill, S.R.; Falush, D.; Blaser, M.J. Natural transformation of Helicobacter pylori involves the Iintegration of short DNA fragments interrupted by gaps of variable size. *PLoS Pathog.* **2009**, *5*, e1000337. [CrossRef] [PubMed]

17. Muller, H.J. The relation of recombination to mutational advance. *Mutat. Res.* **1964**, *1*, 2–9. [CrossRef]

18. Ershova, A.S.; Rusinov, I.S.; Spirin, S.A.; Karyagina, A.S.; Alexeevski, A.V. Role of restriction-modification systems in prokaryotic evolution and ecology. *Biochemistry* **2015**, *80*, 1373–1386. [CrossRef]

19. Roberts, R.J.; Belfort, M.; Bestor, T.; Bhagwat, A.S.; Bickle, T.A.; Bitinaite, J.; Blumenthal, R.M.; Degtyarev, S.K.; Dryden, D.T.F.; Dybvig, K.; et al. A nomenclature for restriction enzymes, DNA methyltransferases, homing endonucleases and their genes. *Nucleic Acids Res.* **2003**, *31*, 1805–1812. [CrossRef]

20. Loenen, W.A.M.; Dryden, D.T.F.; Raleigh, E.A.; Wilson, G.G. Type I restriction enzymes and their relatives. *Nucleic Acids Res.* **2014**, *42*, 20–44. [CrossRef]

21. Liu, Y.-P.; Tang, Q.; Zhang, J.-Z.; Tian, L.-F.; Gao, P.; Yan, X.-X. Structural basis underlying complex assembly and conformational transition of the type I R-M system. *Proc. Natl. Acad. Sci. USA* **2017**, *114*, 11151–11156. [CrossRef] [PubMed]

22. Pingoud, A.; Wilson, G.G.; Wende, W. Type II restriction endonucleases—A historical perspective and more. *Nucleic Acids Res.* **2014**, *42*, 7489–7527. [CrossRef]

23. Morgan, R.D.; Bhatia, T.K.; Lovasco, L.; Davis, T.B. MmeI: A minimal Type II restriction-modification system that only modifies one DNA strand for host protection. *Nucleic Acids Res.* **2008**, *36*, 6558–6570. [CrossRef] [PubMed]

24. Rao, D.N.; Dryden, D.T.F.; Bheemanaik, S. Type III restriction-modification enzymes: A historical perspective. *Nucleic Acids Res.* **2014**, *42*, 45–55. [CrossRef] [PubMed]

25. Czapinska, H.; Kowalska, M.; Zagorskaitė, E.; Manakova, E.; Slyvka, A.; Xu, S.; Siksnys, V.; Sasnauskas, G.; Bochtler, M. Activity and structure of EcoKMcrA. *Nucleic Acids Res.* **2018**, *46*, 9829–9841. [CrossRef]

26. Adhikari, S.; Curtis, P.D. DNA methyltransferases and epigenetic regulation in bacteria. *FEMS Microbiol. Rev.* **2016**, *40*, 575–591. [CrossRef] [PubMed]

27. Messer, W.; Bellekes, U.; Lother, H. Effect of *dam* methylation on the activity of the *E. coli* replication origin, *oriC*. *EMBO J.* **1985**, *4*, 1327–1332. [CrossRef] [PubMed]

28. Kang, S.; Lee, H.; Han, J.S.; Hwang, D.S. Interaction of SeqA and Dam methylase on the hemimethylated origin of Escherichia coli chromosomal DNA replication. *J. Biol. Chem.* **1999**, *274*, 11463–11468. [CrossRef]

29. Sanchez-Romero, M.A.; Busby, S.J.W.; Dyer, N.P.; Ott, S.; Millard, A.D.; Grainger, D.C. Dynamic distribution of SeqA protein across the chromosome of *Escherichia coli* K-12. *MBio* **2010**, *1*. [CrossRef]

30. Welsh, K.M.; Lu, A.L.; Clark, S.; Modrich, P. Isolation and characterization of the *Escherichia coli* mutH gene product. *J. Biol. Chem.* **1987**, *262*, 15624–15629. [PubMed]

31. Au, K.G.; Welsh, K.; Modrich, P. Initiation of methyl-directed mismatch repair. *J. Biol. Chem.* **1992**, *267*, 12142–12148.

32. Putnam, C.D. Evolution of the methyl directed mismatch repair system in *Escherichia coli*. *DNA Repair (Amst).* **2016**, *38*, 32–41. [CrossRef]

33. Zweiger, G.; Marczynski, G.; Shapiro, L. A Caulobacter DNA methyltransferase that functions only in the predivisional cell. *J. Mol. Biol.* **1994**, *235*, 472–485. [CrossRef]

34. Domian, I.J.; Reisenauer, A.; Shapiro, L. Feedback control of a master bacterial cell-cycle regulator. *Proc. Natl. Acad. Sci. USA* **1999**, *96*, 6648–6653. [CrossRef] [PubMed]

35. Gonzalez, D.; Kozdon, J.B.; McAdams, H.H.; Shapiro, L.; Collier, J. The functions of DNA methylation by CcrM in *Caulobacter crescentus*: A global approach. *Nucleic Acids Res.* **2014**, *42*, 3720–3735. [CrossRef]

36. Seshasayee, A.S.N.; Singh, P.; Krishna, S. Context-dependent conservation of DNA methyltransferases in bacteria. *Nucleic Acids Res.* **2012**, *40*, 7066–7073. [CrossRef] [PubMed]

37. Blow, M.J.; Clark, T.A.; Daum, C.G.; Deutschbauer, A.M.; Fomenkov, A.; Fries, R.; Froula, J.; Kang, D.D.; Malmstrom, R.R.; Morgan, R.D.; et al. The epigenomic landscape of prokaryotes. *PLOS Genet.* **2016**, *12*, e1005854. [CrossRef] [PubMed]

38. Nölling, J.; de Vos, W.M. Identification of the CTAG-recognizing restriction-modification systems MthZI and MthFI from Methanobacterium thermoformicicum and characterization of the plasmid-encoded mthZIM gene. *Nucleic Acids Res.* **1992**, *20*, 5047–5052. [CrossRef] [PubMed]

39. Grogan, D.W. Cytosine methylation by the SuaI restriction-modification system: Implications for genetic fidelity in a hyperthermophilic archaeon. *J. Bacterio.* **2003**, *185*, 4657–4661. [CrossRef]

40. Ishikawa, K.; Watanabe, M.; Kuroita, T.; Uchiyama, I.; Bujnicki, J.M.; Kawakami, B.; Tanokura, M.; Kobayashi, I. Discovery of a novel restriction endonuclease by genome comparison and application of a wheat-germ-based cell-free translation assay: PabI (5′-GTA/C) from the hyperthermophilic archaeon *Pyrococcus abyssi. Nucleic Acids Res.* **2005**, *33*, e112. [CrossRef]

41. Watanabe, M.; Yuzawa, H.; Handa, N.; Kobayashi, I. Hyperthermophilic DNA methyltransferase M.PabI from the archaeon *Pyrococcus abyssi. Appl. Environ. Microbiol.* **2006**, *72*, 5367–5375. [CrossRef] [PubMed]

42. Couturier, M.; Lindås, A.-C. The DNA methylome of the hyperthermoacidophilic crenarchaeon *Sulfolobus acidocaldarius. Front. Microbiol.* **2018**, *9*, 137. [CrossRef] [PubMed]

43. Chimileski, S.; Dolas, K.; Naor, A.; Gophna, U.; Papke, R.T. Extracellular DNA metabolism in *Haloferax volcanii. Front. Microbiol.* **2014**, *5*, 57. [CrossRef] [PubMed]

44. Zerulla, K.; Chimileski, S.; Näther, D.; Gophna, U.; Papke, R.T.; Soppa, J. DNA as a phosphate storage polymer and the alternative advantages of polyploidy for growth or survival. *PLoS ONE* **2014**, *9*, e94819. [CrossRef]

45. Ouellette, M.; Gogarten, J.; Lajoie, J.; Makkay, A.; Papke, R. Characterizing the DNA methyltransferases of *Haloferax volcanii* via bioinformatics, gene deletion, and SMRT sequencing. *Genes* **2018**, *9*, 129. [CrossRef]

46. Ouellette, M.; Jackson, L.; Chimileski, S.; Papke, R.T. Genome-wide DNA methylation analysis of *Haloferax volcanii* H26 and identification of DNA methyltransferase related PD-(D/E)XK nuclease family protein HVO_A0006. *Front. Microbiol.* **2015**, *6*, 251. [CrossRef]

47. Parks, D.H.; Imelfort, M.; Skennerton, C.T.; Hugenholtz, P.; Tyson, G.W. CheckM: Assessing the quality of microbial genomes recovered from isolates, single cells, and metagenomes. *Genome Res.* **2015**, gr-186072. [CrossRef] [PubMed]

48. Roberts, R.J.; Macelis, D. REBASE—Restriction enzymes and methylases. *Nucleic Acids Res.* **2001**, *29*, 268–269. [CrossRef]

49. Roberts, R.J.; Vincze, T.; Posfai, J.; Macelis, D. REBASE—A database for DNA restriction and modification: Enzymes, genes and genomes. *Nucleic Acids Res.* **2015**, *43*, D298–D299. [CrossRef] [PubMed]

50. Edgar, R.C. Search and clustering orders of magnitude faster than BLAST. *Bioinformatics* **2010**, *26*, 2460–2461. [CrossRef] [PubMed]

51. Edgar, R.C. MUSCLE: Multiple sequence alignment with high accuracy and high throughput. *Nucl Acids Res.* **2004**, *32*, 1792–1797. [CrossRef] [PubMed]

52. HMMER. Available online: http://hmmer.org/ (accessed on 18 March 2019).

53. Analyses for RMS paper. Available online: https://github.com/Gogarten-Lab/rms_analysis.

54. Makarova, K.S.; Wolf, Y.I.; Koonin, E.V. Archaeal clusters of orthologous genes (arCOGs): An update and application for analysis of shared features between Thermococcales, Methanococcales, and Methanobacteriales. *Life (Basel, Switzerland)* **2015**, *5*, 818–840. [CrossRef] [PubMed]

55. Altschul, S.F.; Madden, T.L.; Schäffer, A.A.; Zhang, J.; Zhang, Z.; Miller, W.; Lipman, D.J. Gapped BLAST and PSI-BLAST: A new generation of protein database search programs. *Nucleic Acids Res.* **1997**, *25*, 3389–3402. [CrossRef] [PubMed]

56. Csardi, G.; Nepusz, T. The igraph software package for complex network research. *InterJournal Complex Syst.* **2006**, *1695*.

57. Caspi, R.; Altman, T.; Dale, J.M.; Dreher, K.; Fulcher, C.A.; Gilham, F.; Kaipa, P.; Karthikeyan, A.S.; Kothari, A.; Krummenacker, M.; Latendresse, M.; Mueller, L.A.; Paley, S.; Popescu, L.; Pujar, A.; Shearer, A.G.; Zhang, P.; Karp, P.D. The MetaCyc database of metabolic pathways and enzymes and the BioCyc collection of pathway/genome databases. *Nucleic Acids Res.* **2010**, *38*, D473–D479. [CrossRef] [PubMed]

58. Hoang, D.T.; Chernomor, O.; von Haeseler, A.; Minh, B.Q.; Le, S.V. UFBoot2: Improving the ultrafast bootstrap approximation. *Mol. Biol. Evol.*. [CrossRef]

59. Nguyen, L.-T.; Schmidt, H.A.; von Haeseler, A.; Minh, B.Q. IQ-TREE: A fast and effective stochastic algorithm for estimating maximum-likelihood phylogenies. *Mol. Biol. Evol.* **2015**, *32*, 268–274. [CrossRef] [PubMed]

60. Stamatakis, A. RAxML version 8: A tool for phylogenetic analysis and post-analysis of large phylogenies. *Bioinformatics* **2014**, *30*, 1312–1313. [CrossRef] [PubMed]

61. Bansal, M.S.; Wu, Y.-C.; Alm, E.J.; Kellis, M. Improved gene tree error correction in the presence of horizontal gene transfer. *Bioinformatics* **2015**, *31*, 1211–1218. [CrossRef]

62. Bansal, M.S.; Kellis, M.; Kordi, M.; Kundu, S. RANGER-DTL 2.0: Rigorous reconstruction of gene-family evolution by duplication, transfer and loss. *Bioinformatics* **2018**, *34*, 3214–3216. [CrossRef]

63. Yu, G.; Smith, D.K.; Zhu, H.; Guan, Y.; Lam, T.T.-Y. GGTREE: An R package for visualization and annotation of phylogenetic trees with their covariates and other associated data. *Methods Ecol. Evol.* **2017**, *8*, 28–36. [CrossRef]

64. Yu, G.; Lam, T.T.-Y.; Zhu, H.; Guan, Y. Two methods formapping and visualizing associated data on phylogeny using ggtree. *Mol. Biol. Evol.* **2018**, *35*, 3041–3043. [CrossRef]

65. Dixon, P. VEGAN, a package of R functions for community ecology. *J. Veg. Sci.* **2003**, *14*, 927–930. [CrossRef]

66. Wickham, H. *Ggplot2: Elegant Graphics for Data Analysis*; Springer: Berlin/Heidelberg, 2016; ISBN 3319242776.

67. Hmisc - Main - Vanderbilt Biostatistics Wiki Available online:. Available online: http://biostat.mc.vanderbilt.edu/wiki/Main/Hmisc (accessed on 18 March 2019).

68. Schliep, K.P. phangorn: phylogenetic analysis in R. *Bioinformatics* **2011**, *27*, 592–593. [CrossRef]

69. Salichos, L.; Rokas, A. Inferring ancient divergences requires genes with strong phylogenetic signals. *Nature* **2013**, *497*, 327–331. [CrossRef]

70. Drost, H. Philentropy: Information Theory and Distance Quantification with R. *J. Open Source Softw.* **2018**, *3*. [CrossRef]

71. Gogarten, J.P. Perl script to measure frequency and distribution of CTAG and GATC motifs in DNA Available online:. Available online: https://github.com/Gogarten-Lab/CTAG-GATC-frequencies (accessed on 12 January 2019).

72. Conesa, A.; Götz, S.; García-Gómez, J.M.; Terol, J.; Talón, M.; Robles, M. Blast2GO: A universal tool for annotation, visualization and analysis in functional genomics research. *Bioinformatics* **2005**, *21*, 3674–3676. [CrossRef] [PubMed]

73. Consortium, U. UniProt: The universal protein knowledgebase. *Nucleic Acids Res.* **2016**, *45*, D158–D169.

74. Soucy, S.M.; Fullmer, M.S.; Papke, R.T.; Gogarten, J.P. Inteins as indicators of gene flow in the halobacteria. *Front. Microbiol.* **2014**, *5*, 1–14. [CrossRef]

75. Gupta, R.S.; Naushad, S.; Fabros, R.; Adeolu, M. A phylogenomic reappraisal of family-level divisions within the class Halobacteria: Proposal to divide the order Halobacteriales into the families Halobacteriaceae, Haloarculaceae fam. nov., and Halococcaceae fam. nov., and the order Haloferacales into th. *Antonie Van Leeuwenhoek* **2016**, *109*, 565–587. [CrossRef]

76. Gupta, R.S.; Naushad, S.; Baker, S. Phylogenomic analyses and molecular signatures for the class Halobacteria and its two major clades: A proposal for division of the class Halobacteria into an emended order Halobacteriales and two new orders, Haloferacales ord. nov. and Natrialbales ord. n. *Int. J. Syst. Evol. Microbiol.* **2015**, *65*, 1050–1069. [CrossRef] [PubMed]

77. Allers, T.; Barak, S.; Liddell, S.; Wardell, K.; Mevarech, M. Improved strains and plasmid vectors for conditional overexpression of His-tagged proteins in *Haloferax volcanii*. *Appl. Environ. Microbiol.* **2010**, *76*, 1759–1769. [CrossRef]

78. Holmes, M.L.; Nuttall, S.D.; Dyall-Smith, M.L. Construction and use of halobacterial shuttle vectors and further studies on Haloferax DNA gyrase. *J. Bacteriol.* **1991**, *173*, 3807–3813. [CrossRef] [PubMed]

79. Kong, Y.; Ma, J.H.; Warren, K.; Tsang, R.S.W.; Low, D.E.; Jamieson, F.B.; Alexander, D.C.; Hao, W. Homologous recombination drives both sequence diversity and gene content variation in Neisseria meningitidis. *Genome Biol. Evol.* **2013**, *5*, 1611–1627. [CrossRef] [PubMed]

80. Furuta, Y.; Namba-Fukuyo, H.; Shibata, T.F.; Nishiyama, T.; Shigenobu, S.; Suzuki, Y.; Sugano, S.; Hasebe, M.; Kobayashi, I. Methylome diversification through changes in DNA methyltransferase sequence specificity. *PLoS Genet.* **2014**, *10*, e1004272. [CrossRef]

81. Furuta, Y.; Kobayashi, I. Mobility of DNA sequence recognition domains in DNA methyltransferases suggests epigenetics-driven adaptive evolution. *Mob. Genet. Elements* **2012**, *2*, 292–296. [CrossRef]

82. Furuta, Y.; Kawai, M.; Uchiyama, I.; Kobayashi, I. Domain movement within a gene: A novel evolutionary mechanism for protein diversification. *PLoS ONE* **2011**, *6*, e18819. [CrossRef]

83. Gophna, U.; Brodt, A. CRISPR/Cas systems in archaea. *Mob. Genet. Elements* **2012**, *2*, 63–64. [CrossRef]

84. Fullmer, M.S.; Soucy, S.M.; Swithers, K.S.; Makkay, A.M.; Wheeler, R.; Ventosa, A.; Gogarten, J.P.; Papke, R.T. Population and genomic analysis of the genus Halorubrum. *Front. Microbiol.* **2014**, *5*, 140. [CrossRef] [PubMed]

85. Shalev, Y.; Soucy, S.; Papke, R.; Gogarten, J.; Eichler, J.; Gophna, U. Comparative analysis of surface layer glycoproteins and genes involved in protein glycosylation in the genus Haloferax. *Genes* **2018**, *9*, 172. [CrossRef]

86. Kandiba, L.; Aitio, O.; Helin, J.; Guan, Z.; Permi, P.; Bamford, D.H.; Eichler, J.; Roine, E. Diversity in prokaryotic glycosylation: an archaeal-derived N-linked glycan contains legionaminic acid. *Mol. Microbiol.* **2012**, *84*, 578–593. [CrossRef] [PubMed]

87. Rodriguez-Valera, F.; Martin-Cuadrado, A.-B.; Rodriguez-Brito, B.; Pašić, L.; Thingstad, T.F.; Rohwer, F.; Mira, A. Explaining microbial population genomics through phage predation. *Nat. Rev. Microbiol.* **2009**, *7*, 828–836. [CrossRef] [PubMed]

88. Tang, S.-L.; Nuttall, S.; Ngui, K.; Fisher, C.; Lopez, P.; Dyall-Smith, M. HF2: A double-stranded DNA tailed haloarchaeal virus with a mosaic genome. *Mol. Microbiol.* **2002**, *44*, 283–296. [CrossRef] [PubMed]

89. Olendzenski, L.; Gogarten, J.P. Evolution of genes and organisms: The tree/web of life in light of horizontal gene transfer. *Ann. N. Y. Acad. Sci.* **2009**, *1178*, 137–145. [CrossRef]

90. Naor, A.; Altman-Price, N.; Soucy, S.M.; Green, A.G.; Mitiagin, Y.; Turgeman-Grott, I.; Davidovich, N.; Gogarten, J.P.; Gophna, U. Impact of a homing intein on recombination frequency and organismal fitness. *Proc. Natl. Acad. Sci. USA* **2016**, *113*, E4654-61. [CrossRef]

91. Oliveira, P.H.; Touchon, M.; Rocha, E.P.C. The interplay of restriction-modification systems with mobile genetic elements and their prokaryotic hosts. *Nucleic Acids Res.* **2014**, *42*, 10618–10631. [CrossRef] [PubMed]

92. Karlin, S.; Mrázek, J.; Campbell, A.M. Compositional biases of bacterial genomes and evolutionary implications. *J. Bacteriol.* **1997**, *179*, 3899–3913. [CrossRef] [PubMed]

93. Fournier, P.; Paulus, F.; Otten, L. IS870 requires a 5′-CTAG-3′ target sequence to generate the stop codon for its large ORF1. *J. Bacteriol.* **1993**, *175*, 3151–3160. [CrossRef]

94. Otwinowski, Z.; Schevitz, R.W.; Zhang, R.-G.; Lawson, C.L.; Joachimiak, A.; Marmorstein, R.Q.; Luisi, B.F.; Sigler, P.B. Crystal structure of trp represser/operator complex at atomic resolution. *Nature* **1988**, *335*, 321–329. [CrossRef] [PubMed]

95. Burge, C.; Campbell, A.M.; Karlin, S. Over- and under-representation of short oligonucleotides in DNA sequences. *Proc. Natl. Acad. Sci. USA* **1992**, *89*, 1358–1362. [CrossRef]

96. Bath, C.; Cukalac, T.; Porter, K.; Dyall-Smith, M.L. His1 and His2 are distantly related, spindle-shaped haloviruses belonging to the novel virus group, Salterprovirus. *Virology* **2006**, *350*, 228–239. [CrossRef] [PubMed]

97. Porter, K.; Tang, S.-L.; Chen, C.-P.; Chiang, P.-W.; Hong, M.-J.; Dyall-Smith, M. PH1: An archaeovirus of *Haloarcula hispanica* related to SH1 and HHIV-2. *Archaea* **2013**, *2013*, 456318. [CrossRef] [PubMed]

Article

Insights into Xylan Degradation and Haloalkaline Adaptation through Whole-Genome Analysis of *Alkalitalea saponilacus*, an Anaerobic Haloalkaliphilic Bacterium Capable of Secreting Novel Halostable Xylanase

Ziya Liao [1], Mark Holtzapple [2], Yanchun Yan [1], Haisheng Wang [1], Jun Li [3] and Baisuo Zhao [1,3,*]

[1] Graduate School, Chinese Academy of Agricultural Sciences, Beijing 100081, China; ziyaliao@163.com (Z.L.); yanyanchun@caas.cn (Y.Y.); wanghaisheng@caas.cn (H.W.)

[2] Department of Chemical Engineering, Texas A&M University, College Station, TX 77843, USA; m-holtzapple@mail.che.tamu.edu

[3] Institute of Agricultural Resources and Regional Planning, Chinese Academy of Agricultural Sciences, Beijing 100081, China; junli01@caas.cn

* Correspondence: bszhao@live.com

Received: 30 October 2018; Accepted: 13 December 2018; Published: 20 December 2018

Abstract: The obligately anaerobic haloalkaliphilic bacterium *Alkalitalea saponilacus* can use xylan as the sole carbon source and produce propionate as the main fermentation product. Using mixed carbon sources of 0.4% (*w/v*) sucrose and 0.1% (*w/v*) birch xylan, xylanase production from *A. saponilacus* was 3.2-fold greater than that of individual carbon sources of 0.5% (*w/v*) sucrose or 0.5% (*w/v*) birch xylan. The xylanse is halostable and exhibits optimal activity over a broad salt concentration (2–6% NaCl). Its activity increased approximately 1.16-fold by adding 0.2% (*v/v*) Tween 20. To understand the potential genetic mechanisms of xylan degradation and molecular adaptation to saline-alkali extremes, the complete genome sequence of *A. saponilacus* was performed with the pacBio single-molecule real-time (SMRT) and Illumina Misseq platforms. The genome contained one chromosome with a total size of 4,775,573 bps, and a G+C genomic content of 39.27%. Ten genes relating to the pathway for complete xylan degradation were systematically identified. Furthermore, various genes were predicted to be involved in isosmotic cytoplasm via the "compatible-solutes strategy" and cytoplasmic pH homeostasis though the "influx of hydrogen ions". The halostable xylanase from *A. saponilacus* and its genomic sequence information provide some insight for potential applications in industry under double extreme conditions.

Keywords: Genome Sequencing; Haloalkaliphile; Xylanase; *Alkalitalea saponilacus*

---

1. Introduction

Haloalkaliphiles are extremophilic microorganisms that grow optimally above 0.5 mol·L^{-1} salinity (NaCl) and above pH 9.0 (sodium carbonate/sodium bicarbonate) [1,2]. They are naturally found in saline-alkaline environments such as soda lakes and soda deserts in various dry steppes and semi-desert areas around the world. They also are found in human industrial processes, such as those involving mineral ore, petroleum refining, pulp and paper, textile preparation, leather tanneries, food and potato processing units, lime kilns, and detergent manufacture, all of which generate effluents containing NaOH, Ca(OH)$_2$, etc. [1,3,4]. Over the last three decades, there has been increased interest in exploring haloalkaliphiles as a precious resource that produces stable unique exo-enzymes and organic compounds with potential applications in various industrial processes [1,5,6]. However, to date, our knowledge of anaerobic haloalkaliphiles associated with exploitable enzymology and genetic adaptations is still limited. The genome sequences of haloalkaliphiles may enable many new and potentially transformative biotechnological efforts by providing genetic information to meet rapidly growing industrial demands.

The obligately anaerobic haloalkaliphilic xylanolytic bacterium *Alkalitalea saponilacus* SC/BZ-SP2^T, grows optimally at 0.44–0.69 mol·L^{-1} Na$^+$ (equivalent to 2.6–4.0% NaCl) and pH 9.7. It was retrieved from a meromictic soda lake [7]. This microorganism is classified as a species in genus *Alkalitalea*, family *Marinilabiliaceae*, class *Bacteroidia*, order *Bacteroidetes*. The haloalkaliphile *A. saponilacus* is the first identified anaerobic bacterium that uses xylan as the sole carbon and energy source, and simultaneously produces propionic acid as the major product. If this xylanase is excreted into highly saline-alkaline surroundings and is easily recovered, perhaps it can be applied in industry, such as the biobleaching of wood pulp. This may be the first report about the complete genome sequence of *A. saponilacus*, which could be used by industry in the future.

2. Materials and Methods

2.1. Concentration and Characterization of the Xylanase

To achieve more in-depth understanding of the xylanase characteristics, *A. saponilacus* SC/BZ-SP2^T was optimally grown using birch xylan, sucrose, maltose, glucose, and cellobiose as sole carbon sources as described previously by Zhao and Chen [7]. The bacterial culture (20 mL) was mildly ultra-sonicated in an ice bath for 10 min with 3-s intervals while emitting 200 W (Branson digital sonifier 250, Branson Ultrasonics, Danbury, CT, USA). For crude concentration, the xylanse suspended in culture media was precipitated using a 40% saturated solution of ammonium sulfate and centrifuged for 20 min at 9425× g (i.e., 10,000 rpm) at 4 °C.

Xylanase activity was measured using xylose as the standard with the modified 3, 5-dinitrosalicylic acid colorimetric method (DNS method) [8]. One unit (1 U) of purified xylanase activity was defined as the amount of enzyme that released 1 μmol of xylose equivalent per min under the assay conditions. The relative xylanase activity is defined as the percentage of the maximum xylanase activity measured at various experimental conditions. The conditions for optimal xylanase activity were assayed as follows. (1) NaCl concentrations (0–22%, w/v, at intervals of 2%) at pH 7.0 and at 55 °C; (2) temperatures (30–90 °C, at intervals of 5 °C) with 4% NaCl and pH 7.0; and at pH (4.0–10.5 with intervals of 0.5 pH units) using sodium citrate buffer (pH 4.0–6.0), sodium phosphate buffer (pH 6.0–8.0), and glycine–NaOH buffer (pH 8.0–10.5) at 4% NaCl and at 55 °C. In addition, the effects of surfactants (0.2%, v/v) and various metals (5 mM) on xylanase activity were tested at optimal conditions (i.e., 4% NaCl, pH 7.0, and 55 °C).

2.2. Genome Sequencing, Annotation and Analysis Pipelines

To obtain detailed genetic information of xylan degradation and saline-alkali tolerance, the whole genome of *A. saponilacus* was completely sequenced. Genomic DNA (5 μg) was extracted using Itop^{TM} microbial DNA isolation kit (Itop, Beijing, China) according to the manufacturer's instructions (Beijing, China). After clone library construction, with a mean size of 8–10 kb using g-Tubes, genome sequencing was performed on a Pacific Biosciences RS II sequencer (Pacific Biosciences, Melon Park, CA, USA) using the SMRTbell temperate prep kit version 1.0 (Pacific Biosciences, Menlo Park, CA, USA) and loaded onto a single-molecule real-time (SMRT) cell (Pacific Biosciences, Menlo Park, CA, USA). All cleaned reads were de novo assembled using Hierarchical Genome Assembly Process (HGAP 2.0) [9], resulting in a single contiguous sequence. Briefly, single reads were mapped to seed reads, overlapping consensus sequences were created by a Celera assembler 8.0 [10,11], and the remaining indel and base substitution errors were removed. This method can produce highly accurate and complete de novo assemblies for small prokaryotic genomes [12]. Additionally, libraries were constructed using the TruSeq Nano DNA library preparation kit (Illumina, San Diego, CA, USA) and used to generate 150-bp paired-end reads via the Illumina HiSeq 2000 platform (Illumina, San Diego, CA, USA). Remapping quality-filtered Illumina reads was performed by onto the assembly using BWA [13]. The alignment was passed to Pilon [14] to correct for indels and single nucleotide polymorphisms (SNPs). The average read length of the pacBio raw data was ~6.6 kb, with maximum read length of about 41,148 bases (coverage, ~498) and Illumina paired-end sequencing generated 1.1 million with 2×150-bp reads (coverage, ~150). The whole complete genome sequence of *A. saponilacus* SC/BZ-SP2^T has been deposited at DDBJ/ENA/GenBank under the accession number CP021904.

Glimmer 3.02 was used for gene prediction, gene number, gene total length, and so on [15]. Automated gene annotation was obtained though NCBI Prokaryotic Genome Annotation Pipeline (PGAP) [16]. Then, the genome files in GenBank format (gb file) were uploaded to the Integrated Microbial Genomes Expert Review (IMG/ER) tool [17] for functional annotation, followed the registration of IMG Analysis Project ID (Ga0265418), and Submission ID (184155) in the Genomes OnLine Database (Gold) [18]. Clusters of orthologous groups of proteins (COG) analyses were undertaken using COG functions and abundance profile analysis within IMG/ER [17]. Kyoto encyclopedia of genes and genomes (KEGG) pathway was analyzed using the online tool [19,20].

3. Results and Discussion

3.1. Characteristics of Alkalitalea saponilacus Xylanase

Noticeably, *A. saponilacus* can use insoluble unsubstituted xylan as the sole substrate, indicating it can secrete extracellular xylanase into the surroundings [21]. This microorganism produces xylanase when using birch xylan, sucrose, maltose, glucose, and cellobiose as carbon sources [7]. Xylanase production with a mixture of 0.4% (*w/v*) sucrose and 0.1% (*w/v*) birch xylan substrates was 3.2 greater times than individual carbon sources of 0.5% (*w/v*) sucrose or 0.5% (*w/v*) birch xylan. This may reduce production costs of industrial xylanase because sucrose is widely distributed and less expensive [22]. Figure 1 shows the *relative* xylanase activity with respect to temperature of 30–90 °C (a broad optimum temperature of 45–55 °C), NaCl concentration of 0–22% (*w/v*) (a wide optimum range of 2–6%), and pH of 4.0–10 (optimum pH 7.0). This xylanase tolerates high temperature, acidic and alkali conditions with its unique halophilic characteristic. Unfortunately, it is not alkaliphilic, which was unexpected based on previous descriptions. However, it is not uncommon because xylanase produced from alkaliphilic *Bacillus* sp. Strain K-1 also has an optimal activity at acidic pH 5.5 [21]. Xylanase activity tolerates surfactant (0.2% *v/v*) such as Tween 20 and Triton X-100; its activity increased by 1.16 times with addition of Tween 20. In addition, activity is inhibited by 5-mM metal ions of Cu^{2+}, Fe^{3+}, Ni^{2+}, Al^{3+}, Mn^{2+}, Co^{2+}, Zn^{2+}, and Ca^{2+} with no influence from Mg^{2+}.

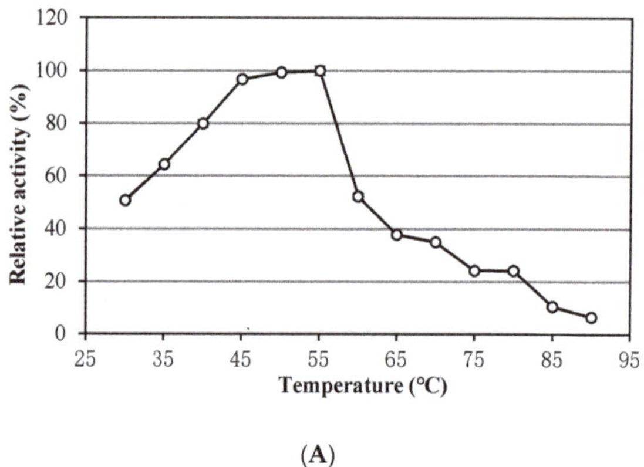

(A)

Figure 1. *Cont.*

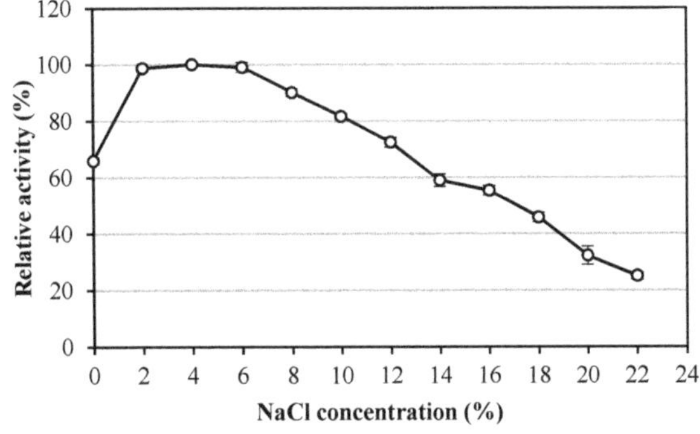

(B)

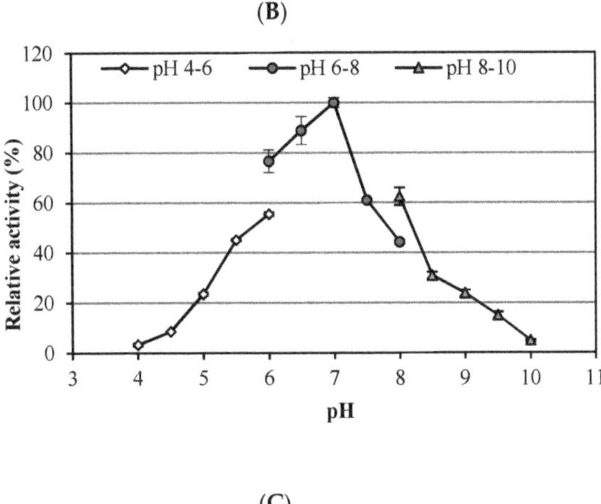

(C)

Figure 1. Characteristics of xylanase produced by *Alkalitalea saponilacus*. (**A**) Influence of temperature on xylanase activity, (**B**) Influence of different NaCl concentration on xylanase activity, (**C**) Influence of pH on xylanase activity. Buffer solutions are citric acid-sodium citrate buffer solution (pH 4–6), sodium hydrogen phosphate-sodium dihydrogen phosphate buffer solution (pH 6–8), glycine-sodium hydroxide buffer solution (pH 8–10), respectively. The error bars indicate standard deviation (SD) of triplicate determination.

3.2. Genome Features of Alkalitalea saponilacus

The complete genome of *A. saponilacus* contained only one chromosome with 4,775,573 bp. The genomic G+C content was 39.27% (Table S1 and Figure S1). Among the 3688 genes predicted, 3626 protein-coding genes (CDS) were predicted that accounted for 98.32% of the whole genome, and 74.02% of which (2684 CDS) were functionally annotated. Also, 12 rRNA genes (four 5S RNAs, four 16S RNA, and four 23S RNA), 48 tRNA genes, and two noncoding RNAs (ncRNAs) were identified.

Of the total 3626 predicted protein-coding genes, 2225 genes (60.33% of the total) were assigned to Clusters of Orthologous Groups (COGs) of proteins (Table S2) and distributed into 23 different categories. The cluster for "cell wall/membrane/envelope biogenesis (COG M)" (200, accounting for 8.99% COGs genes) was the largest group, followed by the classification of "carbohydrate transport and metabolism (COG G)" (181, 8.13%), "translation, ribosomal structure and biogenesis (COG J)" (178, 8.00%) and "amino acid transport and metabolism (COG E)" (159, 7.15%). The categories of "extracellular structures (COG W)" (5, 0.22%) and "cytoskeleton (COG Z)" (1, 0.04%) were the smallest groups.

3.3. The Identified Xylan-Degrading Related Enzymes in Alkalitalea saponilacus

To completely degrade xylans with various substitutions, three major hydrolytic enzymes (endo-β-1,4-xylanases, β-xylosidases, and α-glucuronidase) and several accessory enzymes (e.g., α-L-arabinofuranosidases, α-glucosidases uronidases, and acetyl and feruloyl esterases) might be necessary [23,24]. Genome analysis of *A. saponilacus* revealed one gene of endo-β-1,4-xylanase (XynA), six genes of β-xylosidase (GH43), one gene of α-glucuronidase, and four genes of α-L-arabinofuranosidase (Table 1). Absent were genes encoding acetylxylan esterase, ferulic acid esterase, and *p*-coumaric acid esterase. XynA, which belongs to glycoside hydrolase family 10 (GH10), is a crucial enzyme during xylan degradation and is responsible for fracturing the heteroxylan backbone [25]. XynA of *A. saponilacus* shared the highest sequence identity (55.2%) with the corresponding protein (RefSeq: WP_073173958) of *Tangfeifania diversioriginum* DSM 27063^T, and was 52.4% (WP_074935698) and 52.3% (WP_013549140) identical to that of *Algibacter lectus* DSM 15365^T and *Cellulophaga algicola* DSM 14237^T, respectively. Furthermore, a phylogenetic tree generated from amino acid sequences of 16 xylanase were constructed with neighbor-joining algorithms [26] in MEGA version 7 [27] (Figure 2). The phylogenetic tree showed XynA of *A. saponilacus* formed a single cluster with that of *T. diversioriginum* DSM 27063^T (bootstrap support 76%), and separate branch from those of both *Algibacter lectus* DSM 15365^T and *Cellulophaga algicola* DSM 14237^T, suggesting XynA of *A. saponilacus* may possess distinct characteristics.

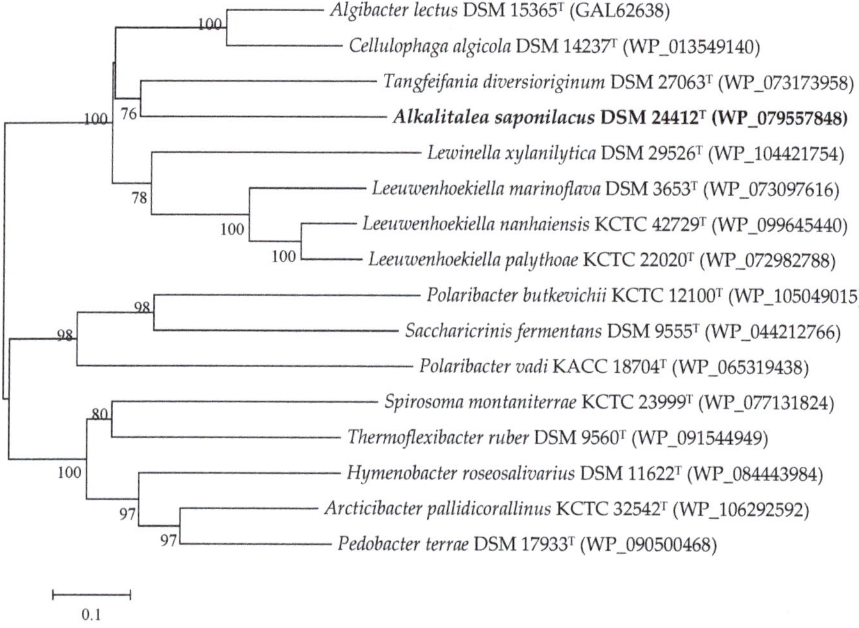

Figure 2. Neighbor-joining phylogenetic tree based on 16 xylanases sequences by using MEGA (Version 6). Numbers on nodes correspond to percentage bootstrap values for 1000 replicates.

Table 1. The identified xylan-degradation-related enzymes. All locus tag numbers are predicted and indicated by IMG/ER. GH: glycoside hydrolase.

Locus Tag	Product Name	GH
CDL62_17705	Endo-β-1,4-xylanase (XynA)	GH10
CDL62_00085	β-xylosidase	GH43
CDL62_06240	β-xylosidase	GH43
CDL62_06275	β-xylosidase	GH43
CDL62_06380	β-xylosidase	GH43
CDL62_15875	β-xylosidase	GH43
CDL62_02285	β-xylosidase	GH43
CDL62_00095	α-glucuronidase	GH67
CDL62_00195	α-L-arabinofuranosidase	GH43
CDL62_00495	α-L-arabinofuranosidase	GH43
CDL62_12950	α-L-arabinofuranosidase	GH43
CDL62_00395	α-L-arabinofuranosidase	GH51

3.4. The Predicted Xylan Degradation Pathways in Alkalitalea saponilacus

To identify xylan degradation pathways in *A. saponilacus*, the 3688 annotated gene sequences to reference canonical pathways in KEGG were mapped and a total of 163 KEGG pathways were obtained. Based on KEGG annotation for genes of potential xylan-degrading enzymes (e.g., endo-β-1,4-xylanase and xylosidase), other potential enzymes involved in xylan degradation (e.g., xylose isomerase and D-xylulokinase) were also identified *A. saponilacus* (Figure 3). Using the genome sequence of *A. saponilacus*, these enzymes can be molecularly cloned to study their inherent characteristics and be modified through genetic engineering technology to be applied into industry.

Under the action of the above enzymes, XynA may possess a putative signal peptide containing 19 amino acids that can guide this enzyme to be secreted outside of cell. *Thermotoga neapolitana*, which was the highest identity sequence, has a similar extracellular enzyme that is secreted into the medium to degrade the macromolecular xylan to xylooligosaccharides, which are transported back into the cell through the specific oligosaccharide transport systems for further assimilation [25]. With the action of XynA, the glycosidic linkage (β-1,4) of xylosides is broken first, and β-1,4-D-xylan oligosaccharides form. Next, xylosidase removes xylose residues from the nonreducing end of β-1,4-D-xylan oligosaccharides, leading to the release of D-xylose, which then converted into D-xylulose using xylose isomerase (XylA). After that, the phosphate and energy from hydrolysis of adenosine triphosphate (ATP) facilitates the conversion of D-xylulose to D-xylulose 5-phosphate, which then is used by the pentose phosphate pathway (PPP). Finally, the D-xylulose 5-phosphate may be transformed into propionic acid using other metabolic pathways [28]. Based on gene annotation, the above pathways are proposed as the way that xylan is metabolized by *A. saponilacus*.

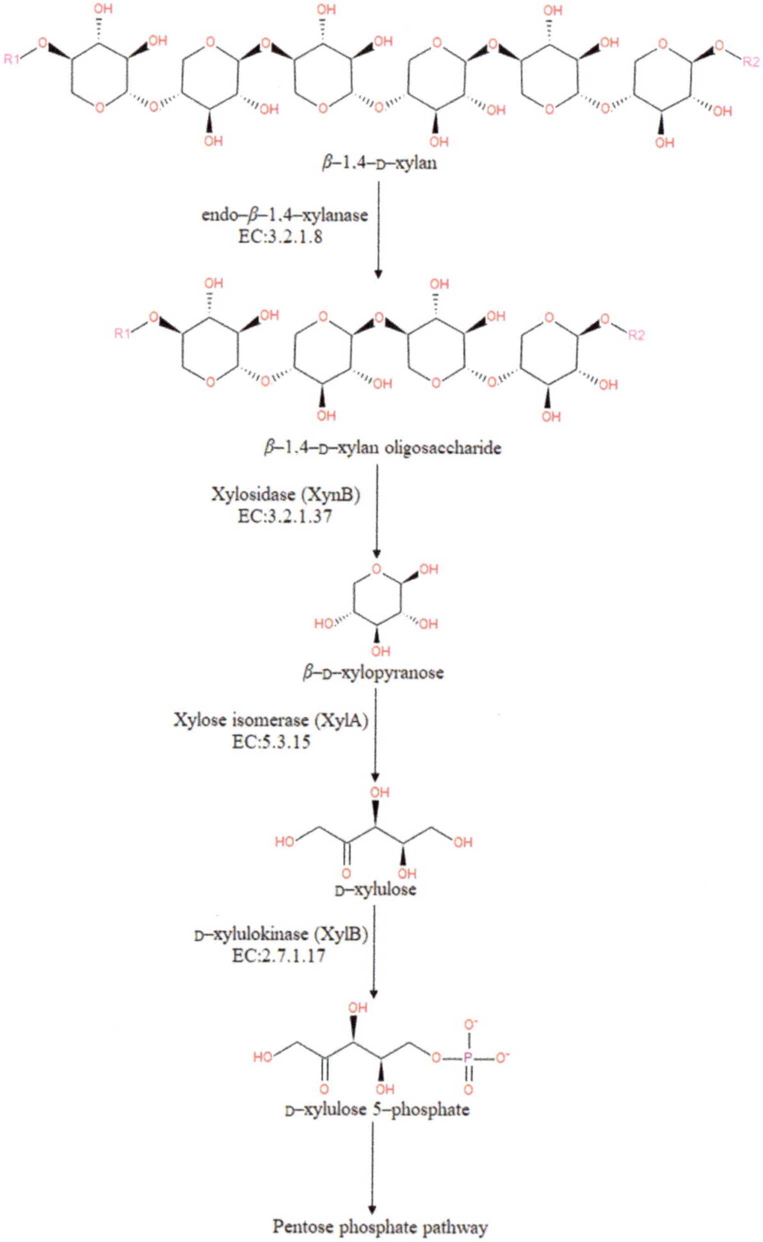

Figure 3. Enzymatic steps of xylan degradation pathways in *A. saponilacus*.

3.5. The Genes Involved in Adaptation to Saline-Alkaline Conditions in Alkalitalea saponilacus

Genome sequence analysis also showed that there are many genes that encode putative proteins potentially associated with the adaptation of *A. saponilacus* to saline-alkaline conditions (Table 2). The presence of *glnA* gene encoding for L-glutamine synthase, one gene coding for choline/glycine/proline betaine transporter (BCCT family), and four genes for Na^+/solute symporter

(SSS family) indicates that *A. saponilacus* maintains osmotic equilibrium across membranes using the "compatible-solutes strategy" when exposed to high salinity [1,29–32]. Additionally, four genes affiliated with the Trk family (two-TrkA-type and two-TrkH-type) responsible for K^+ uptake systems were found, which implies that that *A. saponilacus* might achieve an isosmotic cytoplasm using K^+ to cope rapidly with an osmotic shock [33]. To survive in highly alkaline conditions, *A. saponilacus* has developed genetic adaptations for pH homeostasis via the "influx of hydrogen ions" [34–36]; it harbors seven genes of multisubunit Na^+/H^+ antiporter, four genes of monovalent cation/H^+ antiporter (CPA family, two-CPA1-type and two-CPA2-type), and one gene of Na^+/H^+ antiporter (NhaC family). Eight genes encoding for F_0F_1-ATP synthase and six genes coding for H^+-transporting ATPase (V/A-type) allow *A. saponilacus* to maintain a constant hyper pH cytoplasm using proton gradients [37]. As described above, various predicted genes in *A. saponilacus* offer valuable insights to reveal the adaptive mechanisms of this haloalkaliphile.

Table 2. Genes in *A. saponilacus* involved in adaptation to saline-alkaline environments.

Product Name	Locus Tag
L-glutamine synthesis	
L-glutamine synthetase, GlnA	CDL62_11360
Choline/glycine/proline betaine transporter (BCCT family)	
Choline/glycine/proline betaine transport protein	CDL62_17705
Na^+/solute symporter	
Na^+/solute symporter (SSS family)	CDL62_09935
Na^+/solute symporter (SSS family)	CDL62_06475
Na^+/solute symporter (SSS family)	CDL62_14105
Na^+/solute symporter (SSS family)	CDL62_11075
K^+ transport systems, potassium uptake protein (Trk family)	
Trk system potassium uptake protein, TrkA	CDL62_03510
Trk system potassium uptake protein, TrkH	CDL62_03515
Trk system potassium uptake protein, TrkA	CDL62_03555
Trk system potassium uptake protein, TrkH	CDL62_12070
Na^+/H^+ antiporter (NhaC family)	
H^+/Na^+ antiporter (NhaC family)	CDL62_06020
Multisubunit Na^+/H^+ antiporter	
Multisubunit Na^+/H^+ antiporter, MrpA subunit	CDL62_14320
Multisubunit Na^+/H^+ antiporter, MrpB subunit	CDL62_14325
Multisubunit Na^+/H^+ antiporter, MrpC subunit	CDL62_14330
Multisubunit Na^+/H^+ antiporter, MrpD subunit	CDL62_14335
Multisubunit Na^+/H^+ antiporter, MnhE subunit	CDL62_14340
Multisubunit Na^+/H^+ antiporter, MnhF subunit	CDL62_14345
Multisubunit Na^+/H^+ antiporter, MrpG subunit	CDL62_14350
Monovalent Cation/H^+ antiporter (CPA family)	
K^+/H^+ antiporter (CPA1 family)	CDL62_09425
K^+/H^+ antiporter (CPA1 family)	CDL62_00125
Na^+/H^+ antiporter (CPA2 family)	CDL62_00920
Na^+/H^+ antiporter (CPA2 family)	CDL62_05390
F_0F_1-ATP synthase	
ATP synthase F_1 subcomplex gamma subunit, AtpG	CDL62_07555
ATP synthase F_1 subcomplex alpha subunit, AtpA	CDL62_07560
ATP synthase F_1 subcomplex delta subunit, AtpH	CDL62_07565
ATP synthase F_0 subcomplex B subunit, AtpF	CDL62_07570
ATP synthase F_0 subcomplex C subunit, AtpE	CDL62_07575
ATP synthase F_0 subcomplex A subunit, AtpB	CDL62_07580
ATP synthase F_1 subcomplex epsilon subunit, AtpC	CDL62_07660
ATP synthase F_1 subcomplex beta subunit, AtpD	CDL62_07665
H^+-transporting two-sector ATPase (V-type ATP synthase)	
V/A-type H^+-transporting ATPase subunit E, AtpE	CDL62_11640
V/A-type H^+-transporting ATPase subunit A, AtpA	CDL62_11650
V/A-type H^+-transporting ATPase subunit B, AtpB	CDL62_11655
V/A-type H^+-transporting ATPase subunit D, AtpD	CDL62_11660
V/A-type H^+-transporting ATPase subunit I, AtpI	CDL62_11665
V/A-type H^+-transporting ATPase subunit K, AtpK	CDL62_11670

4. Conclusions

This strain of *A. saponilacus* can grow anaerobically using xylan as the sole carbon source at hypersaline and extremely alkaline conditions. The xylanase activity with the combined substrates of 0.4% sucrose + 0.1% birch xylan substrates was significantly higher than with individual substrates of sucrose or birch xylan. Optimum xylanase activity was obtained at 2–6% NaCl, pH7.0, and 45–55 C. Xylanase activity increased by 1.16 times with addition of Tween 20 whereas it was inhibited by 5-mM Cu^{2+}, Fe^{3+}, Ni^{2+}, Al^{3+}, Mn^{2+}, Co^{2+}, Zn^{2+}, and Ca^{2+}. The genome sequence of *A. saponilacus* has revealed much about the many adaptations of this haloalkaliphile, which allows it to degrade xylan and live in extreme environments. The metabolic enzymes related to xylan degradation, particularly endo-β-1,4-xylanase (XynA), is a new resource of enormous potential value with halophilic characteristics. By synthesizing and transporting compatible solutes, *A. saponilacus* can maintain osmotic equilibrium and survive in hypersaline environments. The chromosome has a wealth of genes that allow Na^+, H^+, and K^+ to be imported and exported, which achieves an isosmotic cytoplasm that adapts to hypersaline environments.

Supplementary Materials: The following are available online at http://www.mdpi.com/2073-4425/10/1/1/s1, Figure S1: Circular genome map of *Alkalitalea saponilacus*. The outermost ring of the circle indicates the size of genome, in which each scale represents 0.5 Mb. The second and third laps illustrate CDSs, colored by COG function classification. The second is forward strand, and the third is backward strand. The fourth circle denotes rRNA and tRNA. The fifth circle is GC content, the red part indicates that GC content is higher than average, whereas the blue is lower. The higher peak value heralds the greater difference in average GC content. The innermost circle states the GC skew (G-C/G+C). The plus-strand is more likely to transcribe CDS when the value is positive, yet minus strand tends to transcribe CDS when the value is negative; Table S1: Genome features of *Alkalitalea saponilacus* SC/BZ-SP2^T; Table S2: Number of genes of *Alkalitalea saponilacus* associated with 23 general COG functional categories.

Author Contributions: Z.L. and B.Z. conceived, designed, collected data and analysis, and wrote the manuscript, M.H. designed and revised the draft paper, Y.Y. and H.W. performed in the sequence alignment and bioinformatics analysis, J.L. participated revised the draft paper. All authors reviewed the final draft.

Funding: This work was supported by grants 31370158 and 31570110 from the National Science Foundation of China (NSFC) and 1610042017001 from the Foundation of Graduate School of Chinese Academy of Agricultural Sciences (CAAS) to Baisuo Zhao.

Conflicts of Interest: The authors declare no conflicts of interest regarding this manuscript.

References

1. Zhao, B.; Yan, Y.; Chen, S. How could haloalkaliphilic microorganisms contribute to biotechnology? *Can. J. Microbiol.* **2014**, *60*, 717–727. [CrossRef] [PubMed]

2. Zhao, B.; Jun, L. Biodiversity of culture-dependent haloalkaliphilic microorganisms. *Acta Microbiol. Sin.* **2017**, *57*, 1409–1420.

3. De Graaff, M.; Bijmans, M.F.; Abbas, B.; Euverink, G.J.; Muyzer, G.; Janssen, A.J. Biological treatment of refinery spent caustics under halo-alkaline conditions. *Bioresour. Technol.* **2011**, *102*, 7257–7264. [CrossRef] [PubMed]

4. Jones, B.E.; Grant, W.D.; Duckworth, A.W.; Owenson, G.G. Microbial diversity of soda lakes. *Extremophiles* **1998**, *2*, 191–200. [CrossRef] [PubMed]

5. Bhatt, H.B.; Gohel, S.D.; Singh, S.P. Phylogeny, novel bacterial lineage and enzymatic potential of haloalkaliphilic bacteria from the saline coastal desert of Little Rann of Kutch, Gujarat, India. *3 Biotech* **2018**, *8*, 53. [CrossRef] [PubMed]

6. Gohel, S.D.; Sharma, A.K.; Dangar, K.G.; Thakrar, F.J.; Singh, S.P. Biology and applications of halophilic and haloalkaliphilic actinobacteria. In *Extremophiles from Biology to Biotechnology*; Durvasula, R.V., Subba Rao, D.V., Eds.; CRC Press: Boca Raton, FL, USA, 2018.

7. Zhao, B.; Chen, S. *Alkalitalea saponilacus* gen. nov. sp. nov. an obligately anaerobic, alkaliphilic, xylanolytic bacterium from a meromictic soda lake. *Int. J. Syst. Evol. Microbiol.* **2012**, *62*, 2618–2623. [CrossRef] [PubMed]

8. Zhao, X.; Luo, K.; Zhang, Y.; Zheng, Z.; Cai, Y.; Wen, B.; Cui, Z.; Wang, X. Improving the methane yield of maize straw: Focus on the effects of pretreatment with fungi and their secreted enzymes combined with sodium hydroxide. *Bioresour. Technol.* **2018**, *250*, 204–213. [CrossRef] [PubMed]

9. Chin, C.S.; Alexander, D.H.; Marks, P.; Klammer, A.A.; Drake, J.; Heiner, C.; Clum, A.; Copeland, A.; Huddleston, J.; Eichler, E.E.; et al. Nonhybrid, finished microbial genome assemblies from long-read SMRT sequencing data. *Nat. Methods* **2013**, *10*, 563–569. [CrossRef]

10. Koren, S.; Schatz, M.C.; Walenz, B.P.; Martin, J.; Howard, J.T.; Ganapathy, G.; Wang, Z.; Rasko, D.A.; McCombie, W.R.; Jarvis, E.D.; et al. Hybrid error correction and de novo assembly of single-molecule sequencing reads. *Nat. Biotechnol.* **2012**, *30*, 693–700. [CrossRef]

11. Myers, E.W.; Sutton, G.G.; Delcher, A.L.; Dew, I.M.; Fasulo, D.P.; Flanigan, M.J.; Kravitz, S.A.; Mobarry, C.M.; Reinert, K.H.; Remington, K.A.; et al. A whole-genome assembly of *Drosophila*. *Science* **2000**, *287*, 2196–2204. [CrossRef]

12. Roberts, R.J.; Carneiro, M.O.; Schatz, M.C. The advantages of SMRT sequencing. *Genome Biol.* **2013**, *14*, 405. [CrossRef] [PubMed]

13. Li, H.; Durbin, R. Fast and accurate long-read alignment with Burrows-Wheeler transform. *Bioinformatics* **2010**, *26*, 589–595. [CrossRef] [PubMed]

14. Walker, B.J.; Abeel, T.; Shea, T.; Priest, M.; Abouelliel, A.; Sakthikumar, S.; Cuomo, C.A.; Zeng, Q.; Wortman, J.; Young, S.K.; et al. Pilon: An integrated tool for comprehensive microbial variant detection and genome assembly improvement. *PLoS ONE* **2014**, *9*, e112963. [CrossRef] [PubMed]

15. Delcher, A.L.; Bratke, K.A.; Powers, E.C.; Salzberg, S.L. Identifying bacterial genes and endosymbiont DNA with Glimmer. *Bioinformatics* **2007**, *23*, 673–679. [CrossRef] [PubMed]

16. Tatusova, T.; DiCuccio, M.; Badretdin, A.; Chetvernin, V.; Nawrocki, E.P.; Zaslavsky, L.; Lomsadze, A.; Pruitt, K.D.; Borodovsky, M.; Ostell, J. NCBI prokaryotic genome annotation pipeline. *Nucleic Acids Res.* **2016**, *44*, 6614–6624. [CrossRef] [PubMed]

17. Markowitz, V.M.; Mavromatis, K.; Ivanova, N.N.; Chen, I.M; Chu, K.; Kyrpides, N.C. IMG ER: a system for microbial genome annotation expert review and curation. *Bioinformatics* **2009**, *25*, 2271–2278. [CrossRef] [PubMed]

18. Mukherjee, S.; Stamatis, D.; Bertsch, J.; Ovchinnikova, G.; Verezemska, O.; Isbandi, M.; Thomas, A.D.; Ali, R.; Sharma, K.; Kyrpides, N.C; et al. Genomes OnLine Database (GOLD) v.6: data updates and feature enhancements. *Nucleic Acids Res.* **2017**, *45*, D446–D456. [CrossRef] [PubMed]

19. Caspi, R.; Foerster, H.; Fulcher, C.A.; Kaipa, P.; Krummenacker, M.; Latendresse, M.; Paley, S.; Rhee, S.Y.; Shearer, A.G.; Tissier, C.; et al. The MetaCyc Database of metabolic pathways and enzymes and the BioCyc collection of Pathway/Genome Databases. *Nucleic Acids Res.* **2008**, *36*, D623–D631. [CrossRef]

20. Kanehisa, M.; Goto, S. KEGG: Kyoto encyclopedia of genes and genomes. *Nucleic Acids Res.* **2000**, *28*, 27–30. [CrossRef] [PubMed]

21. Ratanakhanokchai, K.; Kyu, K.L.; Tanticharoen, M. Purification and properties of a xylan-binding endoxylanase from alkaliphilic *Bacillus* sp. strain K-1. *Appl. Environ. Microbiol.* **1999**, *65*, 694–697.

22. Saleem, M.; Saleem, M.; Jamil, S. Production of xylanase on natural substrates *Bacillus subtilis*. *Int. J. Agric. Biol.* **2013**, *2*, 211–213.

23. Motta, F.L.; Andrade, C.C.P.; Santana, M.H.A. A review of xylanase production by the fermentation of xylan. classification, characterization and applications. In *Sustainable Degradation of Lignocellulosic Biomass-Techniques, Applications and Commercialization*; Chandel, A.K., da Silva, S.S., Eds.; INTECH: Chennai, India, 2013; pp. 251–275.

24. Saha, B.C. Purification and properties of an extracellular beta-xylosidase from newly isolated *Fusarium proliferatum*. *Bioresour. Technol.* **2003**, *90*, 33–38. [CrossRef]

25. Corral, O.L.; Villaseñor-Ortega, F. 14 Xylanases. In *Advances in Agricultural and Food Biotechnology*; Guevara-González, R.G., Torres-Pacheco, I., Eds.; Research Signpost: Kerala, India, 2006; pp. 305–322.

26. Saitou, N.; Nei, M. The neighbor-joining method: A new method for reconstructing phylogenetic trees. *Mol. Biol. Evol.* **1987**, *4*, 406–425.

27. Kumar, S.; Stecher, G.; Tamura, K. MEGA7: Molecular evolutionary genetics analysis version 7.0 for bigger datasets. *Mol. Biol. Evol.* **2016**, *33*, 1870–1874. [CrossRef]

28. Temudo, M.F.; Mato, T.; Kleerebezem, R.; van Loosdrecht, M.C. Xylose anaerobic conversion by open-mixed cultures. *Appl. Microbiol. Biotechnol.* **2009**, *82*, 231–239. [CrossRef]

29. Huang, D.; Liu, J.; Qi, Y.; Yang, K.; Xu, Y.; Feng, L. Synergistic hydrolysis of xylan using novel xylanases, β-xylosidases, and an α-L-arabinofuranosidase from *Geobacillus thermodenitrificans* NG80-2. *Appl. Microbiol. Biotechnol.* **2017**, *101*, 1–15. [CrossRef] [PubMed]
30. Banciu, H.L.; Sorokin, D.Y. Adaptation in haloalkaliphiles and natronophilic bacteria. In *Polyextremophiles: Life Under Multiple Forms of Stress*; Seckbach, J., Oren, A., Stan-Lotter, H., Eds.; Springer: Dordrecht, The Netherlands, 2013; pp. 121–178.
31. Banciu, H.L.; Muntyan, M.S. Adaptive strategies in the double extremophilic prokaryotes inhabiting soda lakes. *Curr. Opin. Microbiol.* **2015**, *25*, 73–79. [CrossRef] [PubMed]
32. Zhao, B.; Mesbah, N.M.; Dalin, E.; Goodwin, L.; Nolan, M.; Pitluck, S.; Chertkov, O.; Brettin, T.S.; Han, J.; Larimer, F.W.; et al. Complete genome sequence of the anaerobic, halophilic alkalithermophile *Natranaerobius thermophilus* JW/NM-WN-LF. *J. Bacteriol.* **2011**, *193*, 4023–4024. [CrossRef] [PubMed]
33. Roberts, M.F. Organic compatible solutes of halotolerant and halophilic microorganisms. *Saline Syst.* **2005**, *1*, 5. [CrossRef]
34. Aono, R.; Ito, M.; Machida, T. Contribution of the cell wall component teichuronopeptide to pH homeostasis and alkaliphily in the alkaliphile *Bacillus lentus* C-125. *J. Bacteriol.* **1999**, *181*, 6600–6606.
35. Krulwich, T.A.; Sachs, G.; Padan, E. Molecular aspects of bacterial pH sensing and homeostasis. *Nat. Rev. Microbiol.* **2011**, *9*, 330–343. [CrossRef] [PubMed]
36. Slonczewski, J.L.; Fujisawa, M.; Dopson, M.; Krulwich, T.A. Cytoplasmic pH measurement and homeostasis in bacteria and archaea. *Adv. Microb. Physiol.* **2009**, *55*, 1–79. [PubMed]
37. Kochegarov, A.A. Modulators of ion-transporting ATPases. *Expert Opin. Ther. Pat.* **2001**, *11*, 825–859. [CrossRef]

Article

Molecular Factors of Hypochlorite Tolerance in the Hypersaline Archaeon *Haloferax volcanii*

Miguel Gomez [1,†], Whinkie Leung [1,†], Swathi Dantuluri [1], Alexander Pillai [1], Zyan Gani [1], Sungmin Hwang [1], Lana J. McMillan [1,2], Saija Kiljunen [3], Harri Savilahti [4] and Julie A. Maupin-Furlow [1,2,*]

[1] Department of Microbiology and Cell Science, Institute of Food and Agricultural Sciences, University of Florida, Gainesville, FL 32611, USA; migmez10@gmail.com (M.G.); whinkie1388@gmail.com (W.L.); swathidantuluri@ufl.edu (S.D.); pillai.alexander@gmail.com (A.P.); z.gani1997@ufl.edu (Z.G.); sungmin.hwang@duke.edu (S.H.); lana.mcmillan@locus-bio.com (L.J.M.)
[2] Genetics Institute, University of Florida, Gainesville, FL 32611, USA
[3] Department of Bacteriology and Immunology, Immunobiology Research Program, University of Helsinki, 00014 Helsinki, Finland; saija.kiljunen@helsinki.fi
[4] Division of Genetics and Physiology, Department of Biology, University of Turku, 20014 Turku, Finland; harri.savilahti@utu.fi
* Correspondence: jmaupin@ufl.edu; Tel.: +1-352-392-4095
† These authors have contributed equally to this work.

Received: 8 October 2018; Accepted: 13 November 2018; Published: 20 November 2018

Abstract: Halophilic archaea thrive in hypersaline conditions associated with desiccation, ultraviolet (UV) irradiation and redox active compounds, and thus are naturally tolerant to a variety of stresses. Here, we identified mutations that promote enhanced tolerance of halophilic archaea to redox-active compounds using *Haloferax volcanii* as a model organism. The strains were isolated from a library of random transposon mutants for growth on high doses of sodium hypochlorite (NaOCl), an agent that forms hypochlorous acid (HOCl) and other redox acid compounds common to aqueous environments of high concentrations of chloride. The transposon insertion site in each of twenty isolated clones was mapped using the following: (i) inverse nested two-step PCR (INT-PCR) and (ii) semi-random two-step PCR (ST-PCR). Genes that were found to be disrupted in hypertolerant strains were associated with lysine deacetylation, proteasomes, transporters, polyamine biosynthesis, electron transfer, and other cellular processes. Further analysis revealed a $\Delta psmA1$ (α1) markerless deletion strain that produces only the α2 and β proteins of 20S proteasomes was hypertolerant to hypochlorite stress compared with wild type, which produces α1, α2, and β proteins. The results of this study provide new insights into archaeal tolerance of redox active compounds such as hypochlorite.

Keywords: archaea; oxidative stress; hypochlorite; redox-active; proteasome

---

1. Introduction

Reactive oxygen species (ROS) and other redox-active compounds can overwhelm the antioxidant mechanisms of a cell and cause damage to most biomolecules including proteins, nucleic acids, lipids, and carbohydrates [1,2]. Oxidation can lead to mutations in DNA by generating single- and double-stranded breaks in the backbone, crosslinks (interstrand and intrastrand), and adducts of bases and sugars [3]. Cell membrane lipids, when oxidized, lose flexibility, which can result in cell lysis [3,4]. Protein oxidation is particularly disruptive, as it leads to protein misfolding, aggregation, breaks in the protein backbone, modified amino acid residues, and loss of catalytic function causing bottlenecks in metabolism [5].

Haloarchaea thrive in hypersaline environments associated with high concentrations of chloride, high doses of ultraviolet (UV) irradiation, oxidative stress, osmotic stress, desiccation, and other extreme conditions [6]. Hypochlorous acid (HOCl) and its derivatives are redox-active compounds commonly encountered in environments of high concentrations of chloride [7,8]. In solution, sodium hypochlorite (NaOCl) forms sodium hydroxide (NaOH) and the strong oxidant HOCl, which can dissociate into hydroxide (OH$^-$) and hypochlorite (OCl$^-$) anions [2].

$$NaOCl + H_2O \Leftrightarrow NaOH + HOCl \Leftrightarrow Na^+ + OH^- + H^+ + OCl^-$$

HOCl interacts with the major classes of biomolecules (i.e., amino acids, proteins, nucleotides, nucleic acids, carbohydrates, and lipids) and inorganic compounds to form free radicals [2]. HOCl exposure commonly damages proteins, DNA, and lipids [2]. Proteins are also reversibly modified by S-thiolation in the presence of HOCl [9]. Particularly destructive is the oxidation of ferrous ion (Fe^{2+}) by HOCl to form hydroxyl radical ($\bullet$OH), chloride ion (Cl$^-$), and ferric ion (Fe^{3+}) [2], with the latter being a catalyst of damaging Fenton chemistry [10].

$$Fe^{2+} + HOCl \Rightarrow Fe^{3+} + \bullet OH + Cl^-$$

Genome-wide en masse insertion mutagenesis is an efficient means to discover gene functions. Recently, the approach was developed for use in *Haloferax volcanii* [11], a model archaeon isolated from the Dead Sea [12]. The strategy is broadly applicable and employs efficient in vitro transposition reaction of phage Mu [13] in combination with random in vivo gene targeting via homologous recombination to generate a mutant library [14].

Our prior work demonstrated that *H. volcanii* responds to hypochlorite stress in a manner that can be quantified at the proteome level by stable isotope labeling in cell culture (SILAC) coupled with tandem mass spectrometry analysis (LC-MS/MS) [15]. To further understand these responses on a global scale, we now report the development of an approach to select for *H. volcanii* mutants that are tolerant of extreme doses of NaOCl on a defined medium. The mutants were selected from a previously described comprehensive random transposon insertion library of *H. volcanii* [11]. The strains were selected for growth on high doses of NaOCl when using glycerol as the carbon/energy source, an organic alcohol common to hypersaline ecosystems [16]. The locations of the transposons on the genome were identified by inverse-nested two-step PCR (INT-PCR) and semi-random two-step PCR (ST-PCR). A selection of markerless deletion (transposon minus) strains, each with a disrupted gene identified in our analysis, was used to further define the cellular mechanisms of hypochlorite tolerance. An isogenic Δ*psmA1* (α1) mutant that produced only the α2 and β proteins of 20S proteasomes was found to be hypertolerant to hypochlorite. Thus, the type of α protein that forms the gate and outermost ring of 20S proteasomes can alter stress responses in this archaeon.

2. Materials and Methods

2.1. Materials

Biochemicals were from Sigma Aldrich (St. Louis, MO, USA). Other inorganic and organic analytical grade chemicals were from Fisher-Scientific (Atlanta, GA, USA). Klenow and other DNA polymerases, restriction endonucleases, and T4 DNA ligase were from New England Biolabs (Ipswich, MA, USA). Agarose for DNA analysis was from Bio-Rad laboratories (Hercules, CA, USA). Desalted oligonucleotide primers were purchased from Integrated DNA Technologies (Coralville, IA, USA). Reagent grade NaOCl solution (available chlorine 10%–15%, 425044–250 mL) was purchased from Sigma Aldrich.

2.2. Strains and Media

Strains and primers used in this study are listed in Table S1. *H. volcanii* strains were grown at 42 °C at 200 rpm orbital shaking in glycerol minimal medium (GMM) with ammonium chloride used as the nitrogen source, as previously described [17]. Uracil was added at a concentration of 50 µg/mL for all Δ*pyrE2* strains. Growth in liquid medium was measured by optical density at 600 nm (OD_{600}). The solid GMM (+uracil) medium was supplemented with 20 g/L agar (Sigma-Aldrich, catalog number: A7002). *H. volcanii* cells were incubated on agar plates in closed zippered bags at 42 °C for 5–10 days in the dark. American Type Culture Collection (ATCC) medium 974 [17] was only used for experiments that compared H26 and markerless deletion strains (not the transposon mutants) by liquid assay (see later section).

2.3. Isolation of Mutants with Enhanced Tolerance to Hypochlorite Stress

To isolate strains with enhanced tolerance to hypochlorite stress, a transposon mutant library of *H. volcanii* H295 [11] was plated on increasing doses of NaOCl using GMM supplemented with uracil (+uracil) to compensate for the Δ*pyrE2* mutation of the parent strain (H295). The medium was devoid of tryptophan to stably maintain the transposon, which carried the tryptophan synthase gene (*trpA*). *H. volcanii* H26 (a *trpA*+ derivative of H295) was included as a control. The *H. volcanii* H295 transposon library was multiplied as previously described [18], stored in 20% glycerol at −80 °C, and thawed upon use. The cell mixture (10 µL) was diluted with 990 µL GMM (+uracil). Aliquots (100 µL) of the diluted cells were spread onto GMM (+uracil) agar plates supplemented with 0 to 1.2 mM NaOCl. Colony forming units (CFUs) per ml of aliquot were determined based on growth at 0 mM NaOCl. The plates were incubated at 42 °C for five days. Transposon mutant strains that grew at 1.2 mM NaOCl were streaked for isolation on GMM (+uracil) and stored at −80 °C in 20% (v/v) glycerol stocks.

2.4. Genomic DNA Isolation

Mutants that grew on GMM (+uracil) agar plates in the presence of 1.2 mM NaOCl were transferred into 3 mL GMM (+uracil) (in 13 × 100 mm tubes) and grown to mid-log phase (OD_{600} 0.6–0.8) at 42 °C with rotary shaking (200 rpm). Cells were pelleted at 14,000× *g* for 10 min (25 °C). The culture broth was removed, and the cell pellets were stored at −80 °C. Genomic DNA was extracted from the pellets by spooling [19].

2.5. Identification of Transposon Insertion Sites

Two PCR-based methods were used to identify transposon insertion sites on the *H. volcanii* genome as outlined in Figure 1.

2.5.1. Inversed Nested Two-Step PCR

Genomic DNA was digested with restriction enzymes, including the following: (i) NdeI and HindIII, (ii) BmtI and BspHI, (iii) NdeI and NheI, or (iv) XhoI and BclI. The restriction enzymes were randomly selected and paired based on optimal activity in a common reaction buffer and temperature, but all were unable to cleave the transposon. The resulting genomic DNA fragments were treated with Klenow DNA polymerase to allow for the fill-in of 5′ overhangs and removal of 3′ overhangs. The Klenow-treated DNA was circularized by blunt-end ligation using T4 DNA ligase. The circularized DNA was used as a template for the INT-PCR approach. In the first (inverse PCR) step, primer M1-1F and M1-1R were designed to anneal to the *trpA* and *cat* genes of the transposon, respectively, and generate a DNA product that carried a portion of the transposon along with the genomic DNA that was adjacent to transposon insertion site. In the second (nested PCR) step, the M1-2F/2R primers were designed to target and amplify the inverse PCR product.

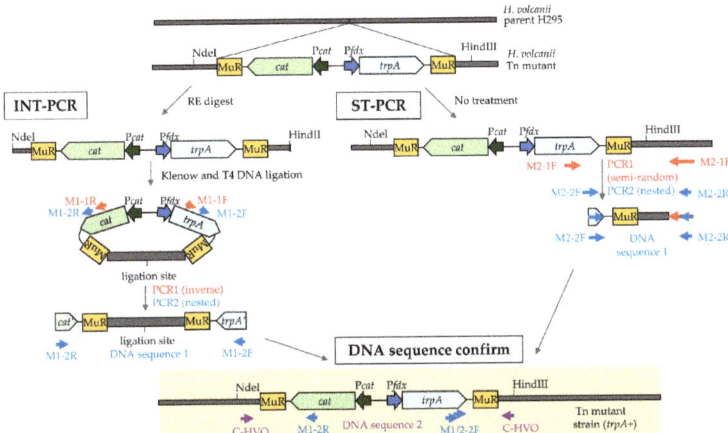

Figure 1. Schematic diagram of the inverted-nested two-step PCR (INT-PCR, left) and the semi-random two-step PCR (ST-PCR, right) strategies to identify the transposon insertion sites in *Haloferax volcanii*. The transposable (Tn) element includes the following: two Mu repeats (MuR), a chloramphenicol acetyltransferase (*cat*) gene, a P*cat* promoter, a tryptophan synthase (*trpA*) gene, and a ferredoxin promoter (P*fdx*). NdeI and HindIII are examples of the restriction enzyme (RE) sites used to cleave the genomic DNA prior to blunt-end ligation to form the circular DNA template used in the INT-PCR method. Primer pairs used for the two PCR steps (PCR1 and PCR2) and the DNA sequencing are color coded (red, blue, and purple) and numbered according to Table S1. See Methods for details.

2.5.2. Semi-Random Two-Step PCR

ST-PCR was performed according to the work of [20] with the following modifications. In the first PCR step, primer M2-F1 was designed to specifically anneal to the *trpA* gene of the transposon, while the 3′ end of the degenerate, primer M2_R1 was used to randomly anneal to the *H. volcanii* genomic DNA including the region adjacent to the transposon insertion site. The DNA product generated by the first PCR was subsequently used as a template for nested PCR, with the M2 2F primer being specific to the 3′ end of the *trpA* gene of the transposon and M2-2R primer being specific to the 5′ end of primer 2.

2.5.3. DNA Sequencing to Identify the Transposon Insertion Sites

The forward and reverse strands of the final DNA products generated by INT-PCR and ST-PCR were sequenced using the nested PCR primers (M1- and M2-2F/R, respectively) by the Sanger method (Eton Bioscience, Inc. San Diego, CA, USA). These DNA sequences (DNA sequence 1, Figure 1) were compared to the *H. volcanii* DS2 genome by NCBI BLAST nucleotide and blastx [21] to determine the transposon insertion site. A confirmation primer (C_HVO locus tag number) that specifically annealed to the genomic region adjacent to the transposon insertion site was also paired with an appropriate nested PCR primer (M1-2F/R or M2-F) to generate a 'confirmation PCR product' using genomic DNA isolated from the mutant strain as a template. The PCR product was excised from the gel and purified with QIAquick Gel Extraction Kit (Qiagen, Germantown, MD, USA), following which each strand of the DNA was analyzed by Sanger sequencing using the primer pairs of the confirmation PCR (DNA sequence 2, Figure 1).

2.6. PCR Conditions

Phusion and OneTaq DNA polymerases were used for INT-PCR and ST-PCR, respectively (New England Biolabs). PCR (50 μL) reactions were mixed on ice with dimethyl sulfoxide (DMSO), buffer, deoxynucleotide triphosphate mix, primers (Table S1), template (spooled genomic DNA or PCR product), and DNA polymerase according to the supplier (New England Biolabs). PCR was performed using Mycyler and Icycler thermal cyclers (Bio-Rad) at the temperatures and times of incubation indicated in Table S2.

2.7. RNA Isolation and Real-time Quantitative Reverse Transcription PCR

Total RNA was isolated from *H. volcanii* cells by using the RNeasy minikit (Qiagen) according to the supplier's instructions. DNA was removed by using a Turbo DNA-free kit (AM1907, Thermo Fisher Scientific, Waltham, MA USA) according to the recommendations of the supplier. The level of contaminating DNA after Turbo DNase digestion was below the limit of detection by PCR. The integrity of the RNA was determined by 2.0% (wt/vol) agarose gel electrophoresis. RNA (50 ng) per reaction mixture volume (50 µL) served as the template. One-step real-time quantitative reverse transcription PCR (qRT-PCR) was performed using the QuantiTect SYBR green RT-PCR kit (Qiagen) following the protocol described in the handbook of the supplier. The qRT-PCR procedure was performed under conditions of 50 °C for 30 min; 95 °C for 15 min; and 40 cycles of 95 °C for 15 s, 51 °C for 30 s, and 72 °C for 30 s, followed by determination of the melting curve using a CFX96 real-time C1000 thermal cycler (Bio-Rad). A single peak revealed by melting curve analysis indicated a single product. The messenger RNA (mRNA) levels were normalized to the internal standard *ribL* (*hvo_1015*). A standard curve was generated by using a QuantiTect SYBR green PCR kit (Qiagen) following the manufacturer's protocol. Purified H26 genomic DNA served as the templates to test different primer pairs for PCR efficiency. Primers with PCR efficiencies between 95% (HVO_2469) and 101% (HVO_1957) were used (Table S1).

2.8. Hypochlorite Stress Plate Assay

H. volcanii strains were streaked from −80 °C glycerol stocks onto GMM (+uracil) agar. The cells were incubated at 42 °C for five days. Isolated colonies were patched on GMM (+uracil) agar supplemented with 0, 0.8, 1.2, or 1.6 mM NaOCl as indicated. Cells were monitored for growth at 42 °C.

2.9. Hypochlorite Stress Liquid Assay

H. volcanii markerless deletion and wild type strains were streaked onto ATCC 974 agar from −80 °C glycerol stocks. The cells were grown for five days at 42 °C. Isolated colonies were inoculated into 25 mL of ATCC 974 medium in 125 mL Erlenmeyer flasks and incubated at 42 °C with rotary shaking (200 rpm). At log phase (OD_{600} of 0.6–0.8), cells were washed twice with GMM (+uracil) by centrifugation ($8600\times g$, 1 min at room temperature) and diluted to an OD_{600} of 0.1 unit in GMM (+uracil) supplemented with 0 or 1.5 mM NaOCl as indicated. The sample (150 µL) was transferred into a 96-well plate. The plate was covered with a lid and sealed with micro-pore tape to protect cells from desiccation. The cells were incubated at 42 °C (807 cycles-per-minute (cpm) shaking) in a micro plate reader (Epoch 2, BioTek, Winooski, VT, USA) with monitoring every 2 h at OD_{600}.

2.10. SDS-PAGE and Immunoblotting Analysis

H. volcanii strains were streaked with a sterile toothpick from −80 °C glycerol stocks onto GMM (+uracil) agar and incubated at 42 °C for five days. Isolated colonies were transferred into 4 mL GMM (+uracil) (in 13 × 100 mm culture tubes) and grown with orbital shaking (at 200 rpm) for two days at 42 °C to an OD_{600} of 0.6. Cells were subcultured and similarly grown to an OD_{600} of 1.2. Cultures (1 mL) were harvested by centrifugation ($16,873\times g$ for 10 min at room temperature). The cell pellets were resuspended to a final OD_{600} of 0.065 per 10 µL by addition of 150–200 µL of 2× reducing SDS loading buffer (50 mM Tris-Cl buffer at pH 6.8 with 2% (w/v) SDS, 10% (v/v) glycerol, 0.3 mg·mL^{-1} bromophenol blue, and 2.5% (v/v) β-mercaptoethanol). Samples were boiled for 10 min. Proteins (10 uL sample) were separated by reducing 10% SDS-PAGE (sodium dodecyl sulfate polyacrylamide gel electrophoresis). Equivalent protein loading was based on OD_{600} of the cell culture (0.065 units per lane) and confirmed by Coomassie blue R-250 staining of parallel gels. Unstained proteins were electroblotted from the gels onto PVDF (polyvinylidene fluoride) membranes (Amersham) as per standard protocol (BioRad). Proteasome α1 subunit was detected using a 1:5000 dilution of an anti-α1 rabbit polyclonal antibody [22] followed by goat anti-rabbit IgG-HRP (horseradish peroxidase) Cruz Marker compatible antibody (SC-2030, Santa Cruz Biotechnology, Dallas, TX, USA) at a 1:1000 dilution. Immunoreactive antigens were detected using the Pierce enhanced chemiluminescence (ECL) Plus Western blotting substrate (Thermo Fisher Scientific) and Amersham Hyperfilm ECL (GE Healthcare Bio-Sciences, Pittsburgh, PA, USA).

2.11. Prediction of Protein Structure and Function

To discern biological mechanisms that may be used by haloarchaea to withstand hypochlorite stress, the function of the genes disrupted in the NaOCl-hypertolerant mutant strains was predicted as follows. Genes encoding proteins that had orthologs with a known function were identified by BlastP [21] and Interpro [23]. Signal peptide (Sec, Tat and lipobox) motifs and transmembrane spanning helices were predicted by SignalP 4.1 [24], TatP 1.0 [25], TatFind 1.4 [26], PSORTb 3.0 [27], and TMHMM 2.0 [28]. A 3D protein structure was modeled using Phyre 2.0 [29]. Conserved active sites/motifs were identified by comparison of the 3D protein models to biochemically characterized proteins using Chimera 1.11 [30]. Operon organization and genome synteny were analyzed using the UCSC (University of California Santa Cruz) Archaeal Genome Browser [31] and SyntTax [32].

3. Results and Discussion

3.1. Haloferax volcanii *Mutants of Enhanced Tolerance to Hypochlorite Stress*

To identify *H. volcanii* mutants of enhanced tolerance to hypochlorite stress, a transposon mutant library was compared to wild type (H26) for growth in the presence of increasing doses of NaOCl. GMM (+uracil) agar plates supplemented with 1.2 mM NaOCl were found to clearly distinguish wild type (H26) from mutant strains when examining 3–5 $\times$ 10^6 CFUs per plate. No growth was observed for the H26 control under these conditions. By contrast, the transposon mutant library yielded ~ 100 CFUs per plate, thus reaching a survival rate of 0.0025%. Individual colonies of the transposon mutant library were further isolated on the selective medium and demonstrated to be tolerant of at least 1.2 mM NaOCl, when compared twith H26 and other strains, which did not survive under these conditions (Figure 2).

0 mM NaOCl   1.2 mM NaOCl

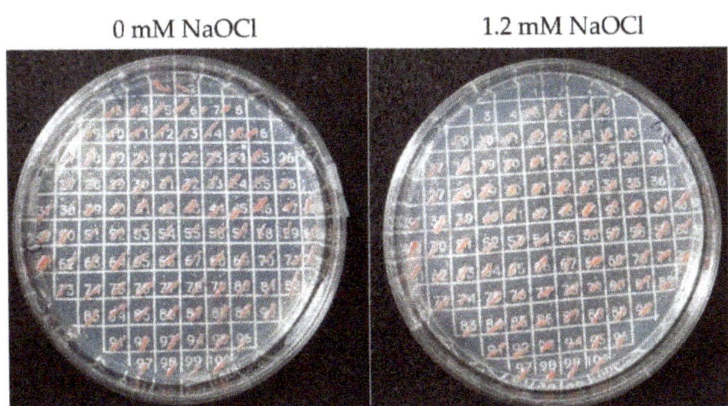

Figure 2. *H. volcanii* transposon mutant strains were found to be hypertolerant to hypochlorite stress. NaOCl-hypertolerant strains (positions 5–100) were compared to H26 (DS70 $\Delta pyrE2$, position 1), HM1041 (H26 $\Delta samp2$, position 2), a transposon library pool (position 3), and HM1052 (H26 $\Delta ubaA$, position 4). Cells were patched on glycerol minimal medium (GMM) (+uracil) with 0 mM NaOCl (left) and 1.2 mM NaOCl (right) for the comparison. See Methods for details.

3.2. Transposon Insertion Sites Mapped on the Genome of Haloferax volcanii *Mutant Strains were Found to be Hypertolerant to Hypochlorite*

Two basic approaches (INT-PCR and ST-PCR) were used to map the transposon insertion sites (Figure 2). These approaches, which differed from the whole genome sequencing method previously reported to map the transposon insertion sites [11,18], were found to be useful in the rapid identification of the transposon insertion site of twenty distinct isolates (Figure 3, Table S3). Fourteen of the sites were identified by INT-PCR, while six sites were identified by ST-PCR. The ST-PCR method was found to be more rapid but less prone to positive identification. Most of the mutant strains (17 of 20 total) had transposons inserted within an open reading frame (ORF), suggesting the loss of gene function. Eleven of these sites were linked to genes that, in wild type (H26) cells, encode proteins detected by SILAC-based LC-MS/MS analysis [15], including three proteins (HVO_2375, HVO_1041, and HVO_1957) of significant differential abundance, after NaOCl stress (see later discussion for details). Three of the isolates had a transposon inserted within an intergenic region upstream (5′) of the predicted TATA box promoter element (isolates 7, 35A, and 36A; Figure S1). This intergenic positioning of the transposon (which has internal promoter elements, P*cat* and P*fdx*) was speculated to have altered the expression of the downstream genes (*hvo_2469* in isolates 7 and 35A and *hvo_1957* in isolate 36A). Thus, the expression of these genes was monitored by qRT-PCR and was found to be significantly altered in response to NaOCl in the mutants compared with wild type (H26) (Figure 4 and later discussion). In addition to the intergenic insertions, several of the clones harbored transposons in the same genomic region as exemplified by two insertions 5′ of *hvo_2469* (isolates 7 and 35A), two insertions in the *hvo_2374–2375* region (isolates 33A and 83A), and two insertions in *hvo_2770* (isolates 30 and 40) (Figure 3).

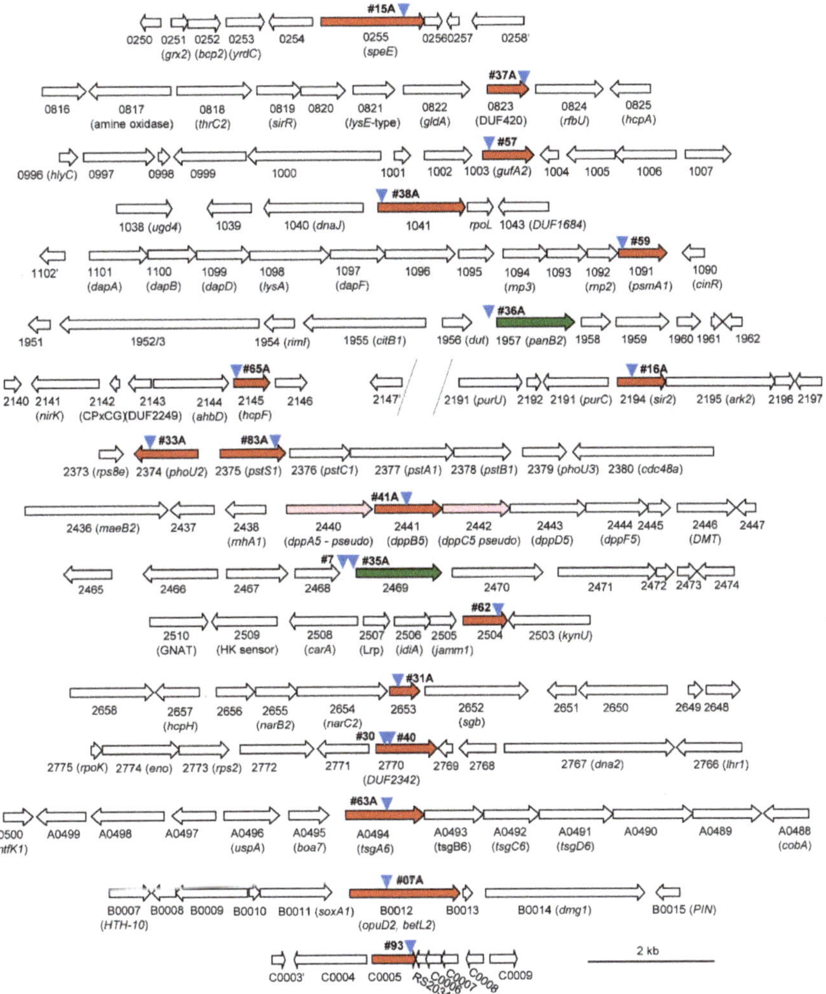

Figure 3. Schematic diagram of the genomic neighborhood of the transposon insertion site of the *H. volcanii* mutants that were hypertolerant to hypochlorite. Down arrowheads (blue) represent the site of transposon insertion with the strain isolate number indicated adjacent to the symbol. Arrows represent open reading frames (ORFs) deduced from the *H. volcanii* genome sequence (UniProt proteome ID: UP000008243). HVO_ gene locus tag number and select gene names are indicated below the ORF. Genes with transposons located within the ORF are in red. Pseudo genes are in pink. Genes in green are downstream of transposons that are inserted within intergenic regions. Scale bar, 2 kb.

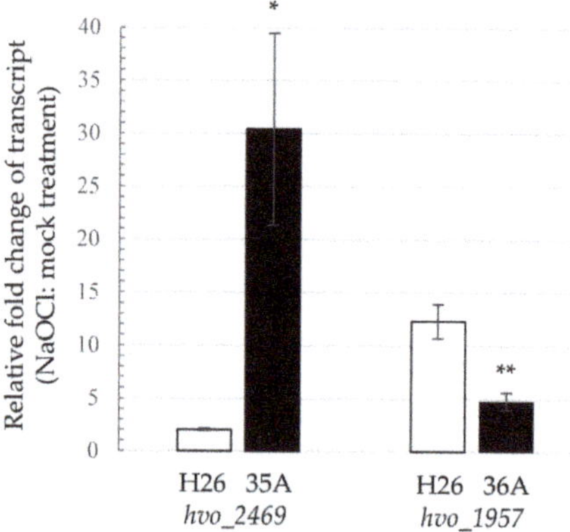

Figure 4. Transcript level responses to NaOCl stress are altered by transposon insertions 5′ of the TATA box of *hvo_2469* and *hvo_1957*. *H. volcanii* strains were grown to exponential phase and treated with 0 and 3 mM NaOCl for 60 min. Total RNA was extracted and used for real-time quantitative reverse transcription PCR (qRT-PCR). Gene expression levels were normalized to the internal reference *ribL* (one-fold). *H. volcanii* strains and the qRT-PCR gene targets are indicated below the *x*-axis. Significant differences between the wild type (H26) and mutant (35A and 36A) strains were calculated by Student's *t*-test analysis with equal variance (p-value < 0.05 * and $p < 0.01$ **). All data are expressed as mean $\pm$ standard error of the mean (SEM) for $n = 3$ technical replicates. See Methods for details.

3.3. Membrane versus Intracellular Functions

More than half (11/20) of the hypertolerant mutant strains had transposon insertions within or adjacent to genes predicted to encode proteins associated with the membrane (Table S3). By comparison, only 24% of the theoretical proteome is estimated to be membrane proteins [33]. Several of the mutant strains were disrupted in ORFs predicted to encode pre-proteins (Figure S2) that would be translocated through the twin arginine translocation (Tat) system, cleaved by a protease to expose an N-terminal cysteine and lipid modified at this cysteine residue [34]. Thus, lipoprotein maturation may be generally sensitive to oxidative stress, as this process occurs within the cell membrane and requires a cysteine thiol group. Of the membrane-associated ORFs that were disrupted by the transposon insertions, most were involved in transport (HVO_1003, HVO_2374, HVO_2375, HVO_2441, HVO_2469, HVO_A0494, HVO_B0012), with others related to redox homeostasis (HVO_0823 and HVO_2145), spermidine synthase (HVO_0255), or unknown functions (HVO_2653) (see later discussion).

3.4. Metal Ion Transport

Gene homologs of metal ion transport were found to be disrupted in cells hypertolerant to hypochlorite stress. HVO_1003, a member of the zinc transport protein (ZIP) family that function in the uptake of zinc and/or other metals [35], was disrupted in isolate 57 (Figure 3). During oxidative stress, zinc can replace the Fe^{2+} released from damaged Fe–S clusters and inactivate these metalloenzymes, thus causing metabolic bottlenecks [36,37]. Reduced levels of intracellular zinc may be the mechanism that enables the *hvo_1003*::Tn mutant to be at a selective advantage over the wild type when challenged with NaOCl. HVO_2441, a homolog of the ATP-binding cassette (ABC) permease DppB that functions in the uptake of dipeptides and heme-iron in bacteria [38], was also found to be disrupted in a NaOCl hypertolerant mutant (Figure 3). Impaired heme-based iron transport is predicted to reduce the intracellular pool of labile iron that causes damaging Fenton chemistry during oxidative stress [39]. Interestingly, *hvo_2441* (*dppB5*) is flanked by two pseudo genes (*dppA5* and *dppC5*) (Figure 3), suggesting the function of this transport system is already reduced in *H. volcanii* DS2 derived strains such as H26 [40].

3.5. Inorganic Phosphate Transport

Gene homologs associated with inorganic phosphate (Pi) regulation (HVO_2374, PhoU2) and transport (HVO_2375, PstS1) were disrupted in two of the NaOCl-hypertolerant strains (Figure 3). Consistent with this finding, PstS1 abundance is down during NaOCl stress in *H. volcanii* [15]. PstS1 is one of the two solute binding protein homologs in haloarchaea that are associated with the ABC-type Pi uptake system operons *pts1* (*pstS1C1A1B1*) and *pts2* (*pstS2C2A2B2*) [41]. Like PstS2, PtsS1 was found to have the conserved residues needed to coordinate and facilitate Pi uptake (Figure S3). In *Halobacterium salinarum*, Δ*pts1* mutants have a higher rate of Pi uptake than Δ*pst2* mutants [41], suggesting the NaOCl-hypertolerant *ptsS1*::Tn mutant had increased intracellular levels of Pi. The *H. volcanii* PhoU2 protein, by contrast, was predicted to be a transcriptional regulator of Pi uptake as it was found to have an N-terminal DNA binding domain and to be in synteny with the *pts1* operon; it also was found to harbor the conserved residues needed to coordinate a multinuclear iron cluster (Figures S4 and S5) that is used in Pi uptake by comparison with characterized PhoU proteins [42]. In bacteria, disruption of *phoU2* increases expression of Pi transport and elevates levels of inorganic polyphosphate (polyP), an intracellular polymer that promotes hypochlorite tolerance [43]. Thus, the mechanism(s) of NaOCl-hypertolerance of the *H. volcanii phoU2*::Tn and *pstS1*::Tn isolates may be related by enhanced levels of intracellular Pi and/or polyP.

3.6. Organic Molecule Transport

Organic molecule transport homologs (HVO_A0494 (TsgA6), HVO_B0012 (BetT), and HVO_2469 (SNF)) were found to be disrupted in several NaOCl-hypertolerant strains (Figure 3). This finding was insightful as the growth medium included only five organic molecule supplements: glycerol (20 mM), thiamine (0.8 µg/mL), biotin (0.1 µg/mL), uracil (50 µg/mL), and tris(hydroxymethyl)aminomethane (Tris, 30 mM).

TsgA6, a predicted Tat lipoprotein of the ABC-type solute binding protein family 1 (IPR006059), was found to be disrupted in isolate 63A (Figure S2). This family includes solute binding proteins that facilitate the uptake of maltose/maltodextrin (MalE/X), oligosaccharide (MsmE), glycerol-3-phosphate (UgpB), and thiamine (TbpA) [44]. While TsgA6 was related to MalE in the 3D structure, it was not predicted to bind maltose (Figure S6). A polar effect on transcription/translation cannot be ruled out, as *tsgA6* was the first gene of the *tsgA6B6C6D6* ABC-transport system operon (Figure 3). One explanation is that the transposon insertion in *tsgA6* rendered cells hypertolerant to hypochlorite by minimizing the synthesis of unnecessary organic molecule transporters that span the membrane, as glycerol was the sole carbon/energy source of the selective conditions.

HVO_B0012, a member of the betaine/carnitine/choline transporter (BCCT) family and the amino acid–polyamine–organocation superfamily (APCS), was found to be disrupted in the NaOCl-hypertolerant strain 67A (Figure 3). BCCT family members mediate the Na^+/H^+-coupled symport or precursor/product antiport of organic molecules with positively charged nitrogen or sulfur head-groups [45,46]. HVO_B0012 was found to be 43% identical to the H^+-driven betaine/choline transporter BetT and in genome synteny with homologs of betaine/choline metabolism and oxidative stress response (Figure S7). While bacteria produce, excrete, and reaccumulate betaine/choline during osmotic stress [47], the salt-in strategy of *H. volcanii* and growth on GMM (+uracil) may have alleviated the need for HVO_B0012 during hypochlorite stress. Betaine/choline metabolism in *H. volcanii* was predicted to require electron transfer and H_2O_2 generation (which may exacerbate ROS damage) (Figure S7).

The third organic molecule transporter identified in the transposon library screen was HVO_2469, a member of the sodium neurotransmitter symporter family (SNF, IPR000175), and close homolog of MhsT, a Na^+-dependent transporter of hydrophobic L-amino acids [48] (Figure S8). Unlike the other transporters that were disrupted in the coding sequence, the site of transposon insertion was 5′ of the BRE (B recognition element) and TATA box promoter elements of *hvo_2469* (Figure S1). Further analysis by qRT-PCR revealed that *hvo_2469* transcripts were up 2-fold in the wild type and over 30-fold in isolate 35A during hypochlorite stress (Figure 4). This result suggested the SNF transporter homolog HVO_2469 was more abundant in the hypertolerant mutant compared with the wild type during hypochlorite stress. One possible explanation is that the transporter facilitated the re-accumulation/uptake of hydrophobic amino acids, known to have strong radical scavenging activities [49].

3.7. Polyamine Synthesis

The polyamine aminopropyltransferase homolog HVO_0255 (SpeE) was found to be disrupted in the hypertolerant mutant 15A (Figure 3). Polyamines are polycationic molecules that interact with negatively charged regions of biomolecules (e.g., nucleic acids, lipids, and proteins) [50] and are considered 'primordial stress molecules' based on their ability to protect cells from oxidative damage or induce oxidative stress [51]. In thermophiles, differences in polyamine ratios are correlated with growth temperature [52]. *H. volcanii* produces the polyamines agmatine and cadaverine [53,54] and has two SpeE homologs (HVO_B0357 and HVO_0255) that are predicted to be integral membrane proteins with conserved active site residues and structural homology to polyamine aminopropyltransferases (Figure S9). Interestingly, the *H. volcanii* SpeEs were found to differ in primary sequence at key regions known to alter polyamine binding specificity in thermophilic polyamine aminopropyltransferases [55]; HVO_0255 had a GG(GA)G(F/Y) motif and long C-terminal extension, while HVO_B0357 had a GGGD(W/Y) motif and short-C-terminal tail (Figures S9 and S10). Thus, the haloarchaeal SpeEs are predicted to have distinct polyamine binding specificities, such that the *hvo_0255* mutation would alter the polyamines ratios that promote NaOCl-hypertolerance.

3.8. Membrane Associated Redox Reactions

Several ORFs associated with redox reactions in the cell membrane were found to be disrupted in the hypertolerant strains, including *hvo_2145* (*hcpF*), *hvo_0823*, and *hvo_2653* (Figure 3). Halocyanins, such as HcpF, are blue (type-1) copper proteins that serve as mobile electron carriers in the peripheral membrane of haloarchaea [56,57]. HcpF is one of eight halocyanins (HcpA-H) predicted to be translocated by the Tat system and one of four (HcpC, HcpD, HcpE, and HcpF) that is a putative Tat lipoprotein (Figure S2). Disruption of *hcpF* may have shifted the cell to halocyanins that are more tolerant of oxidant and/or less prone to transfer electrons to complexes that form oxygen radicals, such as cytochrome oxidase [58,59]. The *hvo_0823*, disrupted in isolate 37A, encodes a cytochrome *c*-oxidase (EC: 1.9.3.1) type helical bundle protein homolog related in structure to the non-heam binding multipass transmembrane domain of *caa3*-type cytochrome oxidases (Figure S11). These results suggest that HVO_0823 could impact the integrity and/or maturation of the *H. volcanii* cytochrome oxidase(s) [60] and, thus, reduce ROS production. The ORF of the multipass transmembrane domain protein HVO_2653 was also found to be disrupted in one of the hypertolerant strains; *hvo_2653* is in genomic synteny with halocyanin (*hcpH*) and nitrate reductase (*narB2* and *narC2*) genes (Figure 3), suggesting it is associated with redox reactions in the cell membrane. Overall, the transposon insertions in *hcpF*, *hvo_0823*, and *hvo_2653* may have shifted electron transfer from enzymes that leak electrons and radicals to systems that avoid metabolic bottlenecks during hypochlorite stress.

3.9. Oxidoreductase and Hydrolase Enzymes

Several soluble oxidoreductase and hydrolase homologs were found to be impaired in hypertolerant mutant strains. Isolate 62 was disrupted in *hvo_2504* encoding a member of the NAD(P)$^+$ -dependent short-chain dehydrogenase/reductase (SDR) family (IPR002347), which was found to be distinct from characterized SDRs based on the absence of conserved catalytic tetrad residues (Figure S12). Isolates 30 and 40 were disrupted in *hvo_2770*, encoding a member of the MA clan of zinicin-like metalloproteases (MEROPS peptidase database, http://www.ebi.ac.uk/merops/) that had the conserved residues to coordinate the catalytic Zn^{2+} ion (Figure S13). While zinicins function in biological processes, such as cell signaling [61], which may explain the enhanced tolerance of isolates 30 and 40 to hypochlorite, the transposon insertions were also positioned at 5′ of *hvo_2771* encoding a glyoxalase-like gene homolog. Enhanced levels of glyoxalase I are associated with tolerance to oxidative/hypochlorite stress [62,63], which could also be the rationale for our findings. The NaOCl-hypertolerant isolate 93 was found to be disrupted in *hvo_c0005*, encoding a member of the six-hairpin glycosidase-like (IPR008928) superfamily, including enzymes that synthesize/break glycosidic bonds damaged by hypochlorite [2]. Isolate 38A was disrupted in *hvo_1041*, encoding a metallo-β-lactamase homolog that had the conserved active site residues needed to coordinate the two catalytic Zn^{2+} ions (Figure S14). HVO_1041 was most closely related to *Thermus thermophilus* TTHA1429 with an unknown function [64], based on 3D structural homolog modeling (Figure S14), and was predicted to cleave RNA based on genomic synteny with *rpoL* (DNA-directed RNA polymerase subunit L) (Figure 3). However, our previous SILAC-based proteomics reveals HVO_1041 to be up in abundance after exposure to NaOCl [15], suggesting that disruption of this gene was not responsible for the observed NaOCl hypertolerance. Instead, we propose that the transposon insertion caused an upregulation of the adjacent gene *hvo_1040*, encoding a DnaJ (Hsp40) chaperone domain protein that facilitated protein folding.

3.10. Protein Lysine Deacetylation

The NAD^{+}-dependent histone deacetylase (HDAC) family protein Sir2 (HVO_2194) of the Gcn5-related N-acetyltransferase (GNAT) subfamily was found to be disrupted in the hypochlorite tolerant strain 16A (Figure 3). *H. volcanii* encodes two HDACs (the class III Sir2 and the class II HdaI) and three histone acetyltransferase (HAT) family homologs (Pat1, Pat2, and Elp3). Of these, HdaI is essential for growth [65], and Pat1/2 are inversely correlated in protein abundance during hypochlorite stress (Pat1 is down 3.1-fold, while Pat2 is up 1.8-fold) [15]. Thus, enzymes that control lysine acetylation are linked to hypochlorite stress. In *Sulfolobus* sp., Sir2 deacetylates the chromatin protein Alba, resulting in Alba binding to the chromosome and transcriptional repression [66]. While Alba-like superfamily (IPR036882) proteins are not conserved in *H. volcanii*, a wide variety of proteins are found to be lysine acetylated in *Haloferax mediterranei* [67]. The *sir2*::Tn mutation of isolate 16A is presumed to stabilize proteins in their acetylated state. The addition of an acetyl group could generally block and protect the amino groups of lysine residues, which are particularly susceptible to free radical formation by HOCl [2]. Alternatively, lysine acetylation may be associated with specific pathway(s), such as chromatin remodeling, DNA replication repair, and/or metabolism, to overcome hypochlorite stress.

3.11. Proteasome Components

Isolate 59 was found to have a transposon insertion in *psmA1* (*hvo_1091*) encoding proteasomal subunit α1, while isolate 36A had a transposon inserted at 5′ of *panB1* (*hvo_1957*) encoding a proteasomal Rpt-like AAA ATPase (Figure 3). Archaea encode different types of 20S proteasomes and AAA ATPases that form a proteasome system network [68,69]. In the case of *H. volcanii*, the α1 and α2 proteins associate with β in combinations of α1β, α2β, and α1α2β to form at least three distinct types of 20S proteasomes [22,70,71]. While 20S proteasomes are essential for growth, the individual type of α subunit is not; in other words, *H. volcanii* requires the β subunit and at least one type of α subunit (α1 or α2) to grow [72]. By contrast, the AAA ATPase complexes of PanA1, PanA1/B2, and PanB2 are not essential for growth of *H. volcanii* [71,72]. Consistent with disruption of the *psmA1* gene, the α1 protein was not detected in the *psmA1*::Tn mutant (isolate 59) by immunoblotting analysis (Figure 5). Based on this finding, the isogenic Δ*psmA1* (*psmA*, α1) strain GZ130 was analyzed for hypertolerance to hypochlorite and was found to grow on GMM (+uracil) agar plates supplemented with 1.6 mM NaOCl; by contrast, the Δ*psmA2* (*psmC*, α2) (GZ114) and parent (H26) strains were unable to grow under these conditions (Figure 6A). In liquid culture, where higher concentrations of NaOCl are required for toxicity, the Δ*psmA1* mutant was found to be more tolerant of NaOCl than the parent (H26) and Δ*samp1* ubiquitin-like mutant (HM1041), with the latter being found to be hypersensitive to NaOCl (Figure 6B), as previously described [15]. The AAA ATPase PanB2 was also associated with NaOCl hypertolerance. After exposure to NaOCl, the *panB2* transcript levels were found to be up by more than 10-fold in the wild type (H26) and up 5-fold in isolate 36A, suggesting the transcript level stress response at this locus was dysregulated in the mutant (Figure 4). This apparent dysregulation may explain the hypertolerance of isolate 36A to hypochlorite stress. A previous study revealed that the levels of α2 and PanB2 were elevated during the stationary phase; whereas α1 and PanA1 were prevalent during log phase growth [71]. In addition, PanB2 was found to be up after exposure to NaOCl [15]. These changes in the composition of proteasome system apparently prepare the cell for stresses, such as hypochlorite and stationary phase, that are encountered in hypersaline environments.

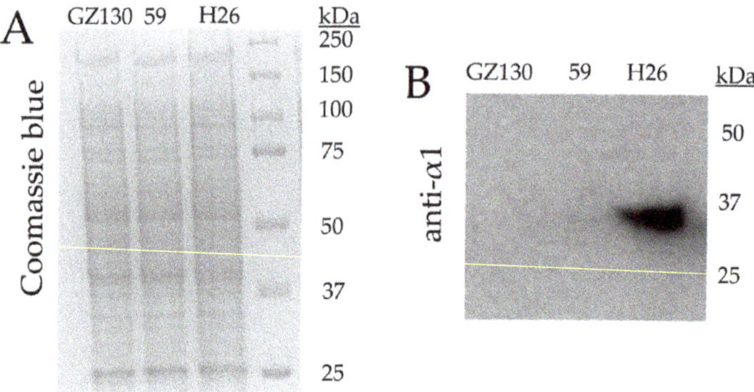

Figure 5. The 20S proteasomal α1 protein is not detected in one of the strains (isolate 59) that was hypertolerant to hypochlorite stress. Whole cells lysate, separated by reducing 10% SDS-PAGE (sodium dodecyl sulfate polyacrylamide gel electrophoresis), was analyzed for total protein by Coomassie Blue staining (**A**) and α1 protein by immunoblotting (**B**). Strains used for analysis included H26 (parent strain), GZ130 (Δ*psmA1*), and isolate 59 (*psmA1*::Tn) as indicated. See Methods for details.

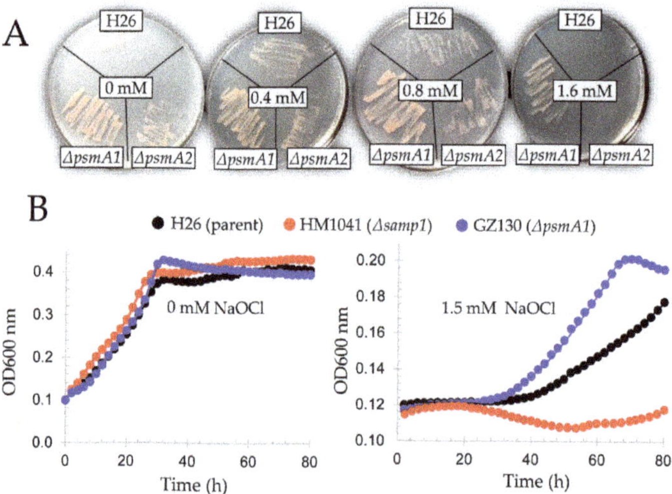

Figure 6. Markerless deletion of *psmA1* renders cells hypertolerant to hypochlorite stress. *H. volcanii* strains H26 (parent), GZ130 (Δ*psmA1*), GZ114 (Δ*psmA2*), and HM1041 (Δ*samp1*) were analyzed for tolerance to hypochlorite by plate (**A**) and/or liquid (**B**) assay. *H. volcanii* cells are more tolerant of NaOCl in liquid culture compared with agar plates. Experiments were performed in experimental and technical triplicate and are presented as an average. See Methods for details.

4. Conclusions

Here, we developed an assay to select and isolate *H. volcanii* mutant strains from a random transposon library that displayed enhanced tolerance to hypochlorite stress when grown on glycerol. We demonstrate that two economical and rapid PCR-based methods (INT-PCR and ST-PCR) can be used to identify the transposon insertion sites on the *H. volcanii* genome. Thus, whole genome sequencing is not required to identify the insertion sites. In the hypochlorite tolerant strains, transposon insertions were found within or upstream of genes associated with lysine deacetylation, proteasome components, transporters, polyamine biosynthesis, electron transfer, and other functions. qRT-PCR analysis revealed that transposons inserted at 5′ of promoter consensus sequences can perturb the transcript abundance of the downstream gene. Subsequent analysis of markerless deletion strains demonstrated that cells with 20S proteasomes composed of α2β are more tolerant of hypochlorite stress than those of α1β subunit composition or a combination thereof. Overall, this approach provided a global view of hypochlorite tolerance in the model archaeon *H. volcanii*.

Supplementary Materials: The following are available online at http://www.mdpi.com/2073-4425/9/11/562/s1, Figure S1: Transposon insertion sites with respect to the promoter region of *hvo_2469* (SNF family transporter) and *hvo_1957* (AAA-type ATPase proteasome associated nucleotidase PanB2); Figure S2: Twin arginine translocation (Tat) signal peptide and lipoprotein motifs of select *H. volcanii* proteins; Figure S3: *H. volcanii* PstS1 (HVO_2375) is structurally related to bacterial phosphate (Pi) solute binding proteins; Figure S4: PhoU domain family proteins of *H. volcanii*; Figure S5: Relationship of *H. volcanii* PhoU2 (HVO_2374) to *Thermotoga maritima* Tm1734; Figure S6: Comparison of *H. volcanii* TsgA6 (HVO_A0494) with the *E. coli* maltose binding protein (MalE); Figure S7: *H. volcanii* HVO_B0012 (BetT) and its predicted function in choline transport; Figure S8: *H. volcanii* SNF family transporter HVO_2469; Figure S9: *H. volcanii* SpeE homologs are related to the polyamine aminopropyltransfease of *Thermus thermophiles*; Figure S10: Multiple amino acid sequence alignment of the active site region of haloarchaeal members of the SpeE family; Figure S11: *H. volcanii* HVO_0823 is a cytochrome *c*-oxidase (EC: 1.9.3.1) type helical bundle protein; Figure S12: HVO_2504 is a putative oxidoreductase of the short-chain dehydrogenase/reductase (SDR, IPR002347) family; Figure S13: *H. volcanii* HVO_2770 is a DUF2342 protein of the zinicin-like metallopeptidase type 2 (IPR018766) and F420 biosynthesis associated (IPR022454) families; Figure S14: *H. volcanii* HVO_1041 is a metallo-β-lactamase superfamily protein; Table S1: Strains, plasmids, and oligonucleotide primers used in this study; Table S2: Cycling conditions for the inverse-nested two-step PCR (INT-PCR) and semi-random two-step PCR (ST-PCR); Table S3: Identification of transposon insertion sites on the genome of *H. volcanii* mutant strains are found to be hypertolerant of hypochlorite stress. Ref: [73–79].

Author Contributions: S.D., L.J.M., and J.A.M.-F. designed the research. M.G., W.L., L.J.M., and J.A.M.-F. wrote the paper. S.K. and H.S. provided the transposon insertion library. M.G., W.L., S.D., A.P., Z.G., S.H., S.M., L.J.M., and J.A.M.-F. performed the research. All authors read and approve the paper.

Funding: Funds awarded to J.A.M.-F. for this project were through the Bilateral BBSRC-NSF/BIO program (NSF 1642283), U.S. Department of Energy, Office of Basic Energy Sciences, Division of Chemical Sciences, Geosciences and Biosciences, Physical Biosciences Program (DOE DE-FG02-05ER15650), and the National Institutes of Health (NIH R01 GM57498).

Conflicts of Interest: The authors do not have a conflict of interest to declare.

References

1. Riley, P.A. Free radicals in biology: Oxidative stress and the effects of ionizing radiation. *Int. J. Radiat. Biol.* **1994**, *65*, 27–33. [CrossRef] [PubMed]

2. Panasenko, O.M.; Gorudko, I.V.; Sokolov, A.V. Hypochlorous acid as a precursor of free radicals in living systems. *Biochemistry (Mosc.)* **2013**, *78*, 1466–1489. [CrossRef] [PubMed]

3. Cabiscol, E.; Tamarit, J.; Ros, J. Oxidative stress in bacteria and protein damage by reactive oxygen species. *Int. Microbiol.* **2000**, *3*, 3–8. [PubMed]

4. Pratt, D.A.; Tallman, K.A.; Porter, N.A. Free radical oxidation of polyunsaturated lipids: New mechanistic insights and the development of peroxyl radical clocks. *Acc. Chem. Res.* **2011**, *44*, 458–467. [CrossRef] [PubMed]

5. Reichmann, D.; Voth, W.; Jakob, U. Maintaining a healthy proteome during oxidative stress. *Mol. Cell.* **2018**, *69*, 203–213. [CrossRef] [PubMed]

6. Jones, D.L.; Baxter, B.K. DNA repair and photoprotection: Mechanisms of overcoming environmental ultraviolet radiation exposure in halophilic archaea. *Front. Microbiol.* **2017**, *8*, 1882. [CrossRef] [PubMed]

7. Ortiz-Bermudez, P.; Srebotnik, E.; Hammel, K.E. Chlorination and cleavage of lignin structures by fungal chloroperoxidases. *Appl. Environ. Microbiol.* **2003**, *69*, 5015–5018. [CrossRef] [PubMed]

8. Wang, G. Chloride flux in phagocytes. *Immunol. Rev.* **2016**, *273*, 219–231. [CrossRef] [PubMed]

9. Loi, V.V.; Rossius, M.; Antelmann, H. Redox regulation by reversible protein S-thiolation in bacteria. *Front. Microbiol.* **2015**, *6*, 187. [CrossRef] [PubMed]

10. Wardman, P.; Candeias, L.P. Fenton chemistry: An introduction. *Radiat. Res.* **1996**, *145*, 523–531. [CrossRef] [PubMed]

11. Kiljunen, S.; Pajunen, M.I.; Dilks, K.; Storf, S.; Pohlschroder, M.; Savilahti, H. Generation of comprehensive transposon insertion mutant library for the model archaeon, *Haloferax volcanii*, and its use for gene discovery. *BMC Biol.* **2014**, *12*, 103. [CrossRef]

12. Mullakhanbhai, M.F.; Larsen, H. *Halobacterium volcanii* spec. nov., a Dead Sea halobacterium with a moderate salt requirement. *Arch. Microbiol.* **1975**, *104*, 207–214. [CrossRef] [PubMed]

13. Haapa, S.; Taira, S.; Heikkinen, E.; Savilahti, H. An efficient and accurate integration of mini-Mu transposons in vitro: A general methodology for functional genetic analysis and molecular biology applications. *Nucleic Acids Res.* **1999**, *27*, 2777–2784. [CrossRef] [PubMed]

14. Kiljunen, S.; Pajunen, M.I.; Savilahti, H. Transposon insertion mutagenesis for archaeal gene discovery. *Methods Mol. Biol.* **2017**, *1498*, 309–320. [CrossRef] [PubMed]

15. McMillan, L.J.; Hwang, S.; Farah, R.E.; Koh, J.; Chen, S.; Maupin-Furlow, J.A. Multiplex quantitative SILAC for analysis of archaeal proteomes: A case study of oxidative stress responses. *Environ. Microbiol.* **2018**, *20*, 385–401. [CrossRef] [PubMed]

16. Nissenbaum, A. The microbiology and biogeochemistry of the Dead Sea. *Microb. Ecol.* **1975**, *2*, 139–161. [CrossRef] [PubMed]

17. Sherwood, K.; Cano, D.; Maupin-Furlow, J. Glycerol-mediated repression of glucose metabolism and glycerol kinase as the sole route of glycerol catabolism in the haloarchaeon *Haloferax volcanii*. *J. Bacteriol.* **2009**, *191*, 4307–4315. [CrossRef] [PubMed]

18. Legerme, G.; Yang, E.; Esquivel, R.N.; Kiljunen, S.; Savilahti, H.; Pohlschroder, M. Screening of a *Haloferax volcanii* transposon library reveals novel motility and adhesion mutants. *Life (Basel)* **2016**, *6*, E41. [CrossRef] [PubMed]

19. Dyall-Smith, M. The Halohandbook: Protocols for Halobacterial Genetics v.7.2. 2009. Available online: http://www.haloarchaea.com/resources/halohandbook/Halohandbook_2009_v7.2mds.pdf (accessed on 19 November 2018).

20. Chun, K.T.; Edenberg, H.J.; Kelley, M.R.; Goebl, M.G. Rapid amplification of uncharacterized transposon-tagged DNA sequences from genomic DNA. *Yeast* **1997**, *13*, 233–240. [CrossRef]

21. Altschul, S.F.; Gish, W.; Miller, W.; Myers, E.W.; Lipman, D.J. Basic local alignment search tool. *J. Mol. Biol.* **1990**, *215*, 403–410. [CrossRef]

22. Kaczowka, S.J.; Maupin-Furlow, J.A. Subunit topology of two 20S proteasomes from *Haloferax volcanii*. *J. Bacteriol.* **2003**, *185*, 165–174. [CrossRef] [PubMed]

23. Finn, R.D.; Attwood, T.K.; Babbitt, P.C.; Bateman, A.; Bork, P.; Bridge, A.J.; Chang, H.Y.; Dosztanyi, Z.; El-Gebali, S.; Fraser, M.; et al. InterPro in 2017-beyond protein family and domain annotations. *Nucleic Acids Res.* **2017**, *45*, D190–D199. [CrossRef] [PubMed]

24. Nielsen, H. Predicting secretory proteins with SignalP. *Methods Mol. Biol.* **2017**, *1611*, 59–73. [CrossRef] [PubMed]

25. Bendtsen, J.D.; Nielsen, H.; Widdick, D.; Palmer, T.; Brunak, S. Prediction of twin-arginine signal peptides. *BMC Bioinform.* **2005**, *6*, 167. [CrossRef] [PubMed]

26. Rose, R.W.; Bruser, T.; Kissinger, J.C.; Pohlschroder, M. Adaptation of protein secretion to extremely high-salt conditions by extensive use of the twin-arginine translocation pathway. *Mol. Microbiol.* **2002**, *45*, 943–950. [CrossRef] [PubMed]

27. Yu, N.Y.; Wagner, J.R.; Laird, M.R.; Melli, G.; Rey, S.; Lo, R.; Dao, P.; Sahinalp, S.C.; Ester, M.; Foster, L.J.; et al. PSORTb 3.0: Improved protein subcellular localization prediction with refined localization subcategories and predictive capabilities for all prokaryotes. *Bioinformatics* **2010**, *26*, 1608–1615. [CrossRef] [PubMed]

28. Krogh, A.; Larsson, B.; von Heijne, G.; Sonnhammer, E.L. Predicting transmembrane protein topology with a hidden Markov model: Application to complete genomes. *J. Mol. Biol.* **2001**, *305*, 567–580. [CrossRef] [PubMed]

29. Kelley, L.A.; Mezulis, S.; Yates, C.M.; Wass, M.N.; Sternberg, M.J. The Phyre2 web portal for protein modeling, prediction and analysis. *Nat. Protoc.* **2015**, *10*, 845–858. [CrossRef] [PubMed]

30. Pettersen, E.F.; Goddard, T.D.; Huang, C.C.; Couch, G.S.; Greenblatt, D.M.; Meng, E.C.; Ferrin, T.E. UCSF Chimera—A visualization system for exploratory research and analysis. *J. Comput. Chem.* **2004**, *25*, 1605–1612. [CrossRef] [PubMed]

31. Schneider, K.L.; Pollard, K.S.; Baertsch, R.; Pohl, A.; Lowe, T.M. The UCSC Archaeal Genome Browser. *Nucleic Acids Res.* **2006**, *34*, D407–D410. [CrossRef] [PubMed]

32. Oberto, J. SyntTax: A web server linking synteny to prokaryotic taxonomy. *BMC Bioinform.* **2013**, *14*, 4. [CrossRef] [PubMed]

33. Kirkland, P.; Humbard, M.; Daniels, C.; Maupin-Furlow, J. Shotgun proteomics of the haloarchaeon *Haloferax volcanii*. *J. Proteome Res.* **2008**, *7*, 5033–5039. [CrossRef] [PubMed]

34. Gimenez, M.I.; Dilks, K.; Pohlschroder, M. *Haloferax volcanii* twin-arginine translocation substates include secreted soluble, C-terminally anchored and lipoproteins. *Mol. Microbiol.* **2007**, *66*, 1597–1606. [CrossRef] [PubMed]

35. Porcheron, G.; Garenaux, A.; Proulx, J.; Sabri, M.; Dozois, C.M. Iron, copper, zinc, and manganese transport and regulation in pathogenic Enterobacteria: Correlations between strains, site of infection and the relative importance of the different metal transport systems for virulence. *Front. Cell. Infect. Microbiol.* **2013**, *3*, 90. [CrossRef] [PubMed]

36. Imlay, J.A. The mismetallation of enzymes during oxidative stress. *J. Biol. Chem.* **2014**, *289*, 28121–28128. [CrossRef] [PubMed]

37. Nieboer, E.; Richardson, D.H.S. The replacement of the nondescript term 'heavy metal' by a biologically significant and chemically significant classification of metal ions. *Environ. Pollut. B* **1980**, *1*, 3–26. [CrossRef]

38. Letoffe, S.; Delepelaire, P.; Wandersman, C. The housekeeping dipeptide permease is the *Escherichia coli* heme transporter and functions with two optional peptide binding proteins. *Proc. Natl. Acad. Sci. USA* **2006**, *103*, 12891–12896. [CrossRef] [PubMed]

39. Frawley, E.R.; Fang, F.C. The ins and outs of bacterial iron metabolism. *Mol. Microbiol.* **2014**, *93*, 609–616. [CrossRef] [PubMed]

40. Hartman, A.; Norais, C.; Badger, J.; Delmas, S.; Haldenby, S.; Madupu, R.; Robinson, J.; Khouri, H.; Ren, Q.; Lowe, T.; et al. The complete genome sequence of *Haloferax volcanii* DS2, a model archaeon. *PLoS ONE* **2010**, *5*, e9605. [CrossRef] [PubMed]

41. Furtwangler, K.; Tarasov, V.; Wende, A.; Schwarz, C.; Oesterhelt, D. Regulation of phosphate uptake via Pst transporters in *Halobacterium salinarum* R1. *Mol. Microbiol.* **2010**, *76*, 378–392. [CrossRef] [PubMed]

42. Liu, J.; Lou, Y.; Yokota, H.; Adams, P.D.; Kim, R.; Kim, S.H. Crystal structure of a PhoU protein homologue: A new class of metalloprotein containing multinuclear iron clusters. *J. Biol. Chem.* **2005**, *280*, 15960–15966. [CrossRef] [PubMed]

43. Wang, X.; Han, H.; Lv, Z.; Lin, Z.; Shang, Y.; Xu, T.; Wu, Y.; Zhang, Y.; Qu, D. PhoU2 but not PhoU1 as an important regulator of biofilm formation and tolerance to multiple stresses by participating in various fundamental metabolic processes in *Staphylococcus epidermidis*. *J. Bacteriol.* **2017**, *199*, e00219-17. [CrossRef] [PubMed]

44. Tam, R.; Saier, M.H., Jr. Structural, functional, and evolutionary relationships among extracellular solute-binding receptors of bacteria. *Microbiol Rev.* **1993**, *57*, 320–346. [PubMed]

45. Ziegler, C.; Bremer, E.; Kramer, R. The BCCT family of carriers: From physiology to crystal structure. *Mol. Microbiol.* **2010**, *78*, 13–34. [CrossRef] [PubMed]

46. Schweikhard, E.S.; Ziegler, C.M. Amino acid secondary transporters: Toward a common transport mechanism. *Curr. Top. Membr.* **2012**, *70*, 1–28. [CrossRef] [PubMed]

47. Lamark, T.; Styrvold, O.B.; Strom, A.R. Efflux of choline and glycine betaine from osmoregulating cells of *Escherichia coli*. *FEMS Microbiol. Lett.* **1992**, *75*, 149–154. [CrossRef] [PubMed]

48. Malinauskaite, L.; Quick, M.; Reinhard, L.; Lyons, J.A.; Yano, H.; Javitch, J.A.; Nissen, P. A mechanism for intracellular release of Na^+ by neurotransmitter/sodium symporters. *Nat. Struct. Mol. Biol.* **2014**, *21*, 1006–1012. [CrossRef] [PubMed]

49. Liu, R.; Xing, L.; Fu, Q.; Zhou, G.H.; Zhang, W.G. A review of antioxidant peptides derived from meat muscle and by-products. *Antioxidants (Basel)* **2016**, *5*, E32. [CrossRef] [PubMed]

50. Agostinelli, E.; Marques, M.P.; Calheiros, R.; Gil, F.P.; Tempera, G.; Viceconte, N.; Battaglia, V.; Grancara, S.; Toninello, A. Polyamines: Fundamental characters in chemistry and biology. *Amino Acids* **2010**, *38*, 393–403. [CrossRef] [PubMed]

51. Rhee, H.J.; Kim, E.J.; Lee, J.K. Physiological polyamines: Simple primordial stress molecules. *J. Cell. Mol. Med.* **2007**, *11*, 685–703. [CrossRef] [PubMed]

52. Hidese, R.; Im, K.H.; Kobayashi, M.; Niitsu, M.; Furuchi, T.; Fujiwara, S. Identification of a novel acetylated form of branched-chain polyamine from a hyperthermophilic archaeon *Thermococcus kodakarensis*. *Biosci. Biotechnol. Biochem.* **2017**, *81*, 1845–1849. [CrossRef] [PubMed]

53. Prunetti, L.; Graf, M.; Blaby, I.K.; Peil, L.; Makkay, A.M.; Starosta, A.L.; Papke, R.T.; Oshima, T.; Wilson, D.N.; de Crecy-Lagard, V. Deciphering the translation initiation factor 5A modification pathway in halophilic archaea. *Archaea* **2016**, *2016*, 7316725. [CrossRef] [PubMed]

54. Hamana, K.; Hamana, H.; Itoh, T. Ubiquitous occurrence of agmatine as the major polyamine within extremely halophilic archaebacteria. *J. Gen. Appl. Microbiol.* **1995**, *41*, 153–158. [CrossRef]

55. Ohnuma, M.; Ganbe, T.; Terui, Y.; Niitsu, M.; Sato, T.; Tanaka, N.; Tamakoshi, M.; Samejima, K.; Kumasaka, T.; Oshima, T. Crystal structures and enzymatic properties of a triamine/agmatine aminopropyltransferase from *Thermus thermophilus*. *J. Mol. Biol.* **2011**, *408*, 971–986. [CrossRef] [PubMed]

56. Mattar, S.; Scharf, B.; Kent, S.B.; Rodewald, K.; Oesterhelt, D.; Engelhard, M. The primary structure of halocyanin, an archaeal blue copper protein, predicts a lipid anchor for membrane fixation. *J. Biol. Chem.* **1994**, *269*, 14939–14945. [PubMed]

57. Scharf, B.; Wittenberg, R.; Engelhard, M. Electron transfer proteins from the haloalkaliphilic archaeon *Natronobacterium pharaonis*: Possible components of the respiratory chain include cytochrome *bc* and a terminal oxidase cytochrome ba_3. *Biochemistry* **1997**, *36*, 4471–4479. [CrossRef] [PubMed]

58. Giro, M.; Ceccoli, R.D.; Poli, H.O.; Carrillo, N.; Lodeyro, A.F. An in vivo system involving co-expression of cyanobacterial flavodoxin and ferredoxin-NADP$^+$ reductase confers increased tolerance to oxidative stress in plants. *FEBS Open Bio* **2011**, *1*, 7–13. [CrossRef] [PubMed]

59. Apel, K.; Hirt, H. Reactive oxygen species: Metabolism, oxidative stress, and signal transduction. *Annu. Rev. Plant. Biol.* **2004**, *55*, 373–399. [CrossRef] [PubMed]

60. Tanaka, M.; Ogawa, N.; Ihara, K.; Sugiyama, Y.; Mukohata, Y. Cytochrome aa_3 in *Haloferax volcanii*. *J. Bacteriol.* **2002**, *184*, 840–845. [CrossRef] [PubMed]

61. Lenart, A.; Dudkiewicz, M.; Grynberg, M.; Pawlowski, K. CLCAs—A family of metalloproteases of intriguing phylogenetic distribution and with cases of substituted catalytic sites. *PLoS ONE* **2013**, *8*, e62272. [CrossRef] [PubMed]

62. Gray, M.J.; Wholey, W.Y.; Parker, B.W.; Kim, M.; Jakob, U. NemR is a bleach-sensing transcription factor. *J. Biol. Chem.* **2013**, *288*, 13789–13798. [CrossRef] [PubMed]

63. Jo-Watanabe, A.; Ohse, T.; Nishimatsu, H.; Takahashi, M.; Ikeda, Y.; Wada, T.; Shirakawa, J.; Nagai, R.; Miyata, T.; Nagano, T.; et al. Glyoxalase I reduces glycative and oxidative stress and prevents age-related endothelial dysfunction through modulation of endothelial nitric oxide synthase phosphorylation. *Aging Cell.* **2014**, *13*, 519–528. [CrossRef] [PubMed]

64. Yamamura, A.; Ohtsuka, J.; Kubota, K.; Agari, Y.; Ebihara, A.; Nakagawa, N.; Nagata, K.; Tanokura, M. Crystal structure of TTHA1429, a novel metallo-beta-lactamase superfamily protein from *Thermus thermophilus* HB8. *Proteins* **2008**, *73*, 1053–1057. [CrossRef] [PubMed]

65. Altman-Price, N.; Mevarech, M. Genetic evidence for the importance of protein acetylation and protein deacetylation in the halophilic archaeon *Haloferax volcanii*. *J. Bacteriol.* **2009**, *191*, 1610–1617. [CrossRef] [PubMed]

66. Bell, S.D.; Botting, C.H.; Wardleworth, B.N.; Jackson, S.P.; White, M.F. The interaction of Alba, a conserved archaeal chromatin protein, with Sir2 and its regulation by acetylation. *Science* **2002**, *296*, 148–151. [CrossRef] [PubMed]

67. Liu, J.; Wang, Q.; Jiang, X.; Yang, H.; Zhao, D.; Han, J.; Luo, Y.; Xiang, H. Systematic analysis of lysine acetylation in the halophilic archaeon *Haloferax mediterranei*. *J. Proteome Res.* **2017**, *16*, 3229–3241. [CrossRef] [PubMed]

68. Fu, X.; Liu, R.; Sanchez, I.; Silva-Sanchez, C.; Hepowit, N.L.; Cao, S.; Chen, S.; Maupin-Furlow, J. Ubiquitin-like proteasome system represents a eukaryotic-like pathway for targeted proteolysis in archaea. *mBio* **2016**, *7*, e00379-16. [CrossRef] [PubMed]

69. Maupin-Furlow, J.A. Archaeal proteasomes and sampylation. *Subcell. Biochem.* **2013**, *66*, 297–327. [CrossRef] [PubMed]

70. Wilson, H.; Aldrich, H.; Maupin-Furlow, J. Halophilic 20S proteasomes of the archaeon *Haloferax volcanii*: Purification, characterization, and gene sequence analysis. *J. Bacteriol.* **1999**, *181*, 5814–5824. [PubMed]

71. Reuter, C.; Kaczowka, S.; Maupin-Furlow, J. Differential regulation of the PanA and PanB proteasome-activating nucleotidase and 20S proteasomal proteins of the haloarchaeon *Haloferax volcanii*. *J. Bacteriol.* **2004**, *186*, 7763–7772. [CrossRef] [PubMed]

72. Zhou, G.; Kowalczyk, D.; Humbard, M.; Rohatgi, S.; Maupin-Furlow, J. Proteasomal components required for cell growth and stress responses in the haloarchaeon *Haloferax volcanii*. *J. Bacteriol.* **2008**, *190*, 8096–8105. [CrossRef] [PubMed]

73. Hanson, A.D.; Pribat, A.; Waller, J.C.; de Crecy-Lagard, V. 'Unknown' proteins and 'orphan' enzymes: The missing half of the engineering parts list—And how to find it. *Biochem. J.* **2009**, *425*, 1–11. [CrossRef] [PubMed]

74. Ortiz de Orue Lucana, D.; Bogel, G.; Zou, P.; Groves, M.R. The oligomeric assembly of the novel haem-degrading protein HbpS is essential for interaction with its cognate two-component sensor kinase. *J. Mol. Biol.* **2009**, *386*, 1108–1122. [CrossRef] [PubMed]

75. Omasits, U.; Ahrens, C.H.; Muller, S.; Wollscheid, B. Protter: Interactive protein feature visualization and integration with experimental proteomic data. *Bioinformatics* **2014**, *30*, 884–886. [CrossRef] [PubMed]

76. Delmas, S.; Shunburne, L.; Ngo, H.P.; Allers, T. Mre11-Rad50 promotes rapid repair of DNA damage in the polyploid archaeon *Haloferax volcanii* by restraining homologous recombination. *PLoS Genet.* **2009**, *5*, e1000552. [CrossRef] [PubMed]

77. Wendoloski, D.; Ferrer, C.; Dyall-Smith, M.L. A new simvastatin (mevinolin)-resistance marker from *Haloarcula hispanica* and a new *Haloferax volcanii* strain cured of plasmid pHV2. *Microbiology* **2001**, *147*, 959–964. [CrossRef] [PubMed]

78. Allers, T.; Ngo, H.P.; Mevarech, M.; Lloyd, R.G. Development of additional selectable markers for the halophilic archaeon *Haloferax volcanii* based on the *leuB* and *trpA* genes. *Appl. Environ. Microbiol.* **2004**, *70*, 943–953. [CrossRef] [PubMed]

79. Miranda, H.; Nembhard, N.; Su, D.; Hepowit, N.; Krause, D.; Pritz, J.; Phillips, C.; Söll, D.; Maupin-Furlow, J. E1- and ubiquitin-like proteins provide a direct link between protein conjugation and sulfur transfer in archaea. *Proc. Natl. Acad. Sci. USA* **2011**, *108*, 4417–4422. [CrossRef] [PubMed]

Article

Several One-Domain Zinc Finger μ-Proteins of *Haloferax volcanii* Are Important for Stress Adaptation, Biofilm Formation, and Swarming

Chantal Nagel, Anja Machulla, Sebastian Zahn and Jörg Soppa *

Department of Biosciences, Institute for Molecular Biosciences, Goethe-University, Max-von-Laue-Str. 9, D-60438 Frankfurt, Germany; ChantalNagel@web.de (C.N.); Anja.Machulla@gmx.de (A.M.); Zahn@bio.uni-frankfurt.de (S.Z.)
* Correspondence: soppa@bio.uni-frankfurt.de

Received: 1 March 2019; Accepted: 30 April 2019; Published: 10 May 2019

Abstract: Zinc finger domains are highly structured and can mediate interactions to DNA, RNA, proteins, lipids, and small molecules. Accordingly, zinc finger proteins are very versatile and involved in many biological functions. Eukaryotes contain a wealth of zinc finger proteins, but zinc finger proteins have also been found in archaea and bacteria. Large zinc finger proteins have been well studied, however, in stark contrast, single domain zinc finger μ-proteins of less than 70 amino acids have not been studied at all, with one single exception. Therefore, 16 zinc finger μ-proteins of the haloarchaeon *Haloferax volcanii* were chosen and in frame deletion mutants of the cognate genes were generated. The phenotypes of mutants and wild-type were compared under eight different conditions, which were chosen to represent various pathways and involve many genes. None of the mutants differed from the wild-type under optimal or near-optimal conditions. However, 12 of the 16 mutants exhibited a phenotypic difference under at least one of the four following conditions: Growth in synthetic medium with glycerol, growth in the presence of bile acids, biofilm formation, and swarming. In total, 16 loss of function and 11 gain of function phenotypes were observed. Five mutants indicated counter-regulation of a sessile versus a motile life style in *H. volcanii*. In conclusion, the generation and analysis of a set of deletion mutants demonstrated the high importance of zinc finger μ-proteins for various biological functions, and it will be the basis for future mechanistic insight.

Keywords: Archaea; *Haloferax volcanii*; zinc finger protein; small protein; μ-protein; deletion mutant; biofilm; swarming; phenotypic analysis

---

1. Introduction

For a long time, very small proteins (μ-proteins) have been nearly totally neglected. Genome annotations typically used a lower limit of 100 codons to include an open reading frame (ORF) as a predicted protein-coding gene. On the one hand, the annotation of a vast number of false-positive genes was prevented, on the other hand, real genes for μ-proteins were also not included and thus they escaped attention. In addition, experimental protein-analytical methods had been previously optimized for normal-sized proteins, which led to the loss of μ-proteins during purification or analyses.

During the last years the awareness has emerged that many μ-proteins do exist in prokaryotes, as well as in eukaryotes, and that they have important biological functions. Several recent reviews summarize the current knowledge about μ-proteins in prokaryotes [1–3] and in eukaryotes [4–6]. For higher eukaryotes, two databases for μ-proteins have been established during the last two years [7,8]. For humans, the database SmProt contains more than 100,000 small proteins of less than 100 aa (http://bioinfo.ibp.ac.cn/SmProt). The field is so new, that generally accepted definitions and a generally accepted terminology do not yet exist. Upper limits of 50 amino acids (aa), 70 aa, or 100 aa are used to define the group of very small proteins. The terms "small proteins", "peptides", "microproteins", "micropeptides", and "sORFs-encoded peptides" are in use. In this contribution we will use an upper limit of 70 aa and the term "μ-proteins", to distinguish between "small proteins", which could be understood as proteins smaller than an average protein of about 300 aa, and also between "peptides", which could include very small peptides of 10 to 30 aa, which are not encoded by distinct ORFs.

The emerging field of μ-protein research is mainly driven by improvements in two experimental approaches. Ribosomal profiling allows the highly parallel determination of the positions of all ribosomes of a cell on all transcripts [4,9]. All detected transcripts are proven to be translated into proteins, including small transcripts that encode μ-proteins. Another approach is the detection of μ-proteins by mass spectrometry (MS). Because standard proteomic approaches failed to detect μ-proteins, different steps had to be optimized, and the optimized procedure for small proteins is often called peptidomics [10]. The Oesterhelt group used optimized procedures to characterize the low molecular weight proteome of the halophilic archaeon *Halobacterium salinarum* already more than ten years ago [11]. Three hundred and eighty small proteins could be experimentally verified. The majority of these proteins (62%) had no assigned function, underscoring that the small proteome is understudied also in haloarchaea. It was noted that the group of small proteins with unknown function included 20 proteins that contained two C(P)XCG motifs. It was predicted that the four cysteines might complex a zinc ion and the proteins might thus contain zinc fingers [11].

Zinc finger proteins were first discovered in eukaryotes and were long thought to be confined to this domain [12]. They are abundant in eukaryotes and are involved in numerous processes, e.g., regulation of transcription, protein degradation, signal transduction, and many others [13]. Zinc finger domains are interaction modules that can enable the binding of zinc finger proteins to DNA, RNA, other proteins, lipids, and small molecules [14,15]. Therefore, the identification of zinc finger motifs is only indicative for zinc binding, but does not allow to predict protein function or association partners. The "classical" and most studied zinc finger domain contains two cysteines and two histidines (C2H2 zinc fingers). However, C3H and C4 zinc finger domains also exist. The four amino acids that coordinate the zinc ion are found in two motifs. The classical C2H2 zinc finger contains the motifs CXXC and HXXH (X = any amino acid). Many C4 zinc fingers contain two C(P)XCG motifs (P in brackets, because only one of the two motifs contain a proline). However, also many variants of these motifs exist, e.g., with slight differences in the distance of the two cysteines or histidines. Accordingly, zinc finger domains exhibit high structural variability, and they have been divided into eight structural groups [16]. A zinc finger is often used as a generic term for all subclasses, which, however, also have specific names like zinc knuckle, zinc ribbon, or treble clef [16].

Zinc finger proteins have also been found in bacteria, the first one was the Ros protein from *Agrobacterium tumefaciens* [17]. Ros is a transcription factor that regulates the *ipt* gene encoding isopentenyl transferase. Ros binds to an inverted repeat in the *ipt* promoter and represses transcription. Ros contains a classical C2H2 zinc finger, and homologues have been identified in additional bacteria and in archaea [18,19]. In addition, also a few examples of C4 zinc finger proteins have been found in bacteria and in archaea [20,21].

Most zinc finger motifs are parts of larger proteins, and the number of small zinc finger proteins is rather low. A bioinformatics search in archaeal and bacterial genomes with very relaxed patterns (e.g., CXXCX, HXXCX, HXXHX, etc.) revealed that only 1.5% of bacterial small proteins of less than 100 aa are potential zinc finger proteins, while, in contrast, this is true for 8% of archaeal small proteins [22]. The functions of very few small zinc finger proteins are known, e.g., they are small subunits of the RNA polymerase or the ribosome. However, most of them do not have an annotated function.

One of experimentally identified 20 C(P)XCG proteins of *H. salinarum* (see above) [11] was studied in detail. It was shown that it was a positive regulator of its adjacent gene, *bop* (bacterioopsin), and of *crtB1* (phytoene synthase) [22]. The expression of both genes was severely diminished in mutants of this zinc finger gene, which was named *brz* (bacterioopsin-regulating zinc finger protein). To our knowledge, this is the only experimental study of a small putative zinc finger protein in all three domains of life, which is not a subunit of a larger protein complex. Therefore, we decided to study more systematically whether small zinc finger proteins have important biological functions. We used the haloarchaeal model species *Haloferax volcanii*, because genetic (and other) tools are highly developed [23], and, very recently it was nominated as the "Microbe of the month" [24]. Its genome annotation is of very high quality, because it is constantly manually curated and many other haloarchaeal genomes are available for comparisons [25]. Sixteen genes for putative zinc-finger μ-proteins were chosen, and in frame deletion mutants were generated. The phenotypes of the mutants were compared to that of the wild-type under eight different conditions, which were chosen to test many different genes and pathways. For 12 of the mutants, phenotypic differences could be observed under at least one condition, revealing that many of these proteins fulfill important functions in *H. volcanii*.

2. Materials and Methods

2.1. Strains and Media

The *H. volcanii* strain H26 [26] was used in this study as a wild-type for construction of the 16 deletion mutants. It contains a deletion in the *pyrE2* gene, which enables two selection steps during mutant construction, and, thereby, facilitates and accelerates the procedure. It was grown in a complex medium with 2.1 M NaCl at 42 °C and good aeration (220 rpm) [27].

The *Escherichia coli* strain XL1-blue MRF′ (Agilent Technologies, Waldbronn, Germany) was used for cloning and was grown in standard media [28].

2.2. Generation of in Frame Deletion Mutants

The 16 in frame deletion mutants were generated by the so called Pop-In-Pop-Out method as described previously [26,29]. In short, for each deletion strain two PCR fragments of about 500 bpwere generated that contained (1) the upstream region and the first codons of the respective gene, and (2) the last codons and the downstream region. The two PCR fragments had an overlap, so that they could be fused by a third PCR reaction, and the resulting PCR fragment was inserted into the plasmid pMH101 by restriction selection cloning [30]. All oligonucleotide sequences are summarized in Supplementary Table S1. The sequences of the 16 plasmids were verified by sequencing. The plasmids were used to transform *H. volcanii* H26, and the Pop-In clones, which had integrated the plasmid into the genome, were selected by growth in the absence of uracil (the plasmids contain the *pyrE2* gene, which had been deleted from the genome of H26). Subsequently, the Pop-In variants were grown in medium with uracil and 5′-FOA, which selects for the absence of the *pyrE2* gene and thus for the Pop-Out variants. The initial identification of Pop-In clones and Pop-Out clones was performed by colony PCR. *H. volcanii* is polyploid [31], therefore, great care has to be taken that the Pop-Out clones are homozygous and do not contain one or a few remaining copies of the wild-type genome. Therefore, genomic DNA was isolated from candidate clones, and PCR with a large number of cycles, as well as Southern blot analyses, were used for the identification of homozygous deletion mutants. The deleted genomic regions of the 16 in frame deletion mutants are listed in Supplementary Table S2.

2.3. Growth Analyses

H. volcanii can be grown in microtiter plates, which enables parallel growth of many cultures and greatly facilitates phenotypic analyses of mutant collections under different conditions [32]. The complex medium, as well as a synthetic medium with different carbon sources, were used [27], as described in the text. Genomic DNA as phosphate source was added at a final concentration of 250 μg/mL, as described [33]. For each condition, 150 μL medium was inoculated to an OD_{600} of 0.05 from a preculture that had been grown under the respective condition. The cultures were grown on a Heidolph Titramax 1000 rotary shaker (Heidolph, Schwalbach, Germany) with 1100 rpm at 42 °C. The OD_{600} was determined using the microtiter plate photometer Spectramax 340 (Molecular Devices,

Ismaning, Germany) at the time points indicated in the respective Figures. Three biological replicates were performed, and average values and standard deviations were calculated.

2.4. Analysis of the Sensitivity to Bile Acids

Haloarchaea are very sensitive to bile acids [34–36]. A mixture of the sodium salts of 50% cholic acid and 50% deoxycholic acid (Honeywell Fluka; No. 48305) was used to test the sensitivity of *H. volcanii* to bile acids and to optimize the concentration. The wild-type strain H26 was grown in the presence of a various concentration of bile acids in the synthetic medium with casamino acids as carbon and energy source. Concentrations of 0.030 mg/mL and 0.035 mg/mL were chosen for the analysis, because the former still allowed the growth of the wild-type, while the latter completely inhibited growth. The cultures were grown in microtiter plates as described above, except that the precultures did not contain bile acids and thus did not have exactly the same conditions as the test cultures.

2.5. Swarm Plate Assays

The swarming assays were performed in six well plates (Sarstedt, Nümbrecht, Germany). Each well was filled with 5 mL synthetic medium with glucose as carbon and energy source and a reduced agar concentration of 0.3% (*w/v*) one day prior to their usage. Cultures were grown in glucose medium to the mid-exponential growth phase. The OD_{600} was determined, aliquots were pelleted by centrifugation, and the cell pellets were suspended in basal salts (medium without carbon source) to yield an OD_{600} of 20.2 µL of cell suspension, which was injected deep into the semi-liquid medium, to ensure a reduced oxygen concentration for motility of *H. volcanii*, as the cells do not swarm at the surface. The plates were incubated at 42 °C in a Styrofoam box together with a glass of water to inhibit drying. Every day the plates were analyzed, pictures were taken, and the swarming diameter was determined. Three biological replicates were performed, and averages and standard deviations were calculated. The results were normalized to the wild-type H26.

2.6. Analysis of Biofilm Formation

For the biofilm assay, cultures were grown in synthetic medium with glucose to the mid-exponential growth phase. The OD_{600} was measured, cells were pelleted by centrifugation and resuspended in fresh medium to an OD_{600} of 0.5. For biofilm formation, 96 well flat base microtiter plates were used (Sarstedt, Nümbrecht, Germany). The biofilm assay consists of several steps, i.e., formation of a biofilm, removal of planktonic cells, fixation and staining of adherent cells, and destaining and photometric quantification of the supernatant. The assay has been performed as described previously by Legerme et al. [37], with a few modifications. To this end, 150 µL of cell suspensions were pipetted in each well, and the plates were incubated without shaking at 42 °C for 24 h or 48 h. After that, the supernatant was removed, and 200 µL of fixing solution (2% (*w/v*) acetic acid) was given into each well, and the plate was incubated for 10 min at room temperature. The supernatant was removed, and the plate was dried for 10 min at 37 °C. After that, 200 µL staining solution (0.1% (*w/v*) crystal violet) was given into each well, and it was incubated for 10 min at room temperature. The supernatant was removed, and the wells were washed three times with 200 µL distilled water, respectively. After that, 200 µL of destaining solution (10% (*v/v*) acetic acid, 30% (*v/v*) methanol) was given into each well, and the plate was incubated for 10 min at room temperature. The supernatant was transferred into a new microtiter plate, and the OD_{600} was recorded with a microtiter plate photometer (Spectramax 340, Molecular Devices, San Jose, CA, USA). Three biological replicates with six technical replicates each were performed, and average values and standard deviations were calculated. A negative control (medium without cells) was included in the assay, and its value (about 0.05) was subtracted from the values of all tested strains.

2.7. Databases and Bioinformatics Analyses

Bioinformatic analyses of the *H. volcanii* genome were performed at the website Halolex [38]. The Halolex database is freely available, but currently usage is restricted to registered users. To request access, send a mail to halolex@rzg.mpg.de. The Integrated Genome Browser [39] was used to visualize the genome annotation, as well as the results of the dRNA-Seq study [40] and a recent RNA-Seq study [41].

3. Results

3.1. Selection of Genes and Generation of In Frame Deletion Mutants

The annotated proteins of *H. volcanii* were retrieved from the Halolex genome database and sorted according to their predicted size. Table 1 summarizes the numbers of small proteins that are predicted to be present in the size classes from 40 aa to 100 aa, all of which are missed in standard genome annotations.

Table 1. Length distribution of small proteins encoded in the genome of *H. volcanii*.

Length up to aa	All Proteins			C(P)XCG Motif Proteins		
	No.	Annotated Function	(%)	No.	(%)	Annotated Function
40	27	0	0	3	11	0
50	80	3	4	17	21	0
60	167	10	6	36	22	0
70	282	24	9	43	15	0
80	373	30	8	56	15	0
90	468	49	10	61	13	0
100	575	72	13	69	12	0

Only 80 proteins are extremely small with a length of up to 50 aa, however, the number of proteins with a size of up to 100 aa is much higher and reaches 575 proteins. This accounts for about 13% of all annotated proteins of *H. volcanii*, and thus small proteins represent a considerable fraction of the *H. volcanii* proteome.

Similar to most if not all other prokaryotic and eukaryotic species, small proteins are understudied in *H. volcanii*. Only three out of eighty proteins of up to 50 aa, and only 72 of 575 proteins of up to 100 aa have an annotated biological function.

A subgroup of the small proteins contains two C(P)XCG motifs and is thus comprised of putative zinc finger proteins. The fraction is higher in very small proteins of up to 50 aa (21%) as in all small proteins of up to 100 aa (12%). While 69 C(P)XCG zinc finger proteins have a length of up to 100 aa, only 16 are longer than 100 aa. Only four are longer than 150 aa, and the largest has a length of 172 aa. Therefore, all 85 C(P)XCG zinc finger proteins of *H. volcanii* are considerably smaller than the average protein size of about 300 aa. None of the C(P)XCG proteins has a predicted annotated function, and thus these putative single domain zinc finger proteins are of specific novelty. Therefore, we aimed to reveal whether or not C(P)XCG µ-proteins have important or essential roles for *H. volcanii*, and to elucidate their participation at specific biological functions.

43 of the C(P)XCG µ-proteins have a length of up to 70 aa, the maximal length that can be studied in the framework of the German Priority Program 2002 "Small proteins in prokaryotes: An unexplored words" (http://www.spp2002.uni-kiel.de). Therefore, the genes for this study had to be selected from this set of 43 proteins. The experimental design was comprised of the generation of in frame deletion mutants and of the phenotypic comparison of wild-type and mutants under various conditions. Sixteen genes were selected for this study based on the following criteria: (1) The identity of

the neighboring genes, with an emphasis on genes with known functions; (2) the expression level under optimal conditions, derived from the number of reads determined in a recent dRNA-Seq study [40]; (3) the cysteine content (with an emphasis on proteins that do not contain cysteines outside of the two motifs); and (4) the isoelectric point (with the aim to include proteins with acidic, as well as basic pIs). Supplementary Table S2 gives an overview of the characteristics of the 16 selected proteins.

A few years ago we have adapted many steps of the workflow for deletion mutant generation to the application of microtiter plates [29], e.g., PCRs, fusion PCRs, cloning. This parallel approach facilitates and accelerates the construction of a double-digit number of in frame deletion mutants, without the need of any robotics, and it is, therefore, well suited for a small group in a University setting. The oligonucleotides used to construct the 16 deletion plasmids are summarized in Supplementary Table S1. The sequences of the plasmids were verified by sequencing, and the plasmids were used to transform *H. volcanii* strain H26 [26]. The deletion mutants were selected using the so called Pop-In-Pop-Out method [29,42]. *H. volcanii* is polyploid [31], therefore, great care has to be taken to ensure that all genome copies contain the deletion. The homozygocities of the deletion mutants were verified using PCR, as well as Southern blot analyses. Supplementary Table S2 summarizes the genomic coordinates of the 16 deleted regions. All 16 homozygous deletion mutants could readily be generated, showing that none of the 16 proteins is essential for *H. volcanii*. All 16 mutants had the same colony size and colony form as the wild-type. In addition, all colonies developed the same red color, showing that all mutants were proficient in carotenoid biosynthesis. The subsequent phenotypic analyses aimed at testing many different biological functions and pathways.

3.2. Growth Analyses in Media with Different Carbon and Phosphate Sources

At first, the wild-type and the 16 in frame deletion mutants were grown in a complex medium at 42 °C with good aeration, and growth curves were recorded (Figure 1). The growth of all 17 strains was nearly identical, indicating that the 16 CPXCP μ-proteins have no important roles under optimal conditions. Growth of the wild-type and the 16 mutants were also monitored in synthetic medium with casamino acids as carbon source. This condition did not require any amino acid biosynthesis, but catabolism of amino acids and anabolism of sugars. Again, none of the mutants exhibited a growth defect (Supplementary Figure S1). Next, growth in synthetic medium with glucose as the sole carbon and energy source was tested (Figure 2). This condition required the catabolism of glucose and the biosynthesis of all 20 amino acids. In this medium, very slight differences between the mutants and the wild-type could be observed. One mutant grew slightly better than the wild-type (dHVO_2901), two mutants grew indistinguishable from the wild-type (dHVO_A0556), and the remaining 13 mutants had very minor growth defects. The next tested carbon and energy source was glycerol (Figure 3). Glycerol is a preferred carbon source for *H. volcanii* and in fact represses glucose metabolism [43]. Unexpectedly, there was a difference between glycerol and the other carbon sources, and six of the mutants exhibited a severe growth defect (Figure 4). Again, one deletion mutant grew better than the wild-type (dHVO_2400).

H. volcanii can use external, environmental genomic DNA as a phosphate source [33]. Therefore, it was tested whether this ability was affected by any of the deletion mutants, using a synthetic medium with genomic DNA as phosphate source and casamino acids as carbon and energy source (Supplementary Figure S2). It turned out that all 16 deletion mutants grew indistinguishable from the wild-type. Taken together, severe growth defects were observed in only one of five different media tested, i.e., in synthetic medium with glycerol as carbon and energy source.

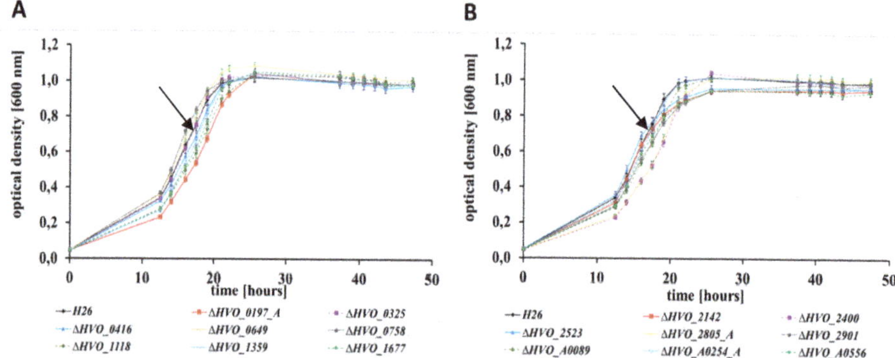

Figure 1. Growth curves of the parent strain *H26* (black) and 16 deletion mutants (in color) grown in a complex medium. (**A**) Parent strain H26 (black) and the deletion mutants Δ*0197_A* (red), Δ*0325* (purple), Δ*0416* (dark blue), Δ*0649* (yellow), Δ*0758* (grey), Δ*1118* (dark green), Δ*1359* (light blue), Δ*1677* (light green). (**B**) Parent strain H26 (black) and the deletion mutants Δ*2142* (red), Δ*2400* (purple), Δ*2523* (dark blue), Δ*2805_A* (yellow), Δ*2901* (grey), Δ*A0089* (dark green), Δ*A0254_A* (light blue), Δ*A0556* (light green). The wild-type is indicated by black arrows.

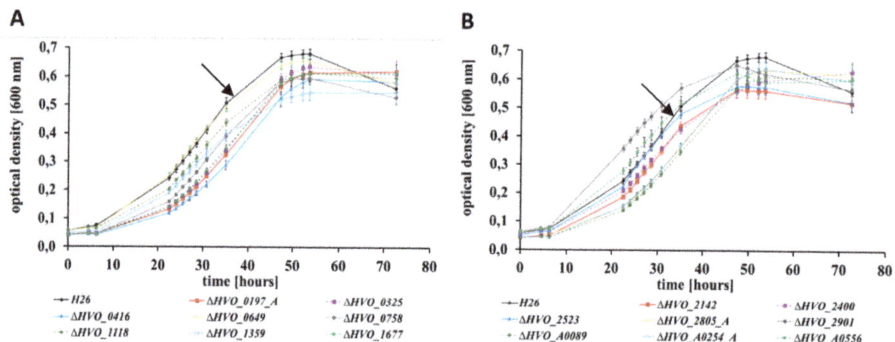

Figure 2. Growth curves of the parent strain *H26* (black) and 16 deletion mutants (in color) grown in synthetic medium with glucose as carbon source. (**A**) Parent strain H26 (black) and deletion mutants Δ*0197_A* (red), Δ*0325* (purple), Δ*0416* (dark blue), Δ*0649* (yellow), Δ*0758* (grey), Δ*1118* (dark green), Δ*1359* (light blue), Δ*1677* (light green). (**B**) Parent strain H26 and deletion mutants Δ*2142* (red), Δ*2400* (purple), Δ*2523* (dark blue), Δ*2805_A* (yellow), Δ*2901* (grey), Δ*A0089* (dark green), Δ*A0254_A* (light blue), Δ*A0556* (light green). The wild-type is indicated by black arrows.

3.3. Sensitivities to Bile Acids

Haloferax and several additional genera of haloarchaea are very sensitive to bile acids, and even moderate concentrations result in cell lysis [34–36]. It can be expected that the degree of sensitivity is correlated with the lipid composition of the membrane, and, thus, it might well be an indicator of lipid metabolism. Therefore, it was tested whether the degree of sensitivity of wild-type and deletion mutants was identical, or whether differences exist. At first, various concentrations were used in casamino acids medium to characterize the sensitivity of the wild-type. It was verified that *H. volcanii* is very sensitive to bile acids, and, it was also found, that very small concentration differences resulted in considerable growth differences. Concentrations of 0.030 mg/mL and 0.035 mg/mL were chosen for further experiments, because the wild-type grew rather well at the former concentration, while it was totally inhibited by the latter concentration. The growth curves of the wild-type and all deletion mutants at both concentrations are shown in Supplementary Figures S3 and S4. The variances between the biological replicates were extremely large, in contrast to all other growth analyses. The reason is most probably that already small differences in evaporation result in concentration changes that affect growth. Therefore, the results have to be treated with care, and only a qualitative analysis seems possible. However, despite the high variances, consistent differences between the wild-type and several mutants were observed at consecutive time points. Four mutants were severely inhibited by 0.030 mg/mL, in contrast to the wild-type and the other mutants (Supplementary Figure S3A, highlighted with a red bar). In contrast, four other mutants showed considerable growth in the presence of 0.035 mg/mL, in contrast to all other strains (Supplementary Figure S4B, highlighted by a red bar). Taken together, the analysis of bile acids sensitivity demonstrated large differences between wild-type and eight deletion mutants. To our knowledge, this is the first time that this assay was used for phenotypic mutant analysis in haloarchaea.

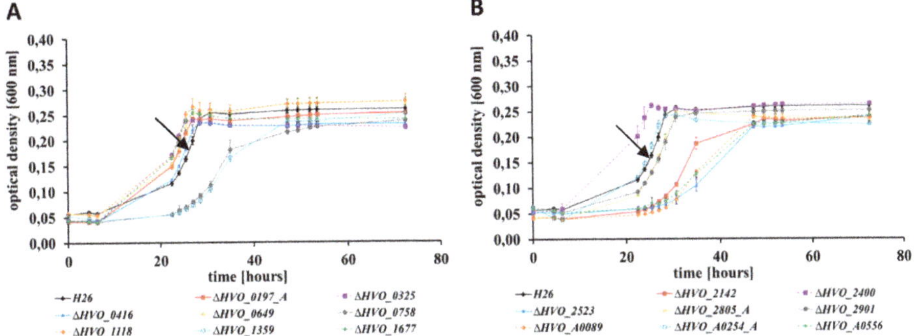

Figure 3. Growth curves of the parent strain *H26* (black) and 16 deletion mutants (in color) grown in synthetic medium with glycerol as carbon source. (**A**) Parent strain H26 (black) and deletion mutants Δ*0197_A* (red), Δ*0325* (purple), Δ*0416* (dark blue), Δ*0649* (yellow), Δ*0758* (grey), Δ*1118* (dark green), Δ*1359* (light blue), Δ*1677* (light green). (**B**) Parent strain H26 (black) and deletion mutants Δ*2142* (red), Δ*2400* (purple), Δ*2523* (dark blue), Δ*2805_A* (yellow), Δ*2901* (grey), Δ*A0089* (dark green), Δ*A0254_A* (light blue), Δ*A0556* (light green). The wild-type is indicated by black arrows.

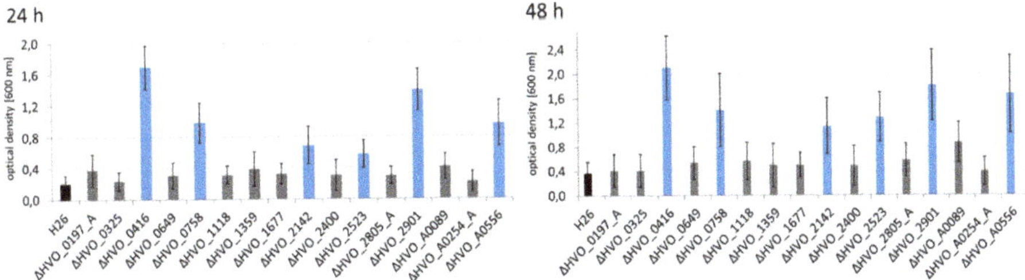

Figure 4. Comparison of the biofilm formation of the parent strain H26 (black) and its 16 deletion mutants after 24 h and 48 h incubation. The mutants with a gain of function phenotypes are shown in blue. Average values of three biological replicates and their standard deviations are shown.

3.4. The Different Life Styles of H. volcanii: Swarming and Biofilm Formation

H. volcanii, like many other haloarchaeal species, can form biofilms [44]. It can be expected that many different gene products are essential for biofilm formation. In fact, more than 10 adhesion mutants were identified using a transposon mutagenesis screen [37]. The principle of the biofilm assay is to incubate a culture to allow the formation of a biofilm, remove planktonic cells, and fix and stain adherent cells [37]. The amount of dye bound to a biofilm correlates with its size, and thus, after destaining, photometric analysis of the dye yields a quantitative value for biofilm formation. The quantification of biofilm formation after 24 h and after 48 h is shown in Figure 4. None of the mutants had a defect in biofilm formation. In stark contrast, six of the mutants exhibited a several fold increase in biofilm formation, from threefold to ninefold (blue columns in Figure 4). Biofilm formation of the remaining ten mutants was very similar to that of the wild-type.

Since its discovery in 1975 *H. volcanii* was regarded to be non-motile [45]. However, a few years ago a swarm plate assay was developed, which revealed that *H. volcanii* can swim and shows chemotaxis under conditions of reduced oxygen concentrations [46]. The assay was applied to quantify swarming velocity of wild-type and deletion mutants after 24 h and 44 h (Figure 5). Six mutants exhibited a null phenotype and had completely lost the ability to swarm (red asterisks in Figure 5. Remarkably, five of the six mutants with a swarm defect concomitantly showed increased biofilm formation (compare Figure 4; Figure 5). The two phenotypes were uncoupled in only two mutants, i.e., mutant dHVO_0416 showed only increased biofilm formation, and mutant dHVO_0649 had solely a swarm defect. Together, the results show a high anti-correlation between a motile and a sessile lifestyle of *H. volcanii*, but in addition, that both life styles can also be affected individually.

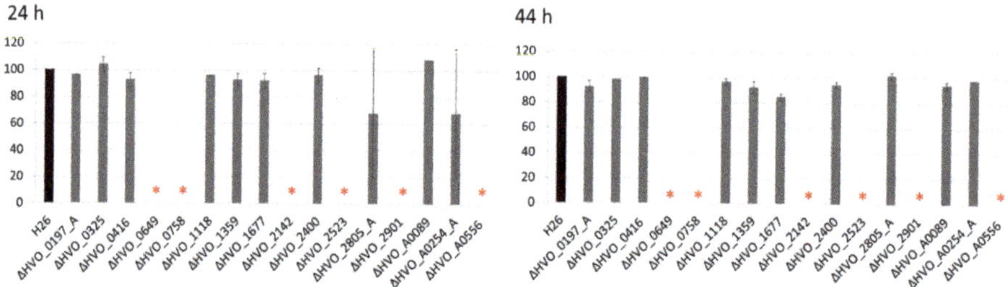

Figure 5. Comparison of the swarming velocity of the parent strain H26 (black) and its 16 deletion mutants after 24 h and 44 h incubation. Average values of three biological replicates and their standard deviations are shown. At both time points, the swarming diameters of the mutants were normalized to that of the parent strain H26 (= 100%). The wild-type had a swarming diameter of 5 mm after 24 h and 23 mm after 48 h. The cultures were injected into the semi-solid agar using a tip with 1 mm diameter. Colonies with no observable swarming with a colony diameter of less than 2 mm after 44 h were regarded to have a null phenotype. The mutants with a null phenotype are indicated by red asterisks.

4. Discussion

Until now, only very few archaeal μ-proteins of less than 70 aa have an annotated function and have been experimentally characterized. Many of these are small subunits of large complexes, e.g., eight subunits of the ribosome belong to this group (Rps14/17e and Rpl20e/24e/29/37e/39e/40e), as well as the subunits RpoK and RpoP of RNA polymerase. Very early, a small DNA binding protein of 64 aa was isolated from cell extracts of *Sulfolobus solfataricus* using classical biochemical approaches [47]. Recently, five μ-proteins from *Methanosarcina mazei* Gö1 were identified by LC-MS/MS, which had sizes from 23 aa to 61 aa [48]. The transcript levels of two of the respective genes were severely decreased in stationary phase, while one was increased 2.5-fold in response to nitrogen limitation. Overproduction of the three proteins, respectively, resulted in changes in transcript levels of 10–20 genes, but, unfortunately, did not result in any phenotypic changes, which could reveal important functions [48]. In *H. volcanii* three "small archaeal modifier proteins (SAMPs)" have been identified, which were first described to modify proteins analogous to the eukaryotic ubiquitin and are thus important for protein degradation in the proteasome [49]. All three are small proteins with less than 100 aa. However, only SAMP2 with 66 aa fulfills the definition of a μ-protein, while SAMP1 has 87 aa and SAMP3 has 92 aa. Later, it was found that the SAMPs are not only involved in protein degradation, but also in sulfur metabolism and oxidative stress response [50,51].

The biological functions of a few additional *H. volcanii* μ-proteins have been studied, however, only 24 of the 282 annotated μ-protein genes have a known function (Table 1). Much more dramatically, none of the 43 putative zinc finger μ-proteins has a known function. Therefore, this group was chosen for this study, with the aim to reveal whether or not they have important biological roles. To our knowledge, prior to our study only one single putative zinc finger μ-protein gene has been studied in any prokaryotic or eukaryotic species, i.e., the *brz* gene of *H. salinarum* [22]. Brz contains a C3H zinc finger with one C(P)XCG motif and one CXXXH motif. Replacement of the second cysteine in the first motif and the histidine in the second motif resulted in the same drastic decrease of the *bop* mRNA level as the deletion of the whole gene, showing that both motifs are indispensable for function.

We chose 16 of the 43 putative zinc finger μ-proteins of *H. volcanii* for analysis, based on the identity of adjacent genes, transcript levels under optimal conditions, pI values, and cysteine content. In frame deletion mutants of all 16 genes could readily be generated, showing that none of the genes is essential. Furthermore, colony size, colony form, and colony color were identical for all mutants and the wild-type, which already indicated that the mutants did not have a severe growth defect in complex medium. The mutants were compared to the wild-type under eight different conditions, which should involve various metabolic pathways and the genes serving them. The results are summarized in Table 2. Notably, none of the mutants exhibited a growth phenotype under optimal or near-optimal conditions. In contrast, growth of eight of the 16 mutants differed from the wild-type in the presence of bile acids, which represent a strong stressor for *H. volcanii*. In total, 12 of the 16 deletion mutants showed a changed phenotype under at least one conditions, underscoring the high importance of the zinc finger μ-proteins. Seven mutants had a phenotype under two or more conditions, and they were thus pleiotropic. About the same number were gain of function and loss of function mutants, indicating that zinc finger μ-proteins are members of regulatory networks that take many environmental conditions into account. The diversity of phenotypes indicates that the functions of the proteins are not uniform or similar. Notably, three deletion mutants had a pleiotropic phenotype under four rather different conditions, which might indicate that the respective zinc finger μ-proteins have a high level in regulatory hierarchies and influence other regulators.

Remarkably, the deletion of each of the disclosed seven genes had an influence on the life styles "biofilm formation" and "swarming" of *H. volcanii*. In five cases the deletions led to an increase in biofilm formation and a concomitant decrease in swarming. Therefore, it seems that these zinc finger μ-proteins promote the motile lifestyle and counteract a sessile lifestyle.

Decision making regulatory circuits between a motile and a sessile lifestyle have also been reported for several bacteria, e.g., *Bacillus subtilis*, *Escherichia coli*, *Pseudomonas aeruginosa*, and *Vibrio cholera* [52]. In bacteria, the signaling molecule c-di-GMP is involved, i.e., high concentrations of c-di-GMP correlate with a high probability of a sessile lifestyle and a low probability of a motile lifestyle [53]. However, archaea do not use c-di-GMP, and thus the molecular mechanism of sessile versus motile decisions must be different in *H. volcanii*.

It seems that also small non-coding RNAs (sRNAs) are involved, because six sRNA gene deletion mutants of *H. volcanii* had a swarming phenotype [29]. Biofilm formation was not addressed in this previous study [29], therefore, it is not known whether the opposite regulation described above for five zinc finger μ-proteins holds true for sRNAs in *H. volcanii*.

The influence of sRNAs on biofilm formation and swarming has also been analyzed in *E. coli* [54]. Ninty-nine sRNAs were overexpressed, and the effects on several traits were quantified. While the overexpression of many sRNAs influenced biofilm formation, swarming, or both, no consistent regulatory pattern could be observed. For example, six sRNAs had an opposing effect on the sessile and motile lifestyle, while, in contrast, eight effected the life styles in the same direction. Biofilm formation and swarming are very complex traits, therefore, complex patterns can be expected, which involve regulators at different levels, as well as structural proteins.

Table 2. Summary of phenotypic analysis of 16 zinc finger gene deletion mutants. "-" indicates that wild-type and deletion mutant are indistinguishable, "loss" indicates a loss of function phenotype of the mutant, "gain" a gain of function phenotype. Three mutants with the same pleiotrophic phenotype are shown in red.

Deletion Mutant	Growth in Four Media [1]	Growth on Glycerol	Bile Acid [2]	Bio-Film	Swarming	Pattern	No. of Phenotypes
HVO_0197_A	-	-	(loss)	-	-		1
HVO_0325	-	-	-	-	-		0
HVO_0416	-	-	-	gain	-		1
HVO_0649	-	-	(loss)	-	loss		2
HVO_0758	-	loss	-	gain	loss	(Yes)	3
HVO_1118	-	-	(loss)	-	-		1
HVO_1359	-	loss	(loss)	-	-		2
HVO_1677	-	-	-	-	-		0
HVO_2142	-	loss	(gain)	gain	loss	Yes	4
HVO_2400	-	gain	-	-	-		1
HVO_2523	-	loss	(gain)	gain	loss	Yes	4
HVO_2805_A	-	-	-	-	-		0
HVO_2901	-	-	(gain)	gain	loss	(Yes)	3
HVO_A0089	-	loss	-	-	-		1
HVO_A0254_A	-	-	-	-	-		0
HVO_A0556	-	loss	(gain)	gain	loss	Yes	4
sum	0	7	8	6	6		

[1] growth in complex medium, synthetic medium with casamino acids and with glucose as carbon source, synthetic medium with genomic DNA as phosphate source [2] the growth in the presence of bile acids had a very high variance. Therefore, the results should be handled with care and they are shown in brackets.

For *H. volcanii*, important components in addition to μ-proteins and sRNAs have been revealed: A transposon mutagenesis screening led to the identification of 17 motility mutants and 11 adhesion mutants, without any overlap between the two groups [37]. Most of the inactivated genes did not have an obvious connection to the motility of biofilm formation, underscoring the complexity of these two traits. Swarming and biofilm formation has already been studied to some extent in *H. volcanii*. For example, it has been shown that flagella are required for swarming, but not for surface adhesion [46]. On the other hand, it was revealed that pili are essential for surface adhesion [55,56]. Confocal scanning microscopy was used to demonstrate the complex morphological development of biofilms over two days [57]. However, much has to be learned about the involved regulatory networks and structural proteins mediating morphological changes. The seven zinc finger μ-proteins described above are unexpected new players in the game.

The phenotypic differences between 12 deletion mutants and the wild-type (Table 2) do not yet yield insight into the regulatory networks and molecular mechanisms of haloarchaeal zinc finger μ-proteins. However, besides the single example of *brz* of *H. salinarum* [22], they represent the first experimental proof in any prokaryotic or eukaryotic species for the high biological importance of zinc finger μ-proteins. The clear phenotypes will pave the way for future in depth characterizations. The next step will be the attempt to complement the mutants with tagged versions of the respective native proteins, which would allow co-affinity purification strategies to reveal interaction networks. This will probably not be easy, because the addition of tags often perturbs the functions of μ-proteins [1], and the number of tags that are compatible with the high salt cytoplasm of haloarchaea is small. In any case, also complementation with the untagged native versions will be informative, because it will allow to test point mutated versions and address the importance of single amino acids for protein function. In conclusion, this study demonstrated that various members of the family of zinc finger μ-proteins, which with one exception has not been studied until now, have important roles in stress adaptation (at least membrane stress) and life style decisions in *H. volcanii*.

Supplementary Materials: The following are available online at http://www.mdpi.com/2073-4425/10/5/361/s1, Table S1: List of primers used for the generation of deletion mutants. Table S2: Overview of characteristics of the analyzed 16 zinc finger proteins. Figures S1–S4: Growth in synthetic medium with casamino acids, growth with genomic DNA as phosphate source, growth in the presence of 0.030 mg/mL and 0.035 mg/mL bile acids.

Author Contributions: Conceptualization, C.N., A.M., S.Z. and J.S.; investigation, C.N., A.M.; analysis, C.N., A.M., S.Z. and J.S.; data curation, S.Z.; preparation of figures, C.N.; writing—original draft preparation, J.S.; writing—review and editing, C.N., A.M., S.Z. and J.S.; supervision, S.Z. and J.S.; project administration, S.Z. and J.S.; funding acquisition, J.S.

Funding: This research was funded by the German Research Council (Deutsche Forschungsgemeinschaft, DFG) through project SO 264/26-1 in the framework of the priority program SPP2002.

Acknowledgments: We thank Friedhelm Pfeiffer (MPI for Biochemistry, Martinsried, Germany) for continuous cooperation on bioinformatics questions, and for keeping Halolex up-to-date as a valuable database for the genomes of *H. volcanii* and other haloarchaea.

Conflicts of Interest: The authors declare no conflict of interest. The funders had no role in the design of the study; in the collection, analyses, or interpretation of data; in the writing of the manuscript, or in the decision to publish the results.

References

1. Duval, M.; Cossart, P. Small bacterial and phagic proteins: An updated view on a rapidly moving field. *Curr. Opin. Microbiol.* **2017**, *39*, 81–88. [CrossRef] [PubMed]

2. Baumgartner, D.; Kopf, M.; Klähn, S.; Steglich, C.; Hess, W.R. Small proteins in cyanobacteria provide a paradigm for the functional analysis of the bacterial micro-proteome. *BMC Microbiol.* **2016**, *16*, 285. [CrossRef] [PubMed]

3. Storz, G.; Wolf, Y.I.; Ramamurthi, K.S. Small proteins can no longer be ignored. *Annu. Rev. Biochem.* **2014**, *83*, 753–777. [CrossRef] [PubMed]

4. Delcourt, V.; Staskevicius, A.; Salzet, M.; Fournier, I.; Roucou, X. Small proteins encoded by unannotated ORFs are rising stars of the proteome, confirming shortcomings in genome annotations and current vision of an mRNA. *Proteomics* **2018**, *18*, e1700058. [CrossRef] [PubMed]

5. Plaza, S.; Menschaert, G.; Payre, F. In Search of lost small peptides. *Annu. Rev. Cell Dev. Biol.* **2017**, *33*, 391–416. [CrossRef]

6. Cabrera-Quio, L.E.; Herberg, S.; Pauli, A. Decoding sORF translation—From small proteins to gene regulation. *RNA Biol.* **2016**, *13*, 1051–1059. [CrossRef]

7. Hao, Y.; Zhang, L.; Niu, Y.; Cai, T.; Luo, J.; He, S.; Zhang, B.; Zhang, D.; Qin, Y.; Yang, F.; et al. SmProt: A database of small proteins encoded by annotated coding and non-coding RNA loci. *Brief. Bioinform.* **2018**, *19*, 636–643. [CrossRef]

8. Olexiouk, V.; van Criekinge, W.; Menschaert, G. An update on sORFs.org: A repository of small ORFs identified by ribosome profiling. *Nucleic Acids Res.* **2018**, *46*, D497–D502. [CrossRef]

9. Mumtaz, M.A.S.; Couso, J.P. Ribosomal profiling adds new coding sequences to the proteome. *Biochem. Soc. Trans.* **2015**, *43*, 1271–1276. [CrossRef]

10. Schrader, M. Origins, technological development, and applications of peptidomics. *Methods Mol. Biol.* **2018**, *1719*, 3–39. [CrossRef]

11. Klein, C.; Aivaliotis, M.; Olsen, J.V.; Falb, M.; Besir, H.; Scheffer, B.; Bisle, B.; Tebbe, A.; Konstantinidis, K.; Siedler, F.; et al. The low molecular weight proteome of *Halobacterium salinarum*. *J. Proteome Res.* **2007**, *6*, 1510–1518. [CrossRef] [PubMed]

12. Maret, W. Zinc biochemistry: From a single zinc enzyme to a key element of life. *Adv. Nutr.* **2013**, *4*, 82–91. [CrossRef]

13. Cassandri, M.; Smirnov, A.; Novelli, F.; Pitolli, C.; Agostini, M.; Malewicz, M.; Melino, G.; Raschellà, G. Zinc-finger proteins in health and disease. *Cell Death Discov.* **2017**, *3*, 17071. [CrossRef] [PubMed]

14. Eom, K.S.; Cheong, J.S.; Lee, S.J. Structural analyses of zinc finger domains for specific interactions with DNA. *J. Microbiol. Biotechnol.* **2016**, *26*, 2019–2029. [CrossRef] [PubMed]

15. Matthews, J.M.; Sunde, M. Zinc fingers—Folds for many occasions. *IUBMB Life* **2002**, *54*, 351–355. [CrossRef] [PubMed]

16. Krishna, S.S.; Majumdar, I.; Grishin, N.V. Structural classification of zinc fingers: Survey and summary. *Nucleic Acids Res.* **2003**, *31*, 532–550. [CrossRef]

17. Chou, A.Y.; Archdeacon, J.; Kado, C.I. *Agrobacterium* transcriptional regulator Ros is a prokaryotic zinc finger protein that regulates the plant oncogene IPT. *Proc. Natl. Acad. Sci. USA* **1998**, *95*, 5293–5298. [CrossRef] [PubMed]

18. Bouhouche, N.; Syvanen, M.; Kado, C.I. The origin of prokaryotic C2H2 zinc finger regulators. *Trends Microbiol.* **2000**, *8*, 77–81. [CrossRef]

19. Malgieri, G.; Palmieri, M.; Russo, L.; Fattorusso, R.; Pedone, P.V.; Isernia, C. The prokaryotic zinc-finger: Structure, function and comparison with the eukaryotic counterpart. *FEBS J.* **2015**, *282*, 4480–4496. [CrossRef] [PubMed]

20. Pereira, L.E.; Tsang, J.; Mrázek, J.; Hoover, T.R. The zinc-ribbon domain of *Helicobacter pylori* HP0958: Requirement for RpoN accumulation and possible roles of homologs in other bacteria. *Microb. Inform. Exp.* **2011**, *1*, 1–10. [CrossRef]

21. Weidenbach, K.; Ehlers, C.; Schmitz, R.A. The transcriptional activator NrpA is crucial for inducing nitrogen fixation in *Methanosarcina mazei* Gö1 under nitrogen-limited conditions. *FEBS J.* **2014**, *281*, 3507–3522. [CrossRef]

22. Tarasov, V.Y.; Besir, H.; Schwaiger, R.; Klee, K.; Furtwängler, K.; Pfeiffer, F.; Oesterhelt, D. A small protein from the *bop-brp* intergenic region of *Halobacterium salinarum* contains a zinc finger motif and regulates *bop* and *crtB1* transcription. *Mol. Microbiol.* **2008**, *67*, 772–780. [CrossRef] [PubMed]

23. Soppa, J. From genomes to function: Haloarchaea as model organisms. *Microbiology* **2006**, *152*, 585–590. [CrossRef] [PubMed]

24. Pohlschroder, M.; Schulze, S. *Haloferax volcanii*. *Trends Microbiol.* **2019**, *27*, 86–87. [CrossRef]

25. Pfeiffer, F.; Oesterhelt, D. A manual curation strategy to improve genome annotation: Application to a set of haloarchael genomes. *Life* **2015**, *5*, 1427–1444. [CrossRef]

26. Allers, T.; Ngo, H.-P.; Mevarech, M.; Lloyd, R.G. Development of additional selectable markers for the halophilic archaeon *Haloferax volcanii* based on the *leuB* and *trpA* genes. *Appl. Environ. Microbiol.* **2004**, *70*, 943–953. [CrossRef] [PubMed]

27. Dambeck, M.; Soppa, J. Characterization of a *Haloferax volcanii* member of the enolase superfamily: Deletion mutant construction, expression analysis, and transcriptome comparison. *Arch. Microbiol.* **2008**, *190*, 341–353. [CrossRef]

28. Green, M.R.; Sambrook, K. *Molecular Cloning: A Laboratory Manual*; Cold Spring Harbor Laboratory Press: Cold Spring Harbor, NY, USA, 2012.

29. Jaschinski, K.; Babski, J.; Lehr, M.; Burmester, A.; Benz, J.; Heyer, R.; Dörr, M.; Marchfelder, A.; Soppa, J. Generation and phenotyping of a collection of sRNA gene deletion mutants of the haloarchaeon *Haloferax volcanii*. *PLoS ONE* **2014**, *9*, e90763. [CrossRef] [PubMed]

30. Hammelmann, M.; Soppa, J. Optimized generation of vectors for the construction of *Haloferax volcanii* deletion mutants. *J. Microbiol. Methods* **2008**, *75*, 201–204. [CrossRef]

31. Breuert, S.; Allers, T.; Spohn, G.; Soppa, J. Regulated polyploidy in halophilic archaea. *PLoS ONE* **2006**, *1*, e92. [CrossRef]

32. Jantzer, K.; Zerulla, K.; Soppa, J. Phenotyping in the archaea: Optimization of growth parameters and analysis of mutants of *Haloferax volcanii*. *FEMS Microbiol. Lett.* **2011**, *322*, 123–130. [CrossRef] [PubMed]

33. Zerulla, K.; Chimileski, S.; Näther, D.; Gophna, U.; Papke, R.T.; Soppa, J. DNA as a phosphate storage polymer and the alternative advantages of polyploidy for growth or survival. *PLoS ONE* **2014**, *9*, e94819. [CrossRef]

34. Kamekura, M.; Oesterhelt, D.; Wallace, R.; Anderson, P.; Kushner, D.J. Lysis of halobacteria in bacto-peptone by bile acids. *Appl. Environ. Microbiol.* **1988**, *54*, 990–995. [PubMed]

35. Elevi Bardavid, R.; Oren, A. Sensitivity of *Haloquadratum* and *Salinibacter* to antibiotics and other inhibitors: Implications for the assessment of the contribution of Archaea and Bacteria to heterotrophic activities in hypersaline environments. *FEMS Microbiol. Ecol.* **2008**, *63*, 309–315. [CrossRef]

36. Kumar, V.; Saxena, J.; Tiwari, S.K. Description of a halocin-producing *Haloferax larsenii* HA1 isolated from Pachpadra salt lake in Rajasthan. *Arch. Microbiol.* **2016**, *198*, 181–192. [CrossRef]

37. Legerme, G.; Yang, E.; Esquivel, R.N.; Kiljunen, S.; Savilahti, H.; Pohlschroder, M. Screening of a *Haloferax volcanii* transposon library reveals novel motility and adhesion mutants. *Life* **2016**, *6*, 41. [CrossRef] [PubMed]

38. Pfeiffer, F.; Broicher, A.; Gillich, T.; Klee, K.; Mejía, J.; Rampp, M.; Oesterhelt, D. Genome information management and integrated data analysis with HaloLex. *Arch. Microbiol.* **2008**, *190*, 281–299. [CrossRef] [PubMed]

39. Freese, N.H.; Norris, D.C.; Loraine, A.E. Integrated genome browser: Visual analytics platform for genomics. *Bioinformatics* **2016**, *32*, 2089–2095. [CrossRef] [PubMed]

40. Babski, J.; Haas, K.A.; Näther-Schindler, D.; Pfeiffer, F.; Förstner, K.U.; Hammelmann, M.; Hilker, R.; Becker, A.; Sharma, C.M.; Marchfelder, A.; et al. Genome-wide identification of transcriptional start sites in the haloarchaeon *Haloferax volcanii* based on differential RNA-Seq (dRNA-Seq). *BMC Genom.* **2016**, *17*, 629. [CrossRef] [PubMed]

41. Laass, S.; Monzon, V.A.; Kliemt, J.; Hammelmann, M.; Pfeiffer, F.; Förstner, K.U.; Soppa, J. Characterization of the transcriptome of *Haloferax volcanii*, grown under four different conditions, with mixed RNA-Seq. *PLoS ONE* **2019**, e0215986. [CrossRef]

42. Bitan-Banin, G.; Ortenberg, R.; Mevarech, M. Development of a gene knockout system for the halophilic archaeon *Haloferax volcanii* by use of the *pyrE* gene. *J. Bacteriol.* **2003**, *185*, 772–778. [CrossRef]

43. Sherwood, K.E.; Cano, D.J.; Maupin-Furlow, J.A. Glycerol-mediated repression of glucose metabolism and glycerol kinase as the sole route of glycerol catabolism in the haloarchaeon *Haloferax volcanii*. *J. Bacteriol.* **2009**, *191*, 4307–4315. [CrossRef] [PubMed]

44. Fröls, S.; Dyall-Smith, M.; Pfeifer, F. Biofilm formation by haloarchaea. *Environ. Microbiol.* **2012**, *14*, 3159–3174. [CrossRef] [PubMed]

45. Mullakhanbhai, M.F.; Larsen, H. *Halobacterium volcanii* spec. nov., a Dead Sea halobacterium with a moderate salt requirement. *Arch. Microbiol.* **1975**, *104*, 207–214. [CrossRef]

46. Tripepi, M.; Imam, S.; Pohlschröder, M. *Haloferax volcanii* flagella are required for motility but are not involved in PibD-dependent surface adhesion. *J. Bacteriol.* **2010**, *192*, 3093–3102. [CrossRef]

47. Kimura, M.; Kimura, J.; Davie, P.; Reinhardt, R.; Dijk, J. The amino acid sequence of a small DNA binding protein from the archaebacterium *Sulfolobus solfataricus*. *FEBS Lett.* **1984**, *176*, 176–178. [CrossRef]

48. Prasse, D.; Thomsen, J.; de Santis, R.; Muntel, J.; Becher, D.; Schmitz, R.A. First description of small proteins encoded by spRNAs in *Methanosarcina mazei* strain Gö1. *Biochimie* **2015**, *117*, 138–148. [CrossRef]

49. Humbard, M.A.; Miranda, H.V.; Lim, J.-M.; Krause, D.J.; Pritz, J.R.; Zhou, G.; Chen, S.; Wells, L.; Maupin-Furlow, J.A. Ubiquitin-like small archaeal modifier proteins (SAMPs) in *Haloferax volcanii*. *Nature* **2010**, *463*, 54–60. [CrossRef] [PubMed]

50. Dantuluri, S.; Wu, Y.; Hepowit, N.L.; Chen, H.; Chen, S.; Maupin-Furlow, J.A. Proteome targets of ubiquitin-like samp1ylation are associated with sulfur metabolism and oxidative stress in *Haloferax volcanii*. *Proteomics* **2016**, *16*, 1100–1110. [CrossRef] [PubMed]

51. Hepowit, N.L.; de Vera, I.M.S.; Cao, S.; Fu, X.; Wu, Y.; Uthandi, S.; Chavarria, N.E.; Englert, M.; Su, D.; Söll, D.; et al. Mechanistic insight into protein modification and sulfur mobilization activities of noncanonical E1 and associated ubiquitin-like proteins of Archaea. *FEBS J.* **2016**, *283*, 3567–3586. [CrossRef]

52. Guttenplan, S.B.; Kearns, D.B. Regulation of flagellar motility during biofilm formation. *FEMS Microbiol. Rev.* **2013**, *37*, 849–871. [CrossRef]

53. Verstraeten, N.; Braeken, K.; Debkumari, B.; Fauvart, M.; Fransaer, J.; Vermant, J.; Michiels, J. Living on a surface: Swarming and biofilm formation. *Trends Microbiol.* **2008**, *16*, 496–506. [CrossRef] [PubMed]

54. Bak, G.; Lee, J.; Suk, S.; Kim, D.; Young Lee, J.; Kim, K.-S.; Choi, B.-S.; Lee, Y. Identification of novel sRNAs involved in biofilm formation, motility, and fimbriae formation in *Escherichia coli*. *Sci. Rep.* **2015**, *5*, 15287. [CrossRef]

55. Esquivel, R.N.; Schulze, S.; Xu, R.; Hippler, M.; Pohlschroder, M. Identification of *Haloferax volcanii* Pilin N-glycans with diverse roles in pilus biosynthesis, adhesion, and microcolony formation. *J. Biol. Chem.* **2016**, *291*, 10602–10614. [CrossRef] [PubMed]

56. Pohlschroder, M.; Esquivel, R.N. Archaeal type IV pili and their involvement in biofilm formation. *Front. Microbiol.* **2015**, *6*, 190. [CrossRef] [PubMed]
57. Chimileski, S.; Franklin, M.J.; Papke, R.T. Biofilms formed by the archaeon *Haloferax volcanii* exhibit cellular differentiation and social motility, and facilitate horizontal gene transfer. *BMC Biol.* **2014**, *12*, 65. [CrossRef] [PubMed]

MDPI

St. Alban-Anlage 66

4052 Basel

Switzerland

Tel. +41 61 683 77 34

Fax +41 61 302 89 18

www.mdpi.com

Genes Editorial Office

E-mail: genes@mdpi.com

www.mdpi.com/journal/genes

9 783039 289554